DEVELOPMENTS IN POLYMER FRACTURE—1

THE DEVELOPMENTS SERIES

Developments in many fields of science and technology occur at such a pace that frequently there is a long delay before information about them becomes available and usually it is inconveniently scattered among several journals.

Developments Series books overcome these disadvantages by bringing together within one cover papers dealing with the latest trends and developments in a specific field of study and publishing them within *six months* of their being written.

Many subjects are covered by the series including food science and technology, polymer science, civil and public health engineering, pressure vessels, composite materials, concrete, building science, petroleum technology, geology, etc.

Information on other titles in the series will gladly be sent on application to the publisher.

DEVELOPMENTS IN POLYMER FRACTURE—1

Edited by
E. H. ANDREWS
B.Sc., Ph.D., D.Sc., F.Inst.P., F.I.M., C.Eng.
*Department of Materials,
Queen Mary College, University of London, UK*

APPLIED SCIENCE PUBLISHERS LTD
LONDON

APPLIED SCIENCE PUBLISHERS LTD
RIPPLE ROAD, BARKING, ESSEX, ENGLAND

British Library Cataloguing in Publication Data

Developments in polymer fracture. —(Developments series).
1
1. Polymers and polymerization—Fracture
I. Andrews, Edgar Harold II. Series
620.1'92 TA455.P58

ISBN 0-85334-819-7

WITH 19 TABLES AND 113 ILLUSTRATIONS

Printed in Great Britain by Galliard (Printers) Ltd, Great Yarmouth

PREFACE

Any book in which different chapters are contributed by different authors must suffer from the defects of unevenness and selectivity. This is especially the case in a volume devoted to advances or developments in a field of study rather than a comprehensive review of that field.

There is, however, an inevitability about the production of this kind of book in any subject of research which is expanding rapidly. Only by persuading those expert in various branches of the field to review in depth their own relatively narrow area can the literature be kept up to date in a digestible form. Whilst reviews do appear from time to time in the research journals, those directly engaged in research seldom take time off to write such articles unless enlisted to do so for the purpose of some special project such as the present book. I am particularly grateful to the contributors, therefore, for their willingness to divert effort into this activity in the midst of many other pressures and research commitments.

Since I wrote my monograph *Fracture in Polymers*, published in 1968, the field of polymer fracture has grown enormously. Some investigative techniques, such as electron spin resonance, had at that time made virtually no contribution to the subject, whilst even the application of fracture mechanics to the failure of polymers was at an early stage. The phenomena of fatigue and impact failure in polymers had received very little systematic attention and were poorly understood. The number of workers active in the field was rather small.

All that has changed dramatically during the past ten years. Many metallurgists have accepted the challenge of polymeric solids and have made notable contributions to our subject. The increasing use of plastics and rubbers in engineering applications has stimulated much activity in design studies, and design against fracture plays an obviously vital role. Perhaps more important than any other factor is the emergence of materials science as an academic discipline in its own right, which has enabled polymers to take their place alongside metals as materials of great scientific and technological interest.

Although many valuable books have been produced in recent years dealing with the mechanics and materials science of polymers and polymer composites, few of them have dealt in any depth with the subject of fracture. A recent volume of *Advances in Polymer Science* (vol. 27, 1978) has done something to correct this situation, but contains only three contributions and fails therefore to cover a wide area. For these reasons it seems a most suitable time for the present volume to appear.

This book is by no means exhaustive of the developments made in the field in recent years. One notable omission is the subject of fatigue in polymers, for which I failed to obtain a contributor. This topic will, however, be well covered in a book being written currently by Professors John Manson and Richard Herzberg of Lehigh University. Another topic which does not feature here is that of particulate composites, but again this omission is rectified by Clive Bucknall's chapter in *Advances in Polymer Science*, vol. 27. Again, the present book contains no separate chapter on fracture mechanics, but this is because most chapters make their own reference to this approach to fracture. Professor J. G. Williams has a chapter on linear elastic fracture mechanics in *Advances in Polymer Science*, vol. 27, but the important case of non-linear, inelastic fracture mechanics still awaits an up-to-date review, although some mention of it is made in the opening chapter of this book.

In inviting contributions I have sought a coverage of the field which ranges widely, from the molecular aspects of fracture in polymers to the severely practical area of applications. The particular application chosen, that of dental materials, was selected as having some novelty value, but also because it well illustrates the widely ranging demands made upon polymers in use situations. The chapter on impact performance, although it concentrates on impact testing, is also of direct interest to those concerned with engineering applications, whilst the excellent and exhaustive review of fibre reinforcement deals with polymers in the composite form obligatory for most load-bearing purposes.

At the other end of our spectrum of coverage lie those chapters concerned with the molecular environmental and microscopical aspects of the subject. Here we have attempted to review not only the scientific advances of recent years but also the growing understanding of the way these phenomena relate to macroscopic and engineering fracture behaviour.

Finally, by devoting the second part of the book to particular classes of polymeric solid (some of which receive little attention in more limited reviews), the authors have tried to provide something of interest and value to all concerned with the fracture and failure of these important and fascinating materials.

E. H. ANDREWS,
Queen Mary College,
London

CONTENTS

LIST OF CONTRIBUTORS

E. H. ANDREWS

Department of Materials, Queen Mary College, Mile End Road, London E1 4NS, UK.

M. BRADEN

Professor of Materials Science in Dentistry, The London Hosptial Medical College, Turner Street, London E1 2AD, UK.

B. E. CAUSTON

Lecturer, Materials Science in Dental Prosthetics, The London Hospital Medical College, Turner Street, London E1 2AD, UK.

A. N. GENT

Professor of Polymer Physics and Assistant Director, Institute of Polymer Science, The University of Akron, Akron, Ohio 44325, USA.

E. J. KRAMER

Associate Professor of Materials Science and Engineering, Bard Hall, Cornell University, Ithaca, New York 14853, USA.

D. R. MULVILLE

Department of the Navy, Mechanics of Materials Branch (Code 8433), Naval Research Laboratory, Washington, DC 20375, USA.

C. T. R. PULFORD

Institute of Polymer Science, The University of Akron, Akron, Ohio 44325, USA.

P. E. REED

Department of Materials, Queen Mary College, Mile End Road, London E1 4NS, UK.

I. WOLOCK

Department of the Navy, Mechanics of Materials Branch (Code 8433), Naval Research Laboratory, Washington, DC 20375, USA.

R. J. YOUNG

Department of Materials, Queen Mary College, Mile End Road, London E1 4NS, UK.

A NOTE ON FRACTURE MECHANICS NOTATION

Because the theory of fracture mechanics has been developed by different workers and often from different viewpoints, a variety of different notations has also arisen. The following notes are provided at the outset to avoid confusion.

(i) A distinction is always drawn between a continuously variable parameter and its critical value at which some event (usually crack propagation) is observed. Thus K is the stress intensity factor, but K_c is the critical stress intensity factor or 'fracture toughness'. Whilst K is a continuously variable function of stress and geometry, K_c is characteristic of the solid.

(ii) The critical value of a fracture parameter is not itself a property of the solid. There may, however, be a *property* of the solid such that fracture occurs when the fracture parameter reaches a value equal to that of the property in question. This distinction is not trivial and may be illustrated by reference to 'fracture energy' and 'strain energy release rate'.

The strain energy release rate is defined as

$$\mathcal{G} \text{ or } G = -2\frac{\partial \mathcal{E}}{\partial A}$$

where $\mathcal{E}$ is the total potential energy of the system and A the area of crack surface. The factor 2 occurs because there are two surfaces to the crack. The solid may be considered to possess a property called 'fracture energy' which is the energy expended in the creation of unit area of crack surface and which has the general character of a surface energy. This 'fracture energy' is denoted by the symbol $\mathcal{T}$ derived from the term 'tearing energy'.

The criterion for fracture then becomes

$$G \geq G_c = 2\mathcal{T}$$

That is, fracture occurs when G, the energy available, achieves a critical value which equals $2\mathcal{T}$, the energy required (for two surfaces). Obviously G_c and $2\mathcal{T}$ are formally equivalent, but they are defined differently and have different physical significance. Some authors prefer to present their work in terms of G_c and others in terms of $\mathcal{T}$, and the two approaches are, of course, equally valid and equivalent.

Similarly the minimum fracture energy $2\mathcal{T}_0$ required for surface formation in the absence of mechanical losses, is formally equivalent to a lowest observed critical energy release rate G_{0c}.

(iii) Note that the symbol K is also used to denote bulk modulus and the symbol G can also stand for shear modulus or Gibbs free energy. The usage will be apparent from the context.

Chapter 1

MOLECULAR STRUCTURE AND STRENGTH OF POLYMERS

E. H. ANDREWS
Queen Mary College, University of London, UK

SUMMARY

It has proved extremely difficult to establish a clear correlation between strength and molecular structure in solids. This chapter reviews the efforts that have been made, historically, to relate these two properties in the case of polymeric solids. Consideration is given to the kinetic theory of fracture in polymers, which seeks to explain fracture in terms of bond breakage probabilities; to direct observations of molecular fracture by electron spin resonance (ESR) and other techniques; and to fracture mechanics theories. It is shown that recent theoretical developments permit at least a partial synthesis of these differing viewpoints. This holds out hope for the emergence of a general and quantitative description of the way strength in polymers depends upon the molecular constitution.

1 INTRODUCTION

The fracture of polymeric solids is now the subject of an extensive literature and this body of information will undoubtedly continue to grow as polymers find increasing application as structural materials in engineering. In an engineering context, of course, the word 'strength' is used to express a variety of properties, including stiffness, plastic yield strength, ultimate tensile strength and fatigue resistance. Each of these properties sets limits to the usefulness of the solid in load-bearing applications.

In this chapter we shall use the word 'strength' to indicate the resistance of the material to fracture, i.e. to the creation of new surfaces within the body.[1] The question we shall consider is how far the strengths of polymeric solids can be related to, or predicted from, their molecular structure.

It is obvious, from a vast number of experimental observations, that strength in polymers is not controlled solely by molecular structure. In a given polymer, with a fixed molecular structure, the strength will generally depend upon the state (crystalline or amorphous), the physical microstructure, the distribution of molecular mass, the temperature, the rate of testing, the state of stress in the specimen, the specimen size and the chemical environment. Even this list does not exhaust the potential for variation in the strength of a solid. Thus one may well ask whether the molecular structure has *any* recognisable role in controlling the ultimate properties of polymers. Certainly this role has proved extremely difficult to define or even detect.

Nevertheless, it is intuitively obvious that the nature of the chemical bonding within a solid must, in some way, influence its cohesive properties. It is the purpose of this chapter to explore the manner in which the bonding does exert such an influence and to explain how the variations of strength listed earlier come about.

Historically, there have been advanced a variety of theories of mechanical strength, each theory approaching the subject from a different viewpoint. One obvious approach is to seek a quantitative correlation between the strength of a polymer molecule (e.g. of the C–C bonds forming its backbone) and the tensile strength of the solid. As we shall see, such correlations have been found experimentally. A modification of this approach recognises that bond-breakage can be represented as a chemically activated process analogous to thermal decomposition. This leads to what is known as the 'kinetic theory of fracture', which is successful in explaining the time- and temperature-dependence of fracture in brittle polymers.

A third approach, largely experimental in character so far, is to use electron spin resonance (ESR) or infra-red spectroscopic techniques to observe directly the fracture of molecules in stressed polymers. The number of fracture events can be determined and their contribution to macroscopic deformation and fracture phenomena assessed. A rather complex picture emerges from such studies, as we shall see in Chapter 2.

A fourth, and more recent, approach is to use fracture mechanics to determine bulk fracture behaviour and infer the contribution of the molecular structure from a suitable analysis of the data. For fracture mechanics to tell us *anything* about molecular strength, of course, the equations used must contain molecular parameters. The first-ever statement of fracture mechanics, Griffith's theory,[2] does contain such a parameter, namely the solid surface energy. No subsequent theory contained such a parameter until the publication of the author's 'generalised theory of fracture mechanics'.[3] This theory employs a parameter $\mathscr{T}_0$, analogous to the Griffith surface energy, which is half the energy required to fracture unit area of interatomic bonds across the fracture plane. (The fraction $\frac{1}{2}$ occurs because $\mathscr{T}_0$ is referred to unit area of crack surface and two such surfaces are produced from a single fracture plane.)

The areas of study outlined above are detailed more fully in what follows.

2 DIRECT CORRELATIONS

The theoretical strength of a solid can be easily calculated from a knowledge of the potential energy of interaction between its constituent atoms.[4] The result is conveniently stated in the form

$$\sigma_f = \alpha E \tag{1}$$

where σ_f is the theoretical fracture stress in tension, α is a constant depending on the solid but having an average value of about 0·1 and E is the Young's modulus of the solid.

In specially prepared materials (e.g. graphite fibres, whisker crystals) experimental values of α approaching 0·05 can be obtained, agreeing within a factor of two with the theoretical strength. Bulk solids, however, exhibit values of α which are typically of the order of 10^{-3} or less. This is largely attributable to the existence in the solid of stress-concentrating flaws.

The theoretical tensile strengths of polymer molecules can be calculated in terms of the strength of the backbone bonds, normalised to unit area. Mark[5] calculated values for oriented cellulose of 15 000 MN m^{-2}. For a range of polymers with side groups of different sizes, Vincent's data[6] suggest a range of 6000 to 30 000 MN m^{-2} for the ideal strength of polymeric solids. Actual strengths are 200 times smaller (α values of ≈ 0.0005).

Mark[5] recognised that polymeric solids could, however, fail by other processes than main-chain fracture. He calculated that if failure occurred by the breakage of van der Waals (intermolecular) bonds only, the strength of cellulose would be between 900 MN m^{-2} and 270 MN m^{-2} according to whether or not the molecules have to be 'pulled out' from among others. The measured tensile strength is around 500 MN m^{-2}, suggesting, superficially, that secondary bond fracture controls the process of failure in this material. However, thermosetting polymers with fully co-valent molecular networks also have very low strengths compared with theoretical values, so that low strength cannot be attributed to the absence of primary bond fracture.

Vincent[6] defined the 'critical tensile strength' for polymers as the fracture stress at the brittle-to-ductile transition of the material (see Fig. 1).

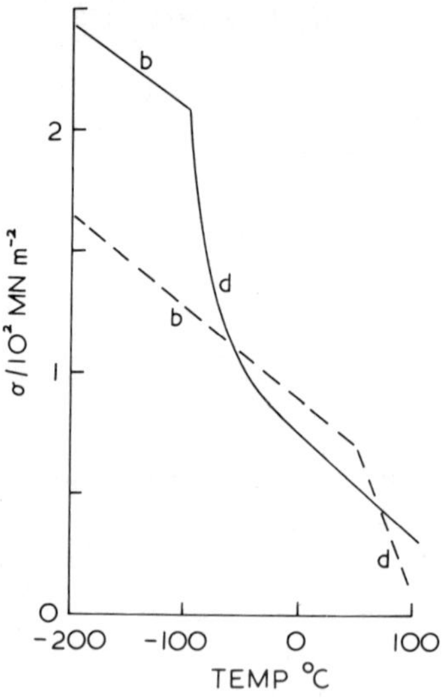

FIG. 1. Brittle fracture stress or yield stress versus temperature for two polymers, polyoxymethylene (——) and poly(methylmethacrylate) (– – –); b, brittle fracture; d, ductile yield. Note break in each curve at the brittle–ductile transition. (After Vincent.[6])

He found a linear correlation between this quantity and the number of backbone bonds per unit area. This number was defined as

$$\text{No. of bonds/area} = \frac{\text{material density} \times \text{length of repeat unit}}{\text{mass of repeat unit}}$$

Vincent's correlation is shown in Fig. 2, the slope of the line being $36{\cdot}8 \times 10^{-18}$ MN. For the theoretical strength to be observed the slope would need to be 61×10^{-16} MN, a factor of 160 greater. Thus although the molecular cross-section obviously exerts an influence on the brittle

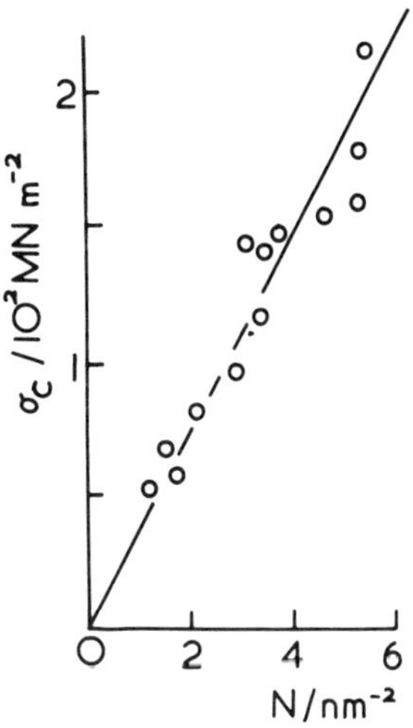

FIG. 2. 'Critical stress' versus number of backbone bonds N per unit area for 14 polymers. (After Vincent.[6])

strength of polymers, the magnitude of the bulk strength is so low as, apparently, to deny molecular strength any direct influence!

This paradox can, of course, be readily explained by appeal to the existence of stress-raising flaws which concentrate the applied stress to such an extent that the theoretical strength is achieved at their tips. However, because *any* discrepancy between real and theoretical strength can be explained by this argument, the explanation is qualitative and unsatisfactory, in that it does not advance our understanding of the matter. It is for this reason that fracture mechanics (FM.) ideas have gained widespread support. Whilst our main discussion of FM. is postponed to the final section of this paper, some reference is appropriate to this discussion.

Fracture mechanics describes the conditions for the propagation of an existing crack. For a small crack in a large specimen, the crack length enters explicitly into the expression for 'fracture energy', and this eliminates uncertainties about such flaws.

The fracture energy is defined as the energy required to create unit area of crack surface and must equal the energy supplied by the system as the crack propagates. The energy supplied is usually expressed in terms of 'energy release rate', $-\partial\mathscr{E}/\partial A$, where $\mathscr{E}$ is the total stored energy and A is the crack surface area.[1] The fracture energy, which we will denote $\mathscr{T}$, is then equal to that value of the energy release rate (its critical value) at which the crack propagates:

$$-(\partial \mathscr{E}/\partial A)_{\text{crit}} = \mathscr{T} \tag{2}$$

The energy release rate at any stage of the experiment can be calculated from the applied stresses or strains and the specimen geometry (including the crack length), so that $\mathscr{T}$ can be measured. If the material is perfectly elastic, and if crack propagation is quasi-static, then the strength parameter $\mathscr{T}$ must be identically equal to the surface energy of the solid (defined as half the energy to break unit area of interatomic bonds). This was Griffith's original criterion for the fracture of brittle solids.

Most solids, however, are not perfectly elastic and $\mathscr{T}$ greatly exceeds the bond fracture energy $\mathscr{T}_0$. The excess energy is dissipated in plastic or visco-elastic deformations. It has been shown recently, by a number of workers,[7,8,9] that a certain class of polymeric solids can be fractured under conditions where $\mathscr{T} \approx \mathscr{T}_0$. Cross-linked gum elastomers become more and more elastic as the temperature is raised and the rate of test lowered. Under sufficiently extreme conditions, a limiting value $\mathscr{T}_0$ of $\mathscr{T}$ is reached which is independent of further changes in rate and temperature and reflects the strength of the molecular network in the absence of visco-elastic loss processes. Typically, this value lies between 20 and 100 J m^{-2}.

Of course, these figures are very high compared with physical-chemical surface energies (20–100 mJ m^{-2}) but it must be remembered that the latter reflect only secondary interactions between atoms whereas $\mathscr{T}_0$ involves the fracture of primary bonds. This, however, does not fully account for the magnitude of $\mathscr{T}_0$, since even fully aligned and perfectly packed polymer molecules would require only between 1 and 10 J m^{-2} to break the C–C backbone bonds.

The answer to this anomaly was given by Lake and Thomas,[10] who pointed out that in a cross-linked network a C–C bond could only be broken if *all* the C–C bonds in that particular network chain (i.e. all bonds between a pair of cross-links) were deformed nearly to their breaking point. When one bond breaks, the energy stored in the remainder of the chain is dissipated irrecoverably. They thus proposed that for cross-linked elastomeric systems $\mathscr{T}_0$ should be given by the energy needed to raise the whole chain of C–C bonds to the breaking point, normalised to unit area of fracture surface. Their theoretical formula is

$$\mathscr{T}_0 = (\tfrac{3}{32})^{1/2} \gamma l U N n^{3/2} \tag{3}$$

where γ is a factor related to the freedom of rotation about C–C bonds, l is the length of a monomer unit, U is the energy required to rupture one C–C bond, N is the number of network chains per unit volume and n is the number of monomer units in the chain.

Very good agreement has been obtained between this theory and the experimentally determined limiting values of $\mathscr{T}_0$ for a variety of cross-linked polymers above their glass-transition temperatures.[7–9] It can thus be said that under conditions of zero mechanical energy loss, the strength of bulk polymers is predictable from their molecular structure.

Although the idea has not yet been tested, it is quite possible that an equation of the form of (3) would explain Vincent's correlation (Fig. 2). Although he considered polymers below their glass-transition temperature, T_g, similar considerations might apply to the limiting strength of plastically deforming molecular assemblies, with entanglements replacing cross-links in the Lake-Thomas theory. Then, by analogy with eqn. (3), $\mathscr{T}_0$ would be governed by N, the number of network chains per unit volume and thus by the number crossing unit area of fracture plane in the plastically oriented solid prior to failure.

Finally, the approach of bulk polymers to theoretical strength has been observed not only in molecular networks at high temperatures and low rates of strain, but also in the large, near-perfect, single crystals that can be obtained by solid-state polymerisation (SSP). Only a few monomers undergo SSP in such a way as to preserve the integrity of large specimens, but this is the case with certain poly-diacetylenes. In these crystals, Baughman *et al.*[11] have observed values of α as high as 0·04 when they were tested in the molecular axis direction. This approach to theoretical strength is attributable to the absence of defects, the perfect alignment and continuity of the molecules in the single crystal and the absence of inelastic deformations for chain-axis tension.

To summarise, therefore, it is possible under some circumstances to observe strengths in polymeric solids which approximate to those predicted directly from molecular structure and bond energies. The circumstances required are: (i) the absence of inelastic deformation processes; and (ii) the absence of flaws or structural imperfections which concentrate stress *or*, alternatively, the use of an FM. approach which eliminates flaw size as an unknown variable. Unless these two conditions are met, the picture is confused firstly by defects which *reduce* the bulk strength and secondly by mechanical loss processes which *increase* the resistance to crack propagation.

3 KINETIC THEORY OF FRACTURE

So far we have considered the bond fracture energy U to be a constant quantity independent of rate and temperature. The kinetic theory of fracture is built around the idea that bond rupture is an activated process analogous to thermal dissociation. Thus the mechanical bond fracture energy U becomes that external energy input which, together with the thermal energy of the atoms, gives rise to a significant rate of bond fracture events. Thus U becomes a variable quantity, dependent on the temperature and the required rate of bond fracture (which in turn can be related to a crack propagation speed or a specimen lifetime under load).

Figure 3 helps us to understand this more easily. State A represents an unbroken interatomic bond and state B the same bond broken. The activation energy G_{AB} must be supplied to obtain the transition A → B. The net rate at which bonds will be broken is given by

$$v = v_0\{[A]\exp(-G_{AB}/kT) - [B]\exp(-G_{BA}/kT)\} \qquad (4)$$

where v_0 is proportional to the thermodynamic temperature T, k is Boltzmann's constant and [A] and [B] are the concentrations of states A and B respectively.

Clearly, as long as the material is unstrained, the free energy, G_B, of

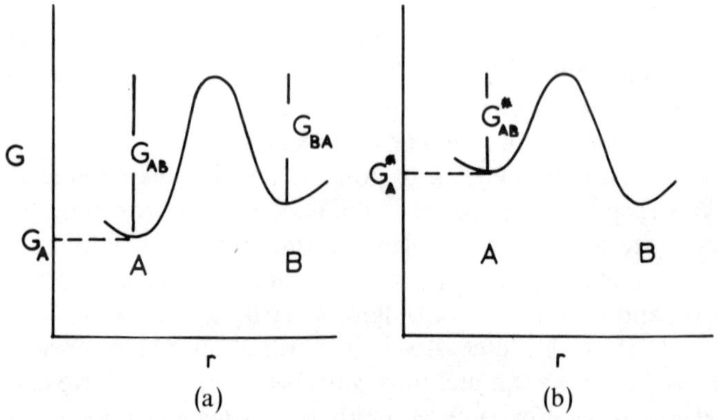

FIG. 3. Free energy diagrams for the activated scission of an interatomic bond; (a) unstressed, (b) stressed. A is the intact state and B the severed state of the bond. r on the horizontal axis represents separation.

state B exceeds that of state A and no net bond breakage will occur. If strain energy is now stored in the material by application of stress, and if this energy can be released by bond breakage, i.e. by the process A → B, the effective activation energy G_{AB} will be reduced by an appropriate amount (see Fig. 3(b)). Mathematically,

$$G^*_{AB} = G_{AB} - \beta\sigma \tag{5}$$

where G^*_{AB} and G_{AB} refer to the stressed and unstressed bonds respectively, σ is the stress on the bond and β is the 'activation volume'.

A minimum stress is required before a net forward reaction A → B can be induced; this stress is given by

$$\sigma_0 = (G_B - G_A)/\beta \tag{6}$$

That is, it is the stress required to make G^*_A equal to G_B. As the stress increases further, G^*_{AB} reduces progressively until the forward reaction predominates. Eventually the stress-induced free energy of state A, G^*_A, will approach the free energy of the activated state and bond fracture will occur by direct mechanical action without the assistance of thermal fluctuations. (This condition can never be precisely attained because of the simultaneous increase in the rate of thermally activated events.)

Thus, to summarise, the mechanical bond fracture energy U, previously treated as a constant, is in fact a variable quantity with lower and upper limits:

$$(G_B - G_A) < U < G_{AB} \tag{7}$$

Between these limits, net bond fracture can occur under the applied stress but will do so at a rate controlled by the temperature. The mechanical strength of a perfectly elastic solid should therefore be time- and temperature-dependent within the limits set by eqn. (7). These ideas can be applied equally well to fracture initiation or to the propagation of a crack by cumulative bond breakage at its tip.

The time- and temperature-dependence of strength in rigid polymers is well documented. If the time to failure, t_f, is measured as a function of stress and temperature, and if t_f is inversely proportional to the bond breakage frequency ν, it follows that,

$$\ln t_f = C + (G_{AB} - \beta\sigma)/kT \tag{8}$$

where C is a constant, and the reverse reaction rate (B → A) is assumed to be negligible. Then a plot of $\ln t_f$ *versus* σ/T gives both β and G_{AB}.

Zhurkov and Tomashevsky[12] obtained values of G_{AB} for a range of polymers and found close agreement with activation energies for thermal dissociation of the materials in question (Table 1).

TABLE 1

ACTIVATION ENERGIES FOR MECHANICAL AND THERMAL DEGRADATION OF SOME SOLID POLYMERS

Polymer	G_{AB} (kcal mol^{-1})	G(*thermal*) (kcal mol^{-1})
PVC	35	32
Polystyrene	54	55
PMMA	54	52–53
Polypropylene	56	55–58
Teflon	75	76–80
Nylon 6	45	43

The weakness of the kinetic theory lies in its limitation to brittle fracture (it takes no account of energy loss processes) and its failure to consider explicitly the propagation phase of fracture. Later we shall see how the ideas of the kinetic theory can be combined with fracture mechanics to provide an extremely comprehensive statement of fracture theory.

4 OBSERVATION OF MOLECULAR FRACTURE IN POLYMERS

We referred earlier to Mark's calculation which suggested superficially that polymeric solids failed by rupture of van der Waals bonds only. That this is not so is amply demonstrated by electron spin resonance studies of polymers under stress. If a co-valently bonded molecular chain breaks, a pair of free radicals is created at the broken ends and these are capable of absorbing microwave energy in the strong magnetic field of an ESR spectrometer.

In practice, the primary radicals (unpaired 'spins') are likely to decay rapidly, either to more stable secondary radicals or by combination with oxygen to form peroxy radicals. Nevertheless, the production of free radicals, of whatever kind, in a stressed polymer is indicative of main-chain fracture.

This subject has been reviewed extensively in recent publications[13,14] and is considered in greater detail in Chapter 2.

Although ESR is the major tool employed to observe molecular fracture, infra-red spectroscopy has also proved valuable. Two distinct experiments have been reported by Russian workers and are discussed fully in Chapter 2. The first approach[15] is to use IR absorption peaks to measure directly the concentration of molecular end-groups before and after loading, thus achieving an absolute determination of fracture events.

The second use of IR spectroscopy[16] estimates the distribution of stress over the population of primary bonds, from the slight shifts of absorption peaks caused by stress. Such experiments show that, typically, a small proportion of the bonds carry a stress much larger than the average, nominal stress. There is a 'cut-off' stress above which the bond population is zero and this stress decreases as the temperature is raised, in harmony with the kinetic theory of fracture. These studies are also discussed further in Chapter 2.

The overall picture given by these studies is that molecular (i.e. main-chain) fracture occurs in solid polymers only when the material's capacity for deformation by uncoiling and flow has been exhausted. (Flow includes crystallographic mechanisms.) While this point is reached, a relatively small proportion of extended molecules carry a disproportionate stress and undergo fracture. If stress or strain continues to increase, such fractures accumulate to form microvoids[17] which may eventually coalesce to give macroscopic fracture. Because of the problems of radical decay, and the contribution to macroscopic failure properties of other deformation mechanisms like plastic flow, it has not so far proved possible to relate free radical production to bulk fracture in any quantitative manner. For further detailed discussion the reader is referred to Chapter 2.

5 FRACTURE MECHANICS AND MOLECULAR STRUCTURE

Fracture mechanics deals with the conditions required for an existing crack to propagate through the solid. There are a variety of fracture mechanics theories, but they all employ some parameter which is supposed to achieve a critical value for crack growth to occur. The theories then relate this parameter to measurable quantities (e.g. the applied stress and crack length) so that its critical value may be determined from the values of stress, strain, crack length, etc. at which the crack is observed to propagate. According to the particular theory employed, the parameter in question may be stress intensity factor (K), energy release rate (G), Rice's contour integral (J), or crack opening displacement (δ). The strength of the solid is then characterised, hopefully, by critical values of such parameters (K_c, G_c, J_c, δ_c).

Most fracture mechanics theories, however, contain no explicit reference to those physical properties of the solid associated with fracture or rupture of the molecular structure (though stiffness and compliance terms do appear). The exceptions are those theories which include the surface energy of the solid, namely Griffith's theory[2] and the generalised theory of Andrews.[3] We shall refer only to Andrews' theory since it reduces to Griffith's theory for perfectly elastic solids.

The relevant equation is that for the surface fracture energy, or 'surface work', denoted here by the symbol $\mathscr{T}$. As previously explained, this quantity is the total energy needed to create unit area of fracture surface by crack propagation, and equals $\frac{1}{2}G_c$, the energy available for this purpose at the moment of crack propagation. (The factor $\frac{1}{2}$ occurs because the crack has two surfaces.)

Obviously, since real materials are not perfectly elastic, part of this energy is used to break interatomic bonds to create the surfaces, and part is dissipated in irreversible processes like visco-elastic and plastic flow throughout the specimen. The generalised theory shows that

$$\mathscr{T} = \mathscr{T}_0 \Phi(\dot{c}, T, \varepsilon_0) \tag{9}$$

where $\mathscr{T}_0$ is the surface energy (defined as half the energy required to break unit area of interatomic bonds across the fracture plane) and Φ is a loss function whose value depends on the material, the crack velocity $\dot{c}$, the temperature, T, and the overall state of strain, ε_0, in the body. Thus $\mathscr{T}$ is not just equal to $\mathscr{T}_0$ plus lost energy, but is *also* related to $\mathscr{T}_0$ in a multiplicative manner.

The loss function Φ reduces to unity for a perfectly elastic solid, but otherwise is given by the theory in an explicit form involving the strain energy distribution in the specimen and the energy loss characteristics of the solid.[3].

One of the most direct tests of the theory has proved to be the study of adhesive failure where a crack is propagated along an interface between two solids rather than cohesively within a single solid. Using eqn. (9) and a series of adhesive systems involving SBR rubber bonded to plastic substrates, Andrews and Kinloch[8,18] were able to derive values for $2\mathscr{T}_0$ which corresponded very closely to the thermodynamic work of adhesion, w_A, calculated quite independently. Since w_A is by definition the energy required to break unit area of interatomic secondary bonds at the interface, the theory was clearly vindicated. When primary bonds were introduced across the interface, they found that $2\mathscr{T}_0 \gg w_A$, but $2\mathscr{T}_0$ could still be explained as the actual energy needed to break bonds across the fracture plane.

Andrews and Fukahori[19] applied eqn. (9) to the cohesive fracture of a series of highly deformable polymers (styrene-butadiene rubber, ethylene-propylene rubber, L.D. polyethylene and plasticised PVC). By measuring $\mathscr{T}$ directly from crack propagation experiments and calculating Φ from the measured strain distribution around propagating cracks, they found that eqn. (9) correctly described the data obtained at different strain rates. Values of $\mathscr{T}_0$ were deduced and found to correspond closely to the minimum fracture energies obtained by other methods at high temperatures and low rates (i.e. under elastic conditions where losses tend to zero). The $\mathscr{T}_0$ values also agreed well with theoretical calculations based on the idea that a whole network chain (i.e. between cross-links) has to be energised to fracture before a single C–C bond in that chain will break.

Since $\mathscr{T}$ can be expressed simply in terms of the applied stresses and specimen geometry, it is clear that eqn. (9) provides a direct link between macroscopic strength and $\mathscr{T}_0$, which is a molecular interaction parameter. Admittedly $\mathscr{T}_0$ is not simply the surface energy of physical chemistry, except in the particular case of a secondary bonded interface. However, $\mathscr{T}_0$ is controlled exclusively by the molecular and supramolecular structure and is normally proportional to the dissociation energy of C–C bonds.

Finally, it is now possible, for the first time, to combine the kinetic theory with fracture mechanics *via* the parameter $\mathscr{T}_0$. Referring back to eqn. (7) we see that the energy that must be supplied mechanically to break bonds is not single-valued, but lies between limits, its actual value depending on rate and temperature according to the kinetic theory. Thus we may write

$$\mathscr{T}_0 = \mathscr{T}_0(v, T) \tag{10}$$

and approximately, assuming $\mathscr{T}_0$ to be proportional to the bond dissociation energy U,

$$\mathscr{T}_0 = \alpha U = \alpha\{G_{AB} - kT \ln(v_0/v)\} \tag{11}$$

where G_{AB} is the bond-breakage activation energy in the absence of stress and v, the frequency of bond fracture, is clearly proportional to crack velocity $\dot{c}$. The limits within which eqn. (11) is valid are, of course, those given in eqn. (7).

The full equation for macroscopic fracture resistance $\mathscr{T}$ is obtained by substituting eqn. (11) into eqn. (9). Thus,

$$\mathscr{T} = \alpha\{G_{AB} - kT \ln(v_0/v)\}\Phi(\dot{c}, T, \varepsilon_0) \qquad (12)$$

This equation has a two-fold rate- and temperature-dependence, namely the kinetic-theory dependence of the first bracket and the loss-function dependence of the second. As long as the behaviour is elastic the observed rate/temperature-dependence will derive from the kinetic theory term. These conditions will tend to apply for very low crack velocities and low applied stresses or strains ($\varepsilon_0 \rightarrow 0$), and at temperatures either very much lower or very much higher than T_g of the polymer. Under conditions of rapid crack growth, high applied stress and intermediate temperature, the loss function is likely to dominate the rate- and temperature-dependence of strength.

The actual variation of $\mathscr{T}_0$ with molecular structure from polymer to polymer has not been studied yet in detail. Since, however, the value of U for C–C or C–O backbone bonds does not vary dramatically from polymer to polymer it is obvious that $\mathscr{T}_0$ will be controlled by two molecular structure factors, namely: (i) the effective cross-sectional area of the molecules, giving rise to Vincent's correlation; and (ii) the number of backbone bonds between cross-links or entanglements, giving rise to the network contribution to $\mathscr{T}_0$ considered by the theory of Lake and Thomas.[10]

6 CONCLUSION

Although no mathematical description of fracture is ever likely to embrace all fracture phenomena, a gratifying degree of completeness is beginning to emerge. This is a result of the synthesis of several different approaches to the explanation and description of fracture in polymers. It is now possible to claim that we possess the necessary theoretical tools to investigate, quantitatively, the relationship between the strength of bulk polymeric solids and their molecular and supramolecular structures.

REFERENCES

1. ANDREWS, E. H. *Fracture in Polymers*, Oliver and Boyd, Edinburgh, 1968, p. 37.
2. GRIFFITH, A. A. *Phil. Trans. Roy. Soc.* (Lond.), **A221** (1920) p. 163.

3. ANDREWS, E. H. *J. Mater. Sci.*, **9** (1974) p. 887.
4. COTTRELL, A. H. *Proc. Roy. Soc.* (Lond.), **A282** (1964) p. 2.
5. MARK, H. In *Cellulose and its Derivatives*, ed. E. Ott, Wiley Interscience, New York, 1943, p. 1001.
6. VINCENT, P. I. *Polymer*, **13** (1972) p. 558.
7. LAKE, G. J. and LINDLEY, P. B. *J. Appl. Polymer Sci.*, **9** (1965) p. 1233.
8. ANDREWS, E. H. and KINLOCH, A. J. *Proc. Roy. Soc.* (Lond.), **A332** (1973) p. 385.
9. AHAGON, A. and GENT, A. N. *J. Polymer Sci.* (*Polym. Phys. Ed.*), **13** (1975) p. 1903.
10. LAKE, G. J. and THOMAS, A. G. *Proc. Roy. Soc.* (Lond.), **A300** (1967) p. 108.
11. BAUGHMAN, R. H., GLEITER, H. and SENDFELD, N. *J. Polymer Sci.* (*Polym. Phys. Ed.*), **13** (1975) p. 1871.
12. ZHURKOV, S. N. and TOMASHEVSKY, E. E. *Physical Basis of Yield and Fracture*, Inst. of Physics, London, 1966, p. 200.
13. KAUSCH, H. H. and DE VRIES, K. L. *Int. J. of Fracture*, **11** (1975) p. 727.
14. ANDREWS, E. H. and REED, P. E. *Advances in Polymer Sci.*, **27** (1978) p. 1.
15. ZHURKOV, S. N. and KORSUKOV, V. E. *J. Polymer Sci.* (*Polym. Phys. Ed.*), **12** (1974) p. 385.
16. ZHURKOV, S. N., VETTEGREN, V. I., KORSOKOV, V. E. and NOVAK, I. I., In *Proc. 2nd. Int. Conf. on Fracture*, Chapman and Hall, London, 1969, p. 545.
17. ZHURKOV, S. N., KUKSENKO, V. S. and SLUTSKER, A. I. *Ibid*, p. 531.
18. ANDREWS, E. H. and KINLOCH, A. J. *Proc. Roy. Soc.* (Lond.), **A332** (1973) p. 401.
19. ANDREWS, E. H. and FUKAHORI, Y. *J. Mater. Sci.*, **12** (1977) p. 1307.

Chapter 2

MOLECULAR FRACTURE

E. H. ANDREWS and P. E. REED
Queen Mary College, University of London, UK

SUMMARY

The breakage of individual molecules during deformation of bulk polymers is considered from an experimental viewpoint. Experimental techniques are described for the detection and measurement of such events and the incidence of molecular fracture is related quantitatively to macroscopic parameters such as stress, strain, strain rate and temperature, as well as to microscopic parameters such as cross-link density and crystallinity. Finally, the relationships between molecular fracture and observable fracture phenomena, such as void formation, crazing and ductile failure are discussed.

1 EXPERIMENTAL STUDY OF MOLECULAR FRACTURE

Although the concepts of molecular fracture date back to the work of Taylor in 1947,[1] it is only in recent years that techniques have been perfected for experimental study. Those techniques fall into three categories.

Firstly, there are indirect techniques in which the consequences of molecular fracture are assessed. They include a wide range of methods including the measurement of molecular weight changes and solubility (gel formation). These have been extensively reviewed by Casale *et al.*[2] Another such technique was employed by Becht and Fischer who observed the polymerisation of a monomer induced by molecular fracture in polymer chains.[3]

Secondly, there are direct techniques, notably electron spin resonance

(ESR) spectroscopy, in which the free radicals produced by the fracture of covalent bonds are directly observed, both in respect of their chemical nature and their number. Much of this review is concerned with the results of ESR studies and this technique is therefore treated at some length below. One little-used technique for the direct assessment of free radicals produced by mechanical means is that of Pazonyi *et al.*[4] and Salloum and Eckert.[5] They chopped various polymers in an ethanolic solution of diphenyl picryl hydrazyl, a chemical indicator, and determined the free radical concentration in the cut surfaces by colorimetric measurements of the colour change. This method is subject to some uncertainty on account of possible side reactions.

Finally, a number of associated techniques have been used to shed light on the phenomenon of molecular fracture. These techniques do not estimate the incidence of fracture events but rather concern phenomena or structures which bear strongly on the subject. Thus infra-red spectroscopy is employed to demonstrate the state of molecular stress prior to fracture, whilst low-angle X-ray studies can be used both to assess the distribution of strain within a semicrystalline polymer and to detect and measure the microvoids which are often associated with molecular fracture in such materials.

1.1 Infra-red (IR) Spectroscopy

Two distinct types of study using IR spectroscopy have been carried out by Zhurkov and co-workers. The first technique[6] involves the determination of chemical end-groups in the polymer by measuring the intensities of their characteristic IR absorption peaks. After mechanical loading, the concentrations of various end-groups were found to increase significantly, suggesting that radicals produced by molecular fracture had reacted to form stable end-groups. Some of the results of Zhurkov *et al.*[6] for polyethylene are given in Table 1.

The increase in end-group concentration, in this case some 35×10^{18} groups cm^{-3}, is assumed to equal twice the number of chain fracture events. It will be seen later that this number is about one to two orders of magnitude larger than the number of free radicals observed by ESR spectroscopy, though it has been argued[7] that ESR reveals only a fraction of the primary radicals actually formed. The IR estimate may therefore be a more accurate method of determining the absolute number of molecular fractures. Zhurkov *et al.* claim that the IR estimate agrees well with molecular weight changes measured on the same specimens after loading.

The second IR technique involves a quite different measurement,

TABLE 1

RESULTS OF INFRA-RED SPECTROSCOPY APPLIED TO POLYETHYLENE

IR Peak frequency (cm^{-1})	*Terminal group*	*Concentration* (cm^{-3}) *Before loading*	*After loading*
910	$RHC{=}CH_2$	18.8×10^{18}	25.6×10^{18}
1 379	$R{-}CH_3$	40.0	60.0
1 710	RCOOH	2.2	3.4
1 735	RCOH	13.4	20.6
	TOTAL	74.4	109.6

namely the shift in main-chain vibrational frequencies induced by stress.[8] If ν_σ and ν_0 are the frequencies of the peak of a given IR absorption in the stressed and unstressed states respectively, Zhurkov *et al.* found that

$$\nu_\sigma = \nu_0 - \alpha\sigma \tag{1}$$

where σ is the applied stress and α is a constant.

This result supports the view that the stress-modified activation energy barrier for bond fracture (see Chapter 1) is linear in stress, since Bernstein[9] has proposed that the shift in dissociation energy D of a bond under stress is related to the corresponding shift in bond stretching frequency, thus:

$$\begin{aligned} c(D_0 - D_\sigma) &= \nu_0^2 - \nu_\sigma^2 \\ &\approx (\nu_0 - \nu_\sigma)2\nu_0 \\ &= 2\nu_0\alpha\sigma \end{aligned} \tag{2}$$

Closer study of the absorption band shifted by stress showed a shape deformation which could be interpreted as a non-uniform stress distribution; i.e. some bonds are more highly stressed than others and consequently suffer a higher-than-average frequency shift. Such bonds were designated 'over-stressed' bonds and their distribution with respect to stress was deduced. An example is shown in Fig. 1 for polypropylene under a load of $0{\cdot}7\,GN\,m^{-2}$. About 50% of the bonds are uniformly stressed, but some appear to carry stresses of up to $1\,GN\,m^{-2}$. Similar results were obtained for PET and oriented polycaproamide (nylon 6). It was further shown, for PET oriented film, that the maximum stress on

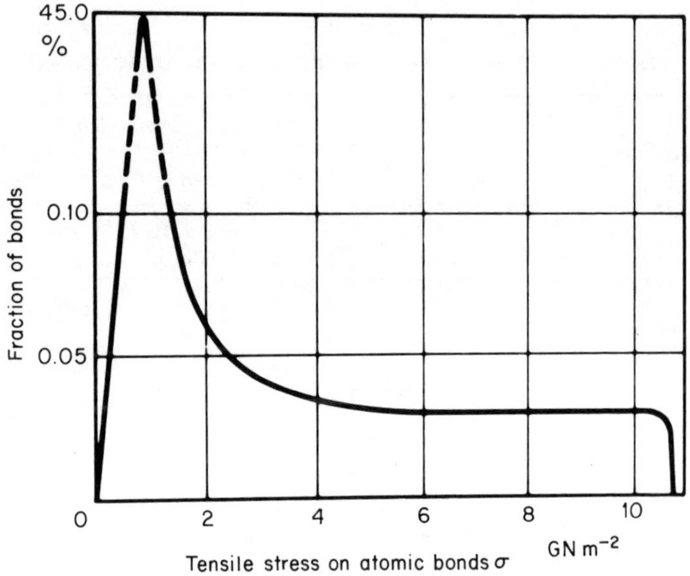

FIG. 1. Stress distribution over a bond population as determined by IR spectroscopy (after reference 8).

bonds is a function of temperature (Fig. 2) in harmony with the kinetic concept of bond rupture.

1.2 Low-angle X-ray Studies

Although low-angle scattering is obtained from semicrystalline polymers as a result of their crystalline-amorphous morphology, and can be used to measure interlamellar spacings, the scattered intensity from cavities or voids is very much stronger. It is therefore possible to use low-sensitivity data to follow the formation and growth of cavities under stress, using the Guinier formula:[10]

$$I(\psi) \approx ANV^2 \exp\{-(4\pi^2/3\lambda^2)L^2\psi^2\} \qquad (3)$$

where I is the diffracted intensity at angle ψ, L is the cavity dimension in the scattering direction, N is the volume concentration of cavities, V is the scattering cavity volume, λ is the wavelength and A a term related to cavity geometry and the electron density difference between cavity and

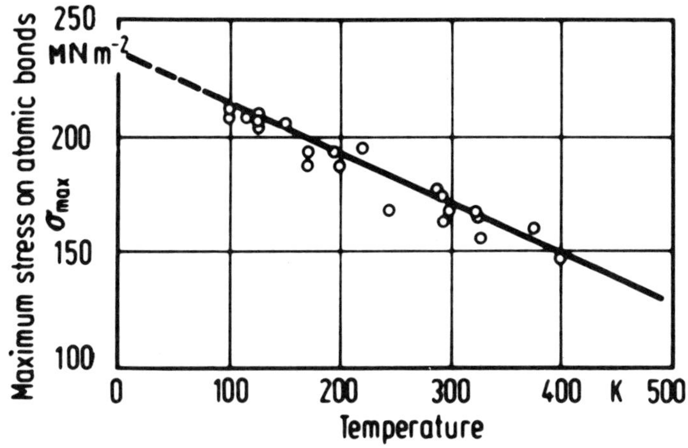

Fig. 2. Dependence of maximum bond stress in PET on temperature (after reference 8).

solid. (For voids this electron density fluctuation is much greater than for crystalline–amorphous structures, hence the much greater intensities for voids.)

Using Cu Kα radiation ($\lambda = 1{\cdot}54$ Å) and scattering angles from several minutes of arc to about 2°, cavity dimensions in the range 10 to 10^3 Å can be measured.[11] Figure 3 shows a typical set of scattering curves for nylon 6 (plotted as log I *versus* ψ^2) showing how the intensity increases with time under load. This is a result of the multiplication and growth of cavities with time.

The curvature of the log I *versus* ψ^2 plots is a consequence of a size distribution, eqn. (3) being derived for an assembly of cavities of uniform size. If a suitable size distribution is chosen, the scattered intensity can be broken down into components which plot linearly and then the cavity size (obtained from the slope) and the number N of cavities (obtained from the intercept $\psi = 0$) can be deduced.

Using this analysis, Zhurkov *et al.*[11] followed the kinetics of cavity growth and correlated their results with ESR measurements of radical formation.

1.3 Electron Spin Resonance: Basic Concepts

Electron spin resonance (ESR) spectroscopy is the branch of spectroscopy which studies the interaction between electronic magnetic

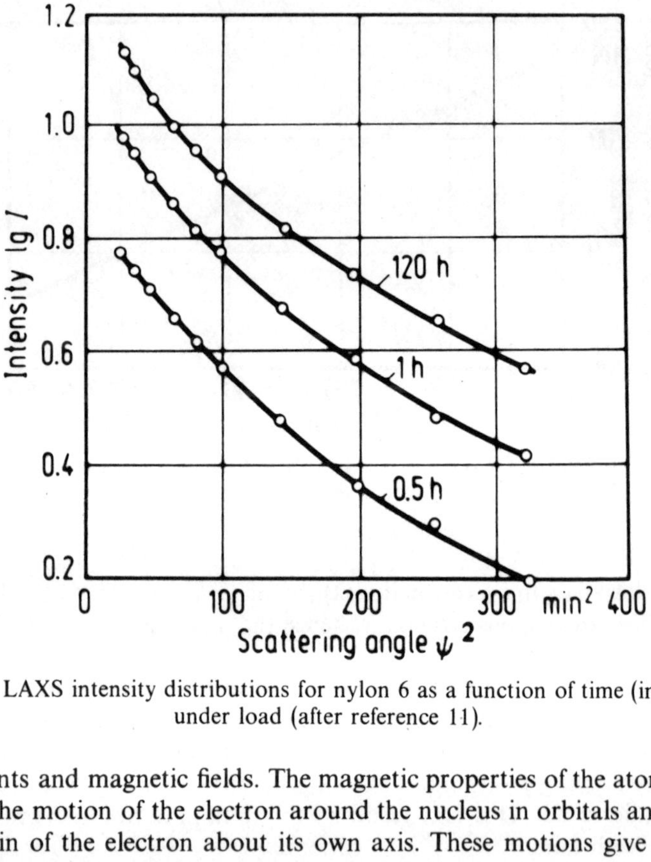

Fig. 3. LAXS intensity distributions for nylon 6 as a function of time (in hours) under load (after reference 11).

moments and magnetic fields. The magnetic properties of the atom arise from the motion of the electron around the nucleus in orbitals and from the spin of the electron about its own axis. These motions give rise to orbital and spin magnetic moments respectively. A net magnetic moment exists in atoms or molecules containing unpaired electrons and only then does the ESR technique become possible. Application of an external magnetic field to a paramagnetic substance causes a splitting of the electronic energy levels, a phenomenon known as the Zeeman effect.

It is not the purpose here to present a detailed account of ESR theory and technique. Many excellent texts cover the subject in considerable depth.[12,13] Our interest here is in the application of ESR to the study of molecular fracture in solid polymers. Generally the orbital and spin components of the magnetic moment are coupled, leading to a rather complex expression for the Zeeman energy levels. However, in solids the orbital and spin motions become decoupled, and the electron then behaves similarly to a free electron. Consequently it is usual to consider only a paramagnetic species containing a single unpaired electron, for which the magnetic quantum number (M_J) takes the values $\pm 1/2$. In this case, the energy difference between the Zeeman energy levels takes the value

$$\Delta E = g\beta H_0 \tag{4}$$

where g is the spectroscopic splitting factor, β the Bohr magneton and H_0 the applied magnetic field. The population difference between the two Zeeman energy levels is given by

$$\Delta N_0 = N_0 g\beta H_0/2kT \tag{5}$$

where N_0 is the number of unpaired electrons per unit mass, k is Boltzmann's constant and T the thermodynamic temperature. Hence ESR provides information on the number of radicals produced as a consequence of a particular process.

Electron spin resonance involving transitions between the Zeeman levels is achieved by applying electromagnetic radiation of frequency ω_0 such that $\Delta E = h\omega_0$, where h is Planck's constant. The resonance condition is then satisfied:

$$h\omega_0 = g\beta H_0 \tag{6}$$

The factor g takes a value close to 2 for the case of a free electron and h and β are true constants. In theory any combination of ω_0 and H_0 which satisfies the resonant condition is satisfactory. In practice ESR employs frequencies of 9–10 GHz with H_0 in the region of 3200 gauss for $g = 2$. The oscillating magnetic field required to induce the transitions between the Zeeman energy levels is therefore in the microwave region. Since the radiation emitted during transitions is polarised in a direction perpendicular to the main magnetic field (H_0), the oscillating magnetic field (H_1) is applied through a waveguide at right angles to the main field.

The main magnetic field is generated by large electromagnets. The dual requirement of a main magnetic field that is strong and truly orthogonal to the oscillating field severely restricts the sample size and the volume of the zone usable for *in-situ* experiments.

Energy is absorbed from the oscillating magnetic field H_1 at resonance. Spectrometers available maintain a constant microwave frequency (ω_0) and vary the frequency of the main magnetic field (H_0) to sweep through resonance. Such an absorption curve would appear as shown in Fig. 4(a). However, the net energy absorption arises from the difference in the upward and downward movements between the energy levels; i.e. the energy absorption results from the population difference between the two

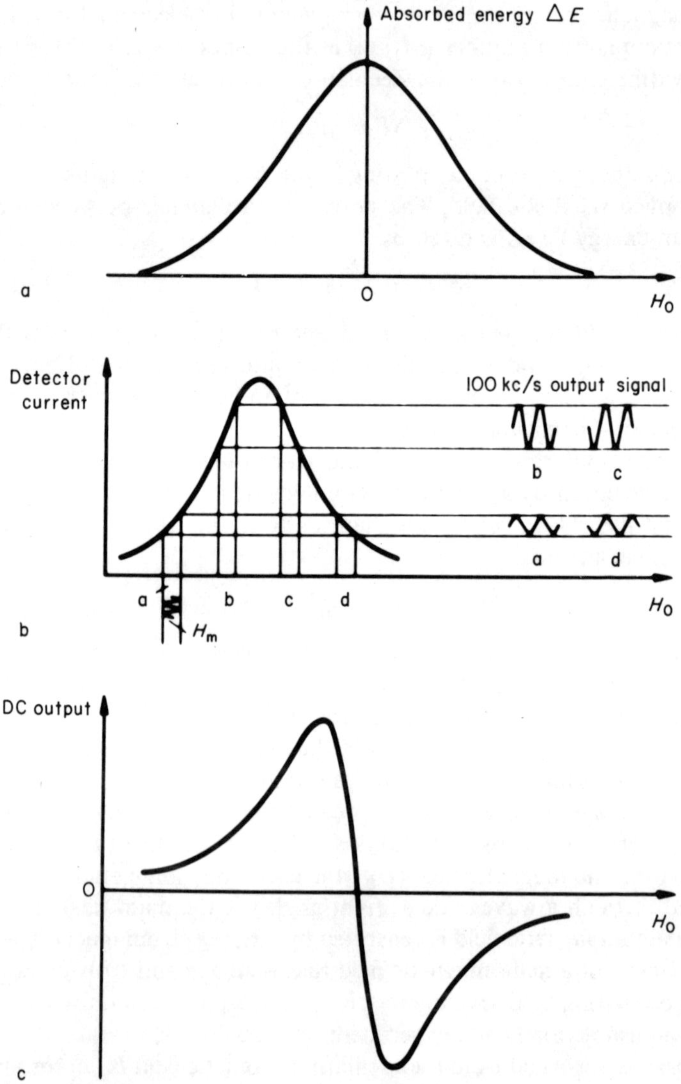

Fig. 4. (a) Singlet absorption curve; (b) Superimposition of modulation on main field to produce phase-sensitive a.c. output corresponding to first derivative of absorption curve; (c) first derivative spectrum for singlet.

energy levels, which is small ($\approx 0{\cdot}07\%$). This problem is overcome by superimposing a small modulating oscillatory field (H_m) on the main field (H_0). The effect, as shown in Fig. 4(b), is that the crystal detector of the microwave bridge responds to the first derivative of the absorption peak, rather than the direct absorption of energy. The a.c. output signal is readily amplified. The usual first-derivative spectrum arising from the absorption curve of Fig. 4(a) is shown in Fig. 4(c) after phase-sensitive detection. Most ESR spectra are presented as first-derivative curves.

An ESR spectrum is characterised by:

(a) the number of lines in the spectrum;
(b) the position of the lines, usually in terms of the g values as computed from the resonance equation;
(c) the separation of the lines (splitting) expressed in gauss;
(d) the relative intensities of different absorption lines;
(e) the line shape.

ESR spectroscopy was developed to study the interaction energies of a paramagnetic atom in a constant magnetic field. It has therefore been mainly used to characterise, rather than identify, paramagnetic species. ESR has been used in polymer studies to identify the radical species generated by various processes, to determine the number of radicals produced and to follow the decay of radicals with time and environment.

1.4 Paramagnetic Centre Formation in Polymers

Since polymers are co-valently bonded molecules formed mainly from elements in the first two periods of the periodic table, all the electrons in the molecule are paired and there is no resultant magnetic moment. It is necessary to break co-valent bonds to obtain unpaired electrons before ESR spectroscopy can be applied. ESR thus provides information on the number and sites of the broken bonds.

Co-valent bonds in polymers are broken during a number of processes, including thermal degradation, degradation by ultra-violet and γ-ray irradiation or by high-energy particle bombardment. Much of the early work on the irradiation of common polymers has been reviewed by Campbell.[14] Irradiation is not highly selective in the choice of initial radical sites and free radicals may be formed either as a result of main-chain cleavage, as in the case of polymethylmethacrylate[15–17] or by the removal of side-groups, as in the case of polyethylene[18] and polypropylene.[19]

More recently attention has been turned to the production of macro-radicals during the mechanical destruction of polymers. Polymers have been ground, milled and crushed in a variety of ways, under inert atmospheres, vacuum or liquid nitrogen.[20–22] Free radicals are also produced in uniaxially stressed fibres, nylon fibres being commonly used for these studies.[23,24] A review and summary of the radicals identified from the mechanical degradation of polymers has been presented by Kausch.[25] Kausch tabulated the assignment of spectra of free radicals formed in a wide range of polymers through degradation by milling and also those formed in stressed fibres and filaments; Campbell reviewed the free radicals formed by irradiation and Kausch those formed from mechanical degradation.

1.5 Limitations on the Application of ESR to Polymer Studies

Since ESR can be used to monitor molecular rupture, it would seem to be a natural technique to use to monitor fracture at the molecular level in a wide range of mechanical testing events. However, it has been applied mainly to study molecular fracture during the rather extraordinary events of grinding or milling plastics and straining highly oriented fibres, often at temperatures around 77 K. De Vries *et al.*[26] sought to apply ESR to the impact fracture of PMMA, but were unable to obtain a spectrum, even though it was obvious that molecules were being ruptured in the process.

The three major limitations on the application of ESR to the general study of fracture in polymers concern:

(a) the sensitivity of existing spectrometers;
(b) the stability of the radicals formed;
(c) the volume of material which can be placed in the cavity of the spectrometer, or the space available within the spectrometer for *in-situ* experiments.

The power absorbed at resonance, and hence the magnitude of the absorption peak, is proportional to the population difference between the two Zeeman energy levels, which is typically only about 0·07% of the total population. Consequently a large population is necessary to obtain a measurable signal. Present spectrometers have sensitivities which require minimum concentrations of 10^{11}–10^{13} spins per gram. While the size of the absorption peak can be increased by increasing the microwave power, this can lead to distortion of the spectrum. Consequently most studies on solid polymers have been carried out at low microwave power (typically 0·01–0·1 mW), which further reduces the sensitivity of the technique. To obtain manageable signals, it is therefore necessary to have a large spin concentration. This requirement is satisfied in irradiated specimens, in ground samples (where the fracture surface area is large) and in tensile-deformed, highly oriented fibres. Hence the extensive use of these techniques in the application of ESR to fracture studies. Backman and De Vries[27] ground and sliced samples of nylon 66, polyethylene and polypropylene at liquid nitrogen temperatures and found that the radical concentration increased proportionally to the fracture area formed. However, it was found that the number of broken bonds per unit area of surface formed was an order of magnitude less than the calculated number of molecules passing through the same area. An 'easy path' fracture mechanism was suggested to explain this deficiency.

Free radicals formed by any technique are highly reactive and seek to react with surrounding matter to eliminate themselves. De Vries *et al.*[28] mention that the radical concentration is essentially independent of temperature up to approximately the glass transition temperature (T_g) of the material, but decreases rapidly above this temperature. Mead and Reed[29,30] have measured the decay rate of radicals in tensile-deformed oriented polybutadiene and found that the radicals were stable for several days at liquid nitrogen temperature, but decayed with increasing rapidity as the temperature was raised through the range from 77 K to T_g. The decay rate curves are shown in Fig. 5. While the decay rate is very small between 77 K and 123 K, at temperatures approaching T_g (approximately 170 K) the radicals decay rapidly. Since it requires considerable time to tune the microwave bridge once the specimen is placed in the cavity, and further time to record the spectrum, it is essential to work with stable free radicals. Hence it is usual to work with polymers below their T_g and often at 77 K, using low-temperature attachments on the ESR spectrometer. Since the population difference between the Zeeman energy levels is inversely proportional to temperature, it is beneficial to operate at low temperature from sensitivity considerations. However, relaxation processes may be adversely affected, necessitating a reduction in microwave power to avoid broadening and a consequential reduction in sensitivity.

The third limitation affecting the application of ESR to fracture studies is the size of the working zone within the spectrometer. Figure 6 shows a dimensionally scaled sectional view of the standard cavity for a Varian E–9 spectrometer fitted with low-temperature attachment glassware. Low temperatures are obtained by passing a stream of nitrogen gas through a

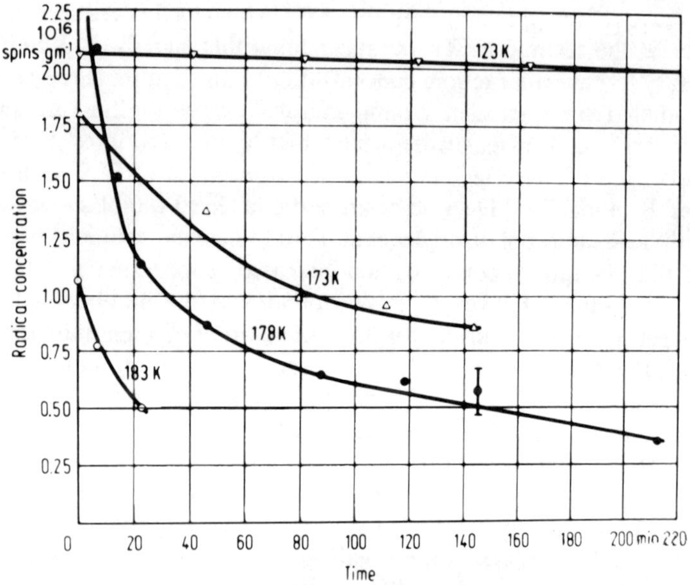

FIG. 5. Decay of radical concentration at four different temperatures in *cis*-1,4-polybutadiene after tensile testing at 83 K.

liquid nitrogen bath and over a thermostatically controlled heater before entering the sample zone. All glassware used in the cavity has to be spectroscopically pure. As can be seen from Fig. 6 the working zone is a mere 4 mm in diameter by 20 mm in length. When the sample has to be placed in a tube, the diameter is reduced to 2 mm. Clearly the type of sample which can be accommodated in the standard cavity is limited. Liquids or powders containing a large radical concentration per unit volume are ideal, which again favours the grinding and milling techniques.

It is important to recognise that all free radicals within the cavity react with the magnetic field, and the spectra and radical concentrations determined are averages for the whole of the specimen within the cavity. ESR spectrometry cannot select areas for investigation as can, say, various types of microscopy. For example ESR has been used to monitor molecular fracture during crazing of polybutadiene at low temperatures.[31] While localised craze bands can be visually observed, the spin concentration can only be averaged over the whole length of the

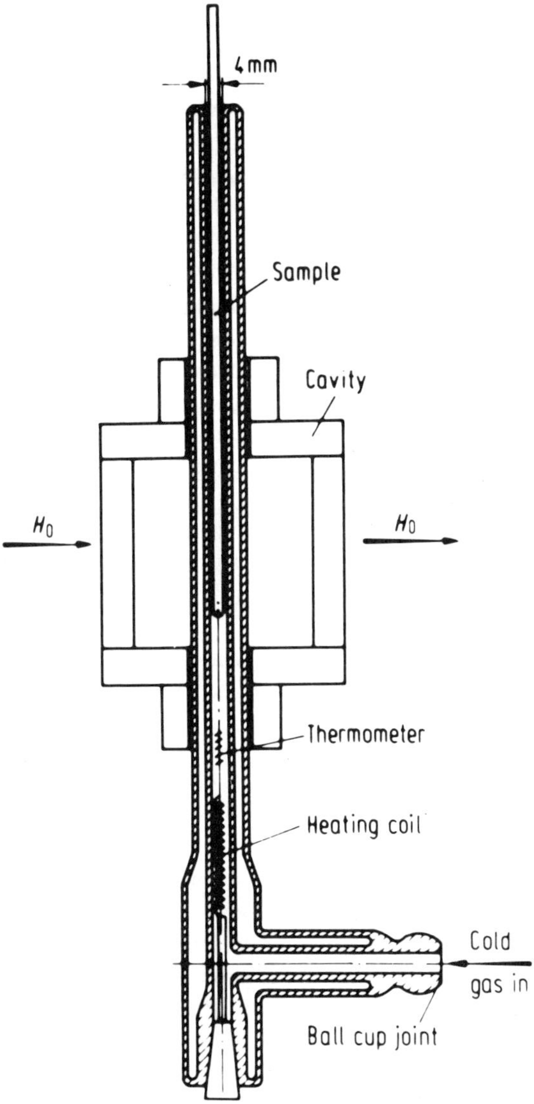

FIG. 6. Glass insert arrangement for low-temperature operation in standard cavity of a Varian E–9 spectrometer.

specimen. It is therefore not possible to determine the radical concentration locally around any particular craze band.

2 MECHANICAL AND PHYSICAL PROCESSES

In this section we examine further some of the results obtained using the techniques reviewed in Section 1. Certain results and their implications have already been indicated for some of the experimental approaches discussed above, and here we shall concentrate on ESR data and the associated mechanical and physical phenomena.

The ultimate purpose of this discussion is to see how far macroscopic deformation and fracture phenomena can be interpreted in molecular terms. The techniques reviewed earlier, and especially that of ESR, provide direct evidence of molecular processes and we wish to see if this evidence is consistent with their playing a dominant role in determining macroscopic properties.

We shall see that the situation is far from simple and that the data on molecular processes cannot be used to make quantitative predictions about macroscopic deformation and failure properties.

2.1 Factors Controlling Molecular Fracture

The approach taken in this section will be to consider in turn the empirical factors which govern the occurrence of molecular fracture and its consequences. The incidence of main-chain fracture in polymers under tensile stress is dependent on a wide range of variables, including polymer chemical structure, physical structure or morphology, additives, environment, time, temperature, orientation and physical state, as well as the more obvious variables of stress and strain.

It is helpful to think of main-chain fracture as a competitive response to deformation; if the deformation can be accommodated more readily by molecular flow or by crystallographic processes, then main-chain fracture will be less evident. This competitive situation underlies many of the phenomena reported in what follows.

2.2 The Importance of Molecular Anchorage

Polymers share with metals the general ability to undergo large deformations before fracturing. This is obvious in the case of elastomers but is equally important in tough engineering plastics which yield and flow under stress. These large deformations involve the uncoiling or straightening of the polymer molecules. In general, this is not a process which depends upon molecular fracture, though molecules may break as a consequence of such deformations. Exceptions to this statement can be found, however, as we shall see.

On the other hand, molecular fracture becomes dominant if molecules are unable to flow because of structural constraints. These constraints may manifest themselves at the outset of a test or may emerge only after a significant deformation has already taken place, and after the molecules have been oriented along the line of action of the applied force.

Any structural feature which prevents mutual molecular flow can be regarded as a molecular anchorage. Experience shows that extensive molecular fracture (i.e. sufficient for detection by ESR spectrometry) can only be induced if such features are present. This does not mean that macroscopic fracture involves no molecular breakage in other cases. Main-chain fracture limited to the tip zone of a single propagating crack, for example, would be undetectable by ESR. Without anchorages, however, it is impossible to induce distributed fracture of molecules throughout a specimen under tensile stress, which alone gives rise to detectable ESR signals.

Molecular anchorages are of two main types: crystalline regions and chemical cross-links (to which physical entanglements may also contribute in polymer networks). Not surprisingly, then, most ESR studies of polymers under tensile stress have been carried out using crystalline fibres (nylon, PET, PE or PP, for example) or cross-linked polymers.[30,32,33]

2.3 ESR Signals from Uncross-linked and Amorphous Polymers

Before considering data from crystalline and cross-linked polymers it should be emphasised that molecular fracture can be observed in such materials as glassy thermoplastics if these are ground or macerated.[20–22] By this means relatively large areas of fracture surface can be generated. Tensile tests on uncross-linked glasses, by contrast, generate only small areas of new surface and do not generally give detectable ESR signals.

Brittle fracture in glassy polymers occurs by the prior formation of crazes[34] which then fracture. In the craze, fibrils of highly oriented polymer are produced and it is the fracture of these fibrils that almost certainly produces the observed radicals. Even in this case, then, the signal derives ultimately from a fibre in which all potential for flow has been exhausted, leaving main-chain fracture as the only accessible deformation mechanism. It is possible that molecular fracture is also involved in the initial cavitation which gives rise to crazing, but this is much less certain for thermoplastic glasses.

Ironically, the results from grinding may have more direct relevance to macroscopic tensile fracture than the ESR data from tensile tests. This is because grinding does produce signals from fracture surfaces whereas tensile testing produces signals from distributable events which may or may not be associated with crack initiation and are unlikely to have any bearing on crack propagation. Unfortunately, grinding experiments are not susceptible of quantitative interpretation since neither the surface generated nor the mechanical forces involved are accurately measured.

Indirect methods of assessing the incidence of molecular fracture in mechano-chemical processes generally have been reviewed by Casale *et al.*[2] One method, for example, is the determination of molecular weight changes in macerated materials. Direct measurements by ESR have been carried out by a number of investigators.[27,32] Backman and De Vries, using several crystalline polymers cut up under liquid nitrogen, estimated that the average fracture surface involves breakage of between 5% and 10% of all molecules intersecting the surface. Salloum and Eckert,[5] using drilled nylon, polyethylene and polypropylene, confirmed these results for nylon and polyethylene but obtained a much stronger ESR signal from polypropylene. But Pazonyi *et al.*, using an indicator to measure the radical concentrations in plasticised PVC and PE cut under ethanol, obtained results which suggest breakage of between 34% and 120% of the chains intersecting a plane in the polymer.[4] Values in excess of 100% are not impossible since molecular fracture occurs not in a plane but in a layer of finite thickness. (Backman and De Vries estimated a layer thickness as large as 6 mm within which 50% of the radicals were produced.) Further work is clearly required to resolve this matter.

2.4 Crystalline Fibre Structures

The most extensive ESR studies of polymers under tensile load have undoubtedly been carried out on drawn crystalline fibres, and this work has been reviewed recently by Kausch and De Vries.[35] It is clear that the morphologies of oriented crystalline fibres, with extensive tie-chain populations and high degrees of molecular uncoiling, strongly favour the incidence of molecular fracture under tensile stress.

In spite of this, and somewhat surprisingly, some of the more obvious experiments have not been carried out. In particular there appear to be no data relating ESR intensities to the degree of crystallinity or other measurable variables by which morphology might be characterised.

Some data reflecting morphological effects have been obtained by annealing experiments on fibres of nylon 6.[36,37]

Lloyd *et al.* found that the histogram of radical concentration *versus* strain was broadened considerably by annealing in a slack condition at 199°C (472 K), and shifted in the direction of higher strain by about 4%. Interpretation is complicated by the fact that annealing produced a shrinkage of 20% but the results do suggest that annealing reduces the homogeneity of the tie-chain length distribution.

The maximum intensity of the ESR signal was also modified by annealing, the results being profoundly affected by the strain under which the specimens were held during annealing. If the maximum radical concentration is plotted against the observed fracture stress for the specimens, Fig. 7 results. The correlation between radical concentration and fracture stress for a given condition may indicate nothing more than that more molecules can be broken if the material sustains higher loads before fracture. The difference between specimens annealed at constant length (0% strain) or while slack, and those annealed under 11·7% tensile strain is, however, more significant. It suggests that specimens annealed under positive strain suffer a loss of breakable tie molecules. This could be due to a number of causes, ranging from thermal fracture of up to three-quarters of the potentially breakable molecules to relaxations of such molecules to produce a 50% reduction in the local strain levels associated

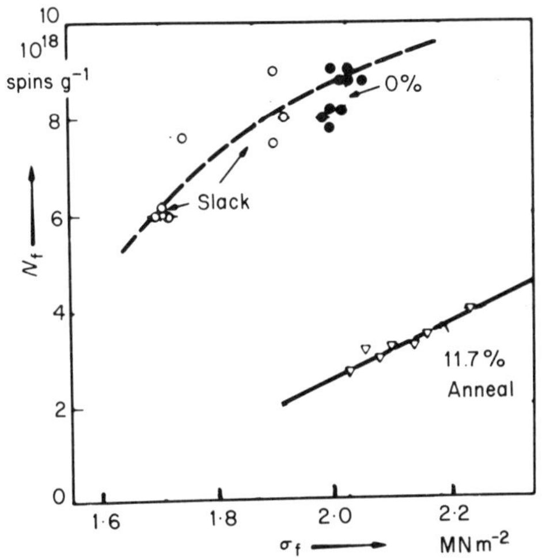

FIG. 7. Concentration of radicals at fracture in nylon 6 as a function of tensile stress for different annealing conditions. Open circles: annealed while slack; closed circles: annealed under 0% strain; triangles: annealed under 11·7% strain.

with a given test strain on the specimen. Whilst it is not possible to draw more precise conclusions, these data do indicate clearly that quite subtle morphological factors can greatly affect the concentration of radicals observed in tensile testing.

2.5 Glassy Polymers and the Role of Cross-links

Highly drawn crystalline fibres produce abundant radicals at strains exceeding about 8% and rising to the breaking strain of, say, 15 to 20%. Such fibres have, of course, already undergone draw ratios of 500% or more from the isotropic condition.

Polymeric glasses, in contrast, have to be deformed beyond the elastic region, i.e. beyond yield, before distributed molecular fracture occurs to a degree detectable by ESR. Cold-drawing polycarbonate gave radical concentrations in the range 0·5 to 1·5 × 10^{16} spins/g, the concentration rising with strain rate in a way that suggests a direct dependence on the flow stress.[38,39]

No detectable signals can be obtained in the brittle fracture of isotropic polymers (even crystalline ones), but once plastic flow is induced (e.g. by small amounts of orientation above T_g before testing below T_g), radicals are readily obtained during deformation.

The radical concentration is strongly dependent on the presence of molecular anchorages, notably cross-links. Mead[29] studied cross-linked polybutadienes deformed plastically below T_g. Pre-strains of different magnitudes were applied above T_g before testing, but comment here is limited to pre-strains below 250% where no strain-induced crystallisation occurs. Although weak signals ($\approx$ 0 to 8 × 10^{14} spins/g) were recorded in uncross-linked material beyond yield, this radical concentration was two orders of magnitude smaller than that obtained on even lightly cross-linked specimens. Figure 8 shows the effect of cross-link density, for specimens pre-extended by 200%, on the radical concentration after a further 200% test strain and at various temperatures of test. The effects of temperature and strain will be discussed later. The point to be noticed here is the major influence of the initial introduction of cross-links; raising the cross-linking from 0 to 10^{-4} mol cm^{-3} of network chains produces a hundredfold increase in radical concentration whilst the further rise from 10^{-4} to 10^{-3} causes only a two- to threefold enhancement of molecular fracture events. This suggests that the major role of cross-linking is to prevent total flow of a molecule relative to its neighbours. This requirement is achieved as soon as there are two cross-links to each molecule, since this is sufficient to establish a continuous network.

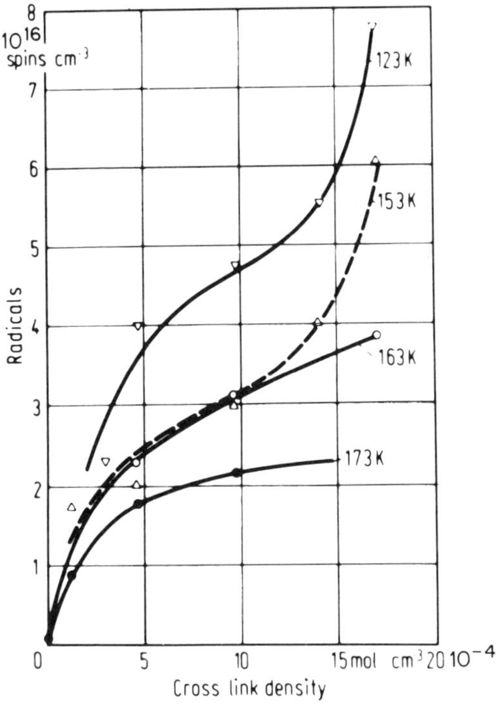

FIG. 8. Effect of temperature and cross-link density (expressed as mol cm^{-3} of network chains) on radical production at 200% test strain after 200% pre-strain above T_g.

Another way to represent an incipient network, or gel point, would be to equate the length of a network chain to the average length of the original molecule, and this would give a network chain concentration of roughly 2×10^{-5} mol cm^{-3}. This is just the range in which rapid increase in radical formation is observed.

At lower temperatures (i.e. more than about 40°C below T_g) a further upturn in radical formation is observed as the cross-link density rises above 10^{-3} mol cm^{-3}. This may reflect the higher stresses required to induce flow in the glassy material at these temperatures, suggesting that molecules are breaking before they become fully straightened, as a consequence of intermolecular friction.

A similar dependence of radical concentration on cross-link density was obtained for *cis*-polyisoprene tested below T_g by Natarajan and Reed.[33]

2.6 The Role of Chemical Structure

The maximum radical concentration observed in cross-linked polybutadiene tested in tension below T_g is of the order of 4×10^{17} spins/g and for *cis*-polyisoprene, which may be more highly crystalline at large pre-extensions, of the order of 11×10^{17} spins/g. Chiang and Sibilia[40] have reported $1{\cdot}5 \times 10^{17}$ spins/g for PET and $0{\cdot}76 \times 10^{17}$ spins/g for nylon 6 at their ultimate elongation. Pazonyi *et al.*[4] recorded a maximum of 8×10^{17} spins/cm^3 for nylon 6 and Becht and Fischer[3] obtained 10^{18} spins/cm^3 for the same material at 17·4% strain. Other polymers commonly give significantly lower maximum radical concentrations, thus:

PMMA, PS	$< 10^{14}$ spins/cm^3	(24,41)
polycarbonate	$1{\cdot}5 \times 10^{16}$ spins/cm^3	(38,39)
PE	5×10^{16} spins/cm^3	(41)
nylon 66	10^{17} spins/cm^3	(42)

Natural silk[41] has a high value of 7×10^{17} spins/cm^3. From these data it is clear that very similar amounts of molecular fracture can be obtained in polymers with widely different chemical structures, values approaching 10^{18} spins/g being recorded for *cis*-PI, *cis*-PBD, nylon 6 and natural silk under appropriate conditions. Secondly, the lower values obtained for other polymers seem to relate more to their morphology or state than to their chemical structure (e.g. the very low figures for glassy polymers).

It appears, therefore, that chemical structure as such has little bearing on whether or not molecules can be induced to fracture under tensile stress. This is determined rather by the amount of strain that can be imposed before fracture, by temperature and by the physical and morphological characteristics of the material.

The chemical structure of the molecules does, of course, profoundly affect the types of radicals which are formed. This subject is dealt with by the present authors elsewhere.[53]

2.7 The Roles of Stress and Strain

Our discussion of the kinetic theory of fracture in Section 1 has already indicated the manner in which applied stress can bring a net accumulation of molecular breakages in a polymeric solid. Since the stress is continuous throughout a specimen loaded in tension, these breakages are distributed throughout the material and can be detected by ESR spectroscopy in terms of a volume concentration of free radicals.

The basic statement of the kinetic theory of fracture (see Chapter 1) is as follows,

$$\nu = \nu_0 \exp\{(-G_{AB} + s\beta\sigma)/kT\} \tag{7}$$

where the applied stress σ accelerates the incidence of main-chain fracture via the term $s\beta\sigma$, s being a stress concentration factor and β an activation volume. This equation describes the probability of fracture of a single bond. It can be applied to the rate of fracture of a population of bonds provided that the reverse reaction (i.e. recombination) has a negligible probability. This will be true if free radicals are separated physically by molecular relaxation or are 'neutralised' by reaction with oxygen or have a reverse activation energy G_{BA} which is large compared with the stress-modified energy barrier $(G_{AB} - s\beta\sigma)$.

How far does eqn. (7) describe accurately the direct observations of ESR? The answer is not completely straightforward.

Equation (7) implies that, for small populations of broken molecules, the rate of radical formation, dN/dt, should be constant at constant stress. This was at first thought to be the case from the data obtained in constant loading-rate tests[41] but it was soon found[43] that stepwise loading tests told a different story. Figure 9 shows data of Becht and Fischer[3] for the stepwise loading of nylon 6. As previously observed by Roylance *et al.*,[43] the rate of radical production under steady strain decreases to zero after a period of minutes (or even seconds) although stress relaxation is only of the order of 10%.

This dilemma is solved by considering in greater detail the actual morphology of the molecular network. Breakable molecular chains are considered to have a distribution of lengths so that the stress is not evenly distributed among load-bearing chains. We have already seen direct evidence of this from the infra-red studies of Zhurkov *et al.*[8] At a given applied stress, the fraction of load-bearing chains under highest local stress do, indeed, fracture according to eqn. (7), but this results in a local redistribution of stress which is more uniform than before. Although the average molecular stress increases as the shorter chains break, the maximum molecular stresses encountered are considerably reduced and the rate of fracture drops accordingly. Significant rates of new radical formation can only be restored by increasing the applied stress, thus repopulating the highest local stress states. The process is then repeated.

Histograms from stepwise straining tests, reflecting as they do the (apparent) length distribution of breakable molecules, are more informative than most other means of displaying ESR data. For example,

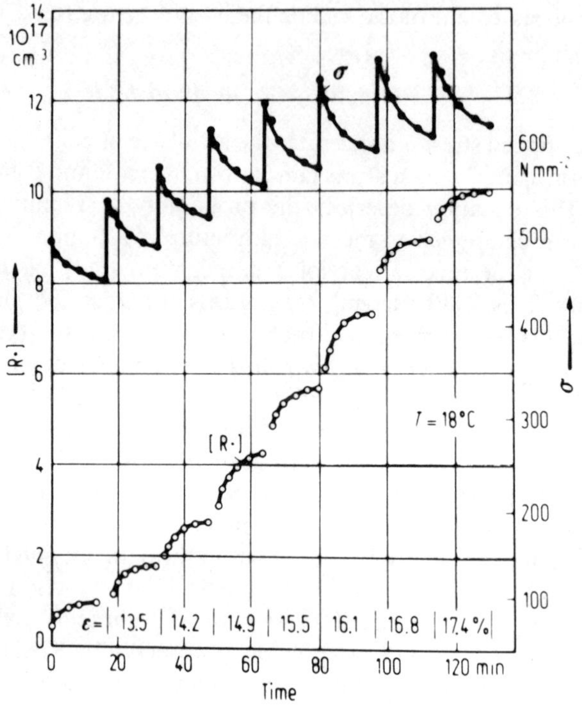

FIG. 9. Radical concentration (○) and stress (●) as functions of time and strain in step-strain tests on nylon 6 (after reference 3).

Hassell[44] has obtained an interesting correlation between the width of the apparent chain-length distribution and the ultimate tensile strength of the fibres concerned. This has not been interpreted[32] as a direct causal relationship, because of the small concentration of molecular fractures relative to load-bearing chains; rather it has been interpreted as reflecting differences in micro-structure, which itself controls macroscopic strength. A more direct interpretation should not, however, be ruled out.

2.8 The Effects of Pre-strain and Plastic Strain

Highly oriented fibres produce ESR signals under tensile deformation in the ranges of 8 to 16% strain and 500 to 900 MN m^{-2} stress. They do not undergo yield and plastic flow for the simple reason that the molecules are already highly aligned in manufacture and their capacity for plastic flow has already been exhausted. Such fibres therefore represent one extreme of the various conditions in which polymeric solids are obtainable.

Cross-linked elastomers (the other main class of polymers studied by ESR) can be pre-strained above T_g to any desired extent (up to fracture) and their orientation stabilised by cooling below T_g before testing. At high pre-strains, strain-induced crystallisation may occur providing a morphology essentially similar to that of synthetic fibres.[45] Cross-linked polymers may therefore be used to explore in a systematic manner the role of strain and orientation in molecular fracture.

The variables to be considered here are pre-orientation, test strain and temperature. The last-named is also considered in a later section but cannot be omitted here because of its interaction with the strain variables. We must also consider the different categories of mechanical behaviour observed in these investigations.

Basically, four types of tensile stress-strain curve are found for cross-linked elastomers deformed below T_g. These are shown schematically in Fig. 10 where pre-orientation is indicated by the point on the abscissa at which the curve begins. For zero or low pre-orientation (curve A), brittle behaviour normally results and no detectable ESR signal is generated. At intermediate pre-orientations (say, 100–300% according to cross-link density, temperature and strain-rate) two types of ductile behaviour can occur. Curve B shows yield and plastic flow accompanied by visible whitening of the specimen and the absorption of environmental gases. Here the primary mode of deformation is crazing which begins in distinct craze bands perpendicular to the strain axis but spreads progressively

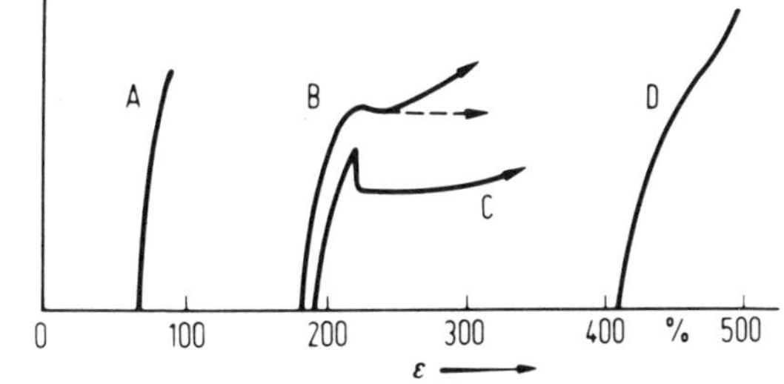

FIG. 10. Types of variation of strain (ε) *versus* stress (σ) observed in cross-linked elastomers below T_g. Origins of curves indicate amount of pre-extension (schematic). Arrows indicate that curves continue.

throughout the whole specimen. In contrast, 'normal' ductile behaviour without crazing (curve C) is also found in this pre-orientation region. Figure 11 shows how the test temperature determines whether crazing or non-crazing ductility occurs in 100% pre-oriented *cis*-polyisoprene. Normal behaviour without crazing is observed at temperatures just below T_g and here the radical concentration is relatively low and short-lived.[33] A ductile-brittle transition occurs as temperature is further reduced but is followed by renewed ductility, now with crazing and strong, long-lived ESR signals.

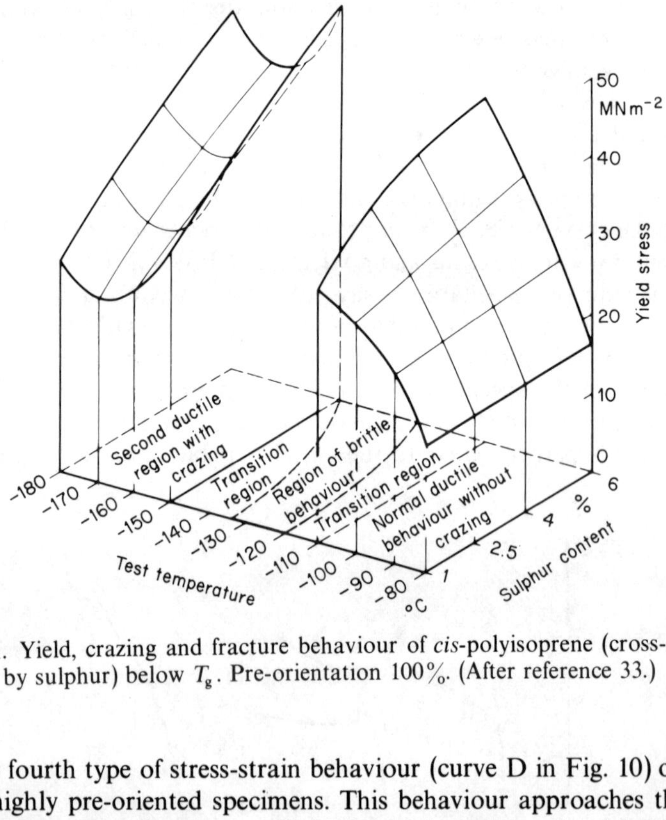

FIG. 11. Yield, crazing and fracture behaviour of *cis*-polyisoprene (cross-linked by sulphur) below T_g. Pre-orientation 100%. (After reference 33.)

The fourth type of stress-strain behaviour (curve D in Fig. 10) occurs with highly pre-oriented specimens. This behaviour approaches that of drawn fibres, though the crystallinity in polyisoprene and polybutadiene is considerably less than in, say, nylon 6. In this region strong ESR signals are generated by tensile testing below T_g but without visual evidence of crazing or voiding and without gas absorption from the environment. It is our view that the phenomena found with specimens of type D are qualitatively similar to those found in drawn crystalline fibres.

The effect of pre-orientation on the radical concentration at rupture in *cis*-PB is illustrated by Fig. 12 (after Mead[29]). It will be seen at once that

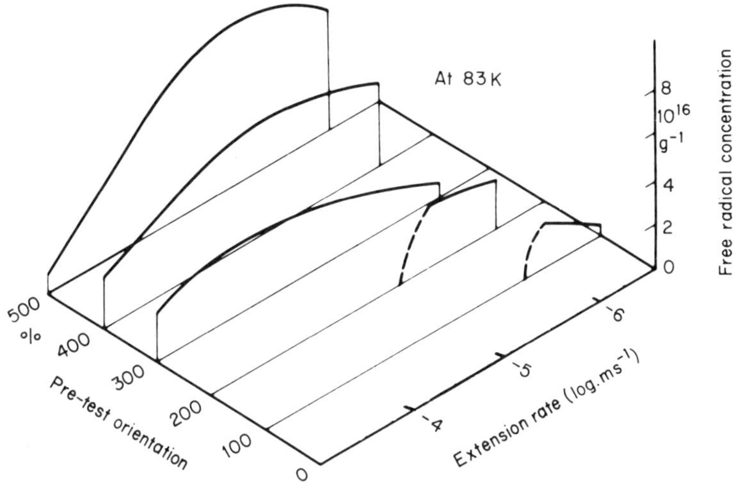

FIG. 12. Effects of pre-orientation and rate of testing on the maximum radical concentration observed in *cis*-polybutadiene tested at 83 K. (After reference 29.)

the picture is complicated by a strain-rate dependence. Thus whilst radical concentration generally increases with pre-orientation as expected, at the highest strain-rates it actually decreases with pre-orientation and this occurs even though the fracture strain is still rising with increasing pre-orientation. These effects can be explained qualitatively by supposing that higher pre-orientations introduce more molecular chains extended to near their breaking point and so, generally, cause an increase in molecular fracture events. If, however, such fracture takes place, not spontaneously, but by a mechanism of microvoid growth (see below) the time for such growth may not be sufficient at high strain rates to effect the rupture of all the potentially breakable chains. This accounts for a decrease in radical formation with increasing strain rate. The decrease with increasing pre-orientation at high rates requires a further supposition that microvoid formation is delayed in highly pre-oriented materials. This may be because the molecular stress distribution is more uniform as a result of shorter chains having already fractured during pre-orientation above T_g. (Evidence of network breakdown in highly strained cross-linked systems is considerable.[46])

The effect of increasing test strain on radical concentration is evident. Figure 9 for crystalline fibres has already indicated an almost linear dependence of radical concentration on test strain, ε, between 10 and 20%, $dN/d\varepsilon$ being 24×10^{18} spins/cm^3. In the plastic deformation of cross-linked glasses, where the test strains are, of course, very much larger, roughly linear plots are again observed though there is a consistent increase of $dN/d\varepsilon$ with increasing strain. The measured values for $dN/d\varepsilon$ for plastic deformation of *cis*-polybutadiene[33] range from about 8×10^{17} for a pre-extension of 400%, a low temperature and a high cross-link density ($1{\cdot}7 \times 10^{-3}$ mol cm^{-3}) to zero for low pre-orientations and higher temperatures. The effect of temperature is illustrated in Fig. 8.

For plastic flow without significant strain hardening, the value of $dN/d\varepsilon$ is two to three orders of magnitude smaller than that for drawn fibres and only begins to approach the latter value at the highest pre-orientations. It is clear, then, that the efficiency of strain in causing molecular fracture is highly dependent on the degree of initial molecular orientation and this can be interpreted in terms of the population of nearly-extended chains at the outset of the test. The effect of cross-link density and the increase in $dN/d\varepsilon$ as plastic flow proceeds to high levels of strain, are consistent with this view. It should also be noted that breaking stresses in the crystalline fibres cited are of the order of 700 MN m^{-2} compared with 200 to 300 MN m^{-2} (true stress) for the plastically deformed polymers.

2.9 The Effects of Strain Rate

The effect of strain rate on maximum radical production in drawn fibres is not great,[47] although we have already noted that under steady load the number of radicals increases with time for periods of up to 20 minutes. In the plastic deformation of polycarbonate also, radical concentration increased only twofold over two decades of strain rate.[38,39] Much larger effects have been observed in plastically deformed, cross-linked glasses, as Fig. 12 shows. Here, for *cis*-polybutadiene at 500% pre-strain, the radical concentration at fracture has a peak value of 7×10^{16} spins/g at intermediate strain rates, and decreases to 5×10^{16} at the lowest and 1×10^{16} at the highest strain rates employed (a range of strain rates of about four decades).[29] Even more dramatic effects were obtained[48] for radical concentration in cross-linked polychloroprene (below T_g) at a fixed test strain (Fig. 13). In this case the number of radicals formed seemed to correlate with the amount of visual crazing. It is likely that the relatively strong

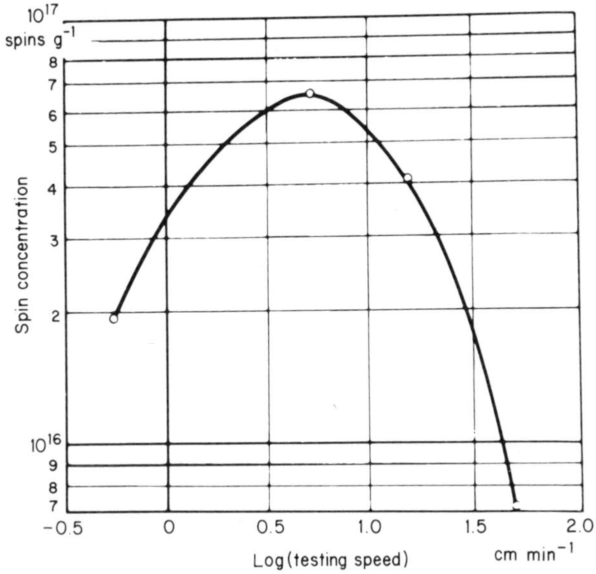

FIG. 13. Effect of testing speed on radical concentration in polychloroprene at 123 K. Pre-orientation 300%. (After reference 48.)

influence of strain rate in plastically deforming specimens is closely related to the growth of crazes or sub-microscopic voids. At very low strain rates molecular fracture may be reduced by molecular flow. As the rate increases so also does the flow or yield stress giving rise to a higher incidence of fracture. As strain rate further increases, however, crazing is suppressed (see Fig. 12) because the necessary time for extensive void growth is not available and molecular fracture associated with such growth is progressively reduced.

2.10 The Effects of Temperature

The kinetic-theory equation for bond breakage, eqn. (7), obviously suggests that radical formation should depend strongly on the temperature of test. Several workers have demonstrated such effects in the tensile loading of drawn fibres. Roylance *et al.*[43] showed that the steady-state radical concentration N_{ss} should be related to the applied stress σ by

$$N_{ss} = \delta \exp \xi\sigma \tag{8}$$

where $\delta = \alpha \exp(-\gamma/kT)$ and $\xi = \beta'/kT$, and α, β' and γ are constants for the material. Plots of ξ *vs* T^{-1} and of $\log \delta$ *vs* T^{-1} for drawn nylon 6 fibres (Fig. 14) confirmed the predicted temperature dependencies.

Johnson and Klinkenberg[49] carried out step-temperature changes on strained polyamide fibres and found the expected increase in radical concentration with increasing temperature. The temperature affects not only the total radical production but also the apparent distribution of breakable chain lengths, as the histograms of Fig. 15 show. It appears that testing at higher temperature permits higher breaking strains and thus a broader distribution of chain lengths available for rupture. At the same time chain population at a given length is greatly diminished. This may be due partly to accelerated radical decay at higher temperatures (see below) which may also offset the expected rise in radical concentration with increasing temperature.

The results on plastic deformation of cross-linked glasses appear at first to contradict those discussed above for crystalline fibres. Fig. 16 shows the effect of temperature on the plot of radical concentration *versus* strain for *cis*-polybutadiene.[29] The number of radicals increases with decreasing temperature, instead of decreasing as might be expected. This effect may be attributable to radical decay during tests, but other evidence indicates that secondary radicals at least are long-lived below 123 K. The more likely explanation is that the flow or yield stress increases rapidly as the temperature falls, so that at 93 K the molecules are subject to much higher shear stresses than at 123 K. The ratio of radical concentration to yield stress appears to decrease with decreasing temperature in harmony with the expectations of the kinetic theory.

2.11 Crazes and Microvoids

Our final topic in this Section concerns the physical process of void growth which is so often associated with the observation of ESR signals in the tensile testing of polymers. We shall also refer at this point to the important effects of the environment on these processes. This topic is particularly important since it is only as we consider the growth of cavities that our discussion of distributed molecular fracture converges with the subject of macroscopic fracture.

Zhurkov *et al.*[11] used light scattering and low angle X-ray scattering (LAXS) to study the time-dependent formation of microcracks or cavities in stressed polymers. The concentrations of microcavities were found to be as high as 10^{15}–10^{16} cm^{-3} uniformly distributed throughout the specimen, though more numerous close to the specimen surface.

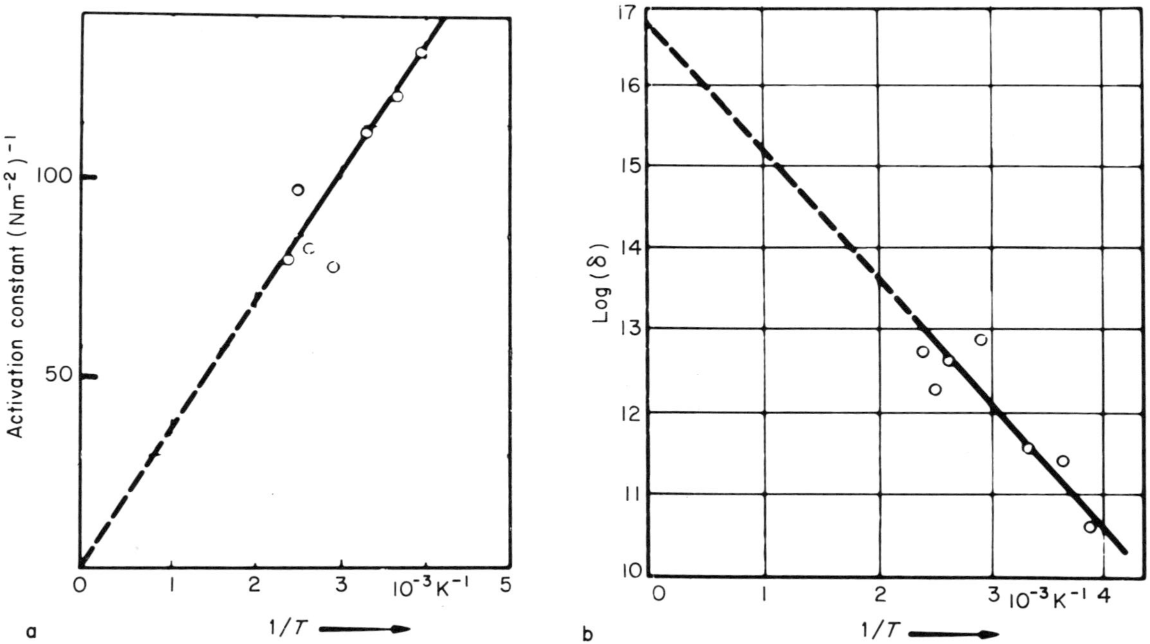

FIG. 14. Effect of temperature on (a) the activation constant ζ and (b) the pre-exponential constant δ of equation (8). (After reference 43.)

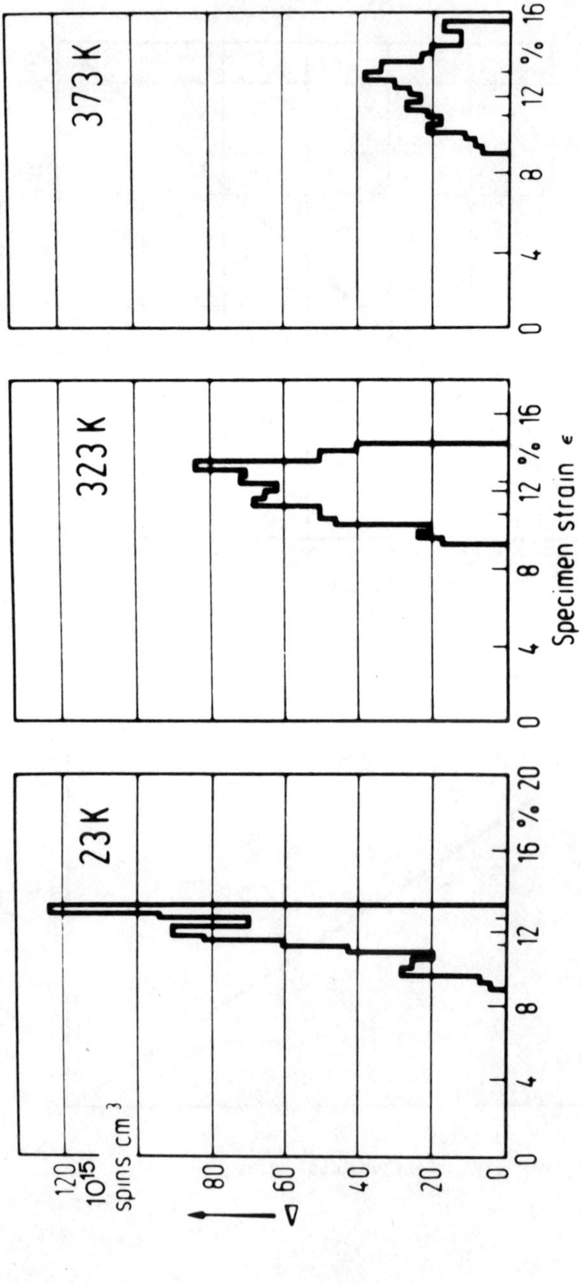

FIG. 15. Histograms of spin increments against strain for nylon 6 fibres, showing effects of temperature (after reference 36.)

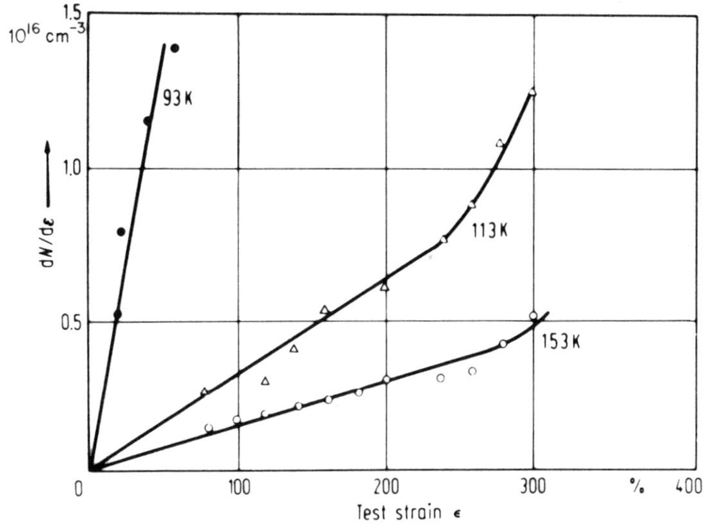

FIG. 16. Effect of temperature of test on the curve of rate of radical formation, dN/dε, *versus* strain for cross-linked *cis*-polybutadiene. Pre-extension 200% (After reference 29.)

Figure 17 shows how the number of microvoids increased with time under load in a PVC film, though a similar behaviour was found in nylon 6. The initial rate of formation of voids, dN/dt, was logarithmic with tensile stress. A further important observation was that the critical concentration of cavities just before macroscopic failure appeared to be constant for specimens whose lifetimes under load varied over four decades of time (according to the applied load). The same critical concentration of microvoids was found close to the tip of a propagating crack.

The microvoid dimensions were such that the ratio of their size (L) to their mean separation (R) was between 0·25 and 0·5. The cavities appeared to be disc-shaped with their planes perpendicular to the stress axis and had an average size around 300 Å. 'Small' and 'large' microvoids could be recognised in the scattering patterns, and whilst most were small, the number of 'large' voids increased with time.

In further studies Zhurkov *et al.*[50] measured microvoid formation and bond rupture on the same specimens. Bond fracture was assessed using the infra-red technique described in Section 1.1 above. They obtained identical behaviour for the kinetics of bond rupture and cavity accumulation, both in respect of time and stress dependence. Their results are summarised in Table 2.

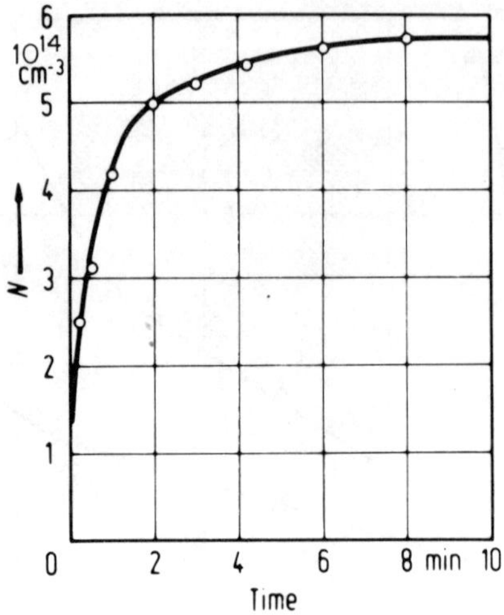

FIG. 17. Growth in number of microvoids with time in PVC under a tensile stress of 6 kg mm^{-2} (After reference 11.)

TABLE 2

BOND RUPTURE AND CAVITY FORMATION COMPARED

Polymer	*F*, cm^{-3}	*N*, cm^{-3}	*D*, Å	R_1	R_2
Nylon 6	27×10^{18}	900×10^{14}	90	300	330
p-Propylene	$1{\cdot}4 \times 10^{18}$	5×10^{14}	320	2 800	2 600
p-Ethylene	6×10^{18}	6×10^{14}	400	10 000	11 000

Here F is the bond rupture concentration, N the microvoid concentration and D the diameter of the disc-shaped microvoid. The two ratios R_1 and R_2 are respectively the number of molecular fracture events per cavity (F/N) and the number of aligned molecules calculated to intersect a plane of the same area as a microvoid. The close agreement was taken to signify that all molecules passing through a microvoid area had been fractured by the growth of the void.

This is a reasonable conclusion for highly oriented crystalline polymers where the amorphous regions are probably composed largely of tie molecules. It is unlikely to be true of microvoids and craze cavities formed in amorphous glasses or poorly oriented crystalline polymers.

The occurrence of crazing in the plastic deformation of cross-linked glasses has already been referred to. Visible crazing in polybutadiene and polyisoprene was obtained under some experimental conditions but not all. For crazing to occur required in general low strain rates, low temperatures and low pre-orientations (though not so low that brittle behaviour ensued). A feature of visible crazing is the absorption of relatively large quantities of the environmental gases. The quantities are too great to allow their presence in vapour form within the voids[30] and it follows that they must be present in a condensed form, possibly as a monolayer within the craze.

This observation, along with the work of Brown and Imai,[51] leads to the conclusion that the crazing observed is environmentally induced, and occurs only at temperatures just below or just above the boiling point of the environmental gases. This hypothesis was confirmed by Mead[29] who deformed polychloroprene and polybutadiene at various temperatures in atmospheres of nitrogen, helium, carbon dioxide and sulphur dioxide. Mead also found corresponding changes in the load-extension curves of his materials for different environments. When conditions were not suitable for environmental crazing, mechanical behaviour reverted to either brittle fracture or shear yielding.

Brown and Imai[51] have shown that in PMMA both environmental and non-environmental ('intrinsic') crazing can occur, and have shown how the incidence of the former depends on the partial pressure of the crazing agent. It is not certain whether intrinsic crazing occurs in the low-temperature deformation of cross-linked glasses or whether craze formation is wholly environmental in these cases.

It is not at all certain that the void formation and crazing discussed in the previous paragraph is directly associated with molecular fracture in the manner proposed by Zhurkov[50] for crystalline fibres and films. ESR signals are certainly produced simultaneously with visible crazing in cross-linked glasses, and the chemical nature of the radicals observed is affected by the penetration of environmental gases. Furthermore, radicals are not always detected in the shear yielding of these materials. On the other hand, even stronger ESR signals are obtained in those deformation regimes (e.g. for polybutadiene) where no visible crazing is observed. It should be remembered, however, that Zhurkov's microvoids do not appear as visible crazes.

In spite of the caution necessary in interpreting the data, it is not unreasonable to suppose that crazing, like microvoid formation, does involve molecular fracture, at least in cross-linked glasses. This could occur both in the initial cavitation event and in the subsequent growth of craze cavities, the latter involving plastic flow in the 'walls' of the cavity. It is unlikely that a one-to-one relationship exists between molecular intersections and void area in crazing, though Zhurkov may be correct in proposing such a relationship for oriented crystalline polymers. Finally, microvoiding may well be taking place in the more highly oriented cross-linked glasses where visible crazing is no longer evident.

The enhancement, or even inducement, of crazing by environmental gases is almost certainly a physical effect rather than a chemical one.[52] That is, it is most unlikely that this process affects the incidence of molecular fracture. What the adsorption of environmental fluid may well do, however, is to make possible the growth of microvoids into visible crazes. The reason for this is that the energetically unfavourable step in the cavitation of a continuous medium is the expansion of the cavity at very small radius against the surface tension of the medium. (The stress required goes as $2\gamma/r$, where γ is surface tension and r is radius.) Adsorption, by reducing γ, favours cavity growth at this critical stage and thus greatly improved the chances of large cavities developing.

2.12 Synopsis of Mechanical and Physical Processes

Finally in this Section we attempt to draw together the evidence provided by the various techniques employed to study molecular fracture. For clarity this synopsis is presented as a brief catalogue of sequential events linking the application of stress to macroscopic deformation and fracture *via* the molecular processes that may occur.

(a) The application of tensile load to a solid polymer elicits various responses at the molecular level, depending on the morphology and state of the solid. Molecules which are randomly coiled tend to uncoil and straighten in the direction of stress. If the material is glassy, this can only occur at high stresses (i.e. above the yield stress) because of the steric hindrance of neighbouring molecules. Molecular chains which are already fully extended (by pre-orientation or fibre formation) will deform elastically to accommodate the load. Some which are close to a fully extended configuration will be brought to full extension as a result of loading.

(b) Molecular chains which are fully extended, being less compliant than their coiled neighbours, will carry a disproportionate fraction of the total load, as in any 'composite' solid. The high stresses on the co-valent bonds of such a molecule may be detectable by shifts in the associated vibrational (infra-red) frequencies. Molecular regions within crystals will share the load uniformly and are unlikely to become 'over-stressed'.

(c) Because of the high dissociation energies of co-valent bonds, intermolecular slippage or flow, involving the rupture of secondary bonds, will normally occur in preference to molecular fracture (i.e. main-chain breakage). Under such circumstances, plastic deformation by flow will be observed without molecular rupture. This flow may be uniform, as in shear yielding, or localised, as in void growth and crazing.

(d) If the molecules are 'anchored' in the structure either by crystalline regions or chemical cross-links (or even by persistent physical entanglement with, for example, long relaxation times) the mutual flow of such molecules will be prevented. They will thus uncoil as the strain increases and eventually become fully extended and elastically loaded. This may occur after small strains (as in oriented fibres) or after large test strains (as in plastically deforming glasses). As already noted, non-uniformities of structure will ensure that different molecular chains achieve full orientation before others, so that the population of fully extended chains will be strain- and pre-strain-dependent. The degree of structural non-uniformity will depend upon morphology and thus upon thermal treatments and other factors.

(e) In the absence of permanent molecular anchorages it is still possible for individual molecules to become highly stressed. This may occur, for example, under the high stress necessary to produce flow at temperatures far below the glass transition. A long molecule may have physical entanglements with its neighbours which relax on a time scale long compared with the deformation rate and thus act as 'permanent' anchorages on this time scale.

(f) Fully extended molecules fracture under the combined influence of mechanical stress ('mechanical excitation') and thermal fluctuations, according to the kinetic theory of fracture. The rate of this process is governed by the molecular stress and the temperature.

(g) As a result of the fracture of fully extended chains, stress is redistributed more evenly and the rate of chain fracture falls at constant overall stress. The process of mechanical degradation can only progress, therefore, if the applied stress or strain is increased to repopulate the fully extended state.

(h) Because stress is redistributed by molecular fracture locally, such fracture events tend to accumulate locally, producing a microvoid.

(i) Microvoids, which are distributed uniformly throughout a specimen, can grow by various methods:

(1) by propagation as a result of further molecular fracture;

(2) by growth as a result of local molecular flow without fracture, possibly aided by environmental adsorption reducing interfacial energies;
(3) by coalescence.

(j) The growth of microvoids can result in either a single large local flaw or crack which propagates to destroy the specimen, or else in a stable array of voids, as in the wholesale crazing sometimes observed. The conversion of a narrow craze, by craze breakdown, into a propagating crack is an intermediate possibility.

(k) The establishment, or nucleation, of a single crack is only the first step towards fracture. For macroscopic failure to be observed, the crack must propagate through the specimen. Propagation involves both the creation of new surfaces (which normally involves molecular fracture) and the dissipation of energy in inelastic deformations (which do not normally involve molecular fracture).

(l) Macroscopic fracture may thus be sub-divided into Regime I, or nucleation, and Regime II, or propagation. For specimens containing pre-existing cracks, Regime II will predominate. How far the fracture parameters of a specimen, such as lifetime under stress and fracture strength, depend on Regime I or Regime II processes will depend on the material's physical properties, the specimen geometry (i.e. whether there are pre-existing cracks) and the conditions of test (rate, temperature, environment).

(m) Molecular fracture is probably important both in Regime I and Regime II, even in polymers where ESR signals cannot be generated by tensile testing. The role of molecular fracture, and its interaction with other physical mechanisms such as mechanical loss processes, is however, significantly different in the two regimes.

REFERENCES

1. Taylor, N. W. *J. Appl. Phys.*, **18** (1947) p. 943.
2. Casale, A., Porter, R. S. and Johnson, J. F. *Rubber Chem. and Tech.*, **44** (1971) p. 534.
3. Becht, J. and Fischer, H. *Kolloid Z.u.Z.Polymere*, **229** (1969) p. 167.
4. Pazonyi, T., Tudos, F. and Dimitrov, M. *Angew. Makrom. Chemie*, **10** (1970) p. 75.
5. Salloum, R. J. and Eckert, R. E. *J. Appl. Polymer Sci.*, **17** (1973) p. 509.
6. Zhurkov, S. N. and Korsukov, V. E. *J. Polymer Sci., Polymer Phys. Ed.*, **12** (1974) p. 385.

7. ZHURKOV, S. N., ZAKREVSKII, V. A., KORSUKOV, V. E. and KUKSENKO, V. S. English translation in *Soviet Phys. Solid State*, **13** (1972) p. 1680.
8. ZHURKOV, S. N., VETTEGREN, V. I., KORSUKOV, V. E. and NOVAK, I. I. In *Proc. 2nd. Int. Conf. on Fracture*, Chapman and Hall, London, 1969, p. 545.
9. BERNSTEIN, H. I. *Spectrochim. Acta*, **18** (1962) p. 161.
10. GUINIER, A. and FOURNET, A. *Small Angle Scattering of X-rays*, John Wiley and Sons, New York, 1955.
11. ZHURKOV, S. N., KUKSENKO, V. S. and SLUTSKER, A. I. *Proc. 2nd. Int. Conf. on Fracture*, 1969, Chapman and Hall, London, 1969, p. 531.
12. AYSCOUGH, P. *Electron Spin Resonance in Chemistry*, Methuen, London, 1967.
13. POOLE, C. P. *Electron Spin Resonance*, John Wiley and Sons, New York, 1967.
14. CAMPBELL, D. *Macrom. Rev.*, **4** (1970) p. 91.
15. SCHNEIDER, E. E. *Disc. Faraday Soc.*, **19** (1955) p. 158, and *J. Chem. Phys.*, **23** (1955) p. 978.
16. ABRAHAM, R. J., MELVILLE, H. W., OVENALL, D. W. and WHIFFEN, D. H. *Trans. Faraday Soc.*, **54** (1958) p. 1133.
17. CHARLESBY, A. and Thomas, D. K. *Proc. Roy. Soc. (London)*, **A269** (1962) p. 104.
18. SMALLER, B. and MATHESON, M. S. *J. Chem. Phys.*, **28** (1958) p. 1169.
19. LIBBY, D., ORMEROD, M. G. and CHARLESBY, A. *Polymer*, **1** (1960) p. 212.
20. BRESLER, S. R., KAZBEKOV, E. N. and SAMINSKII, E. M. *Polymer Sci. (USSR)*, **1** (1959) p. 540.
21. ZHURKOV, S. N., TOMASHEVSKII, E. E. and ZAKREVSKII, V. A. English translation in *Soviet Phys. Solid State*, **3** (1962) p. 2074.
22. RADCIG, V. A. and BUTYAGIN, P. YU. *Polymer Sci. (USSR)*, **A9** (1967) p. 2883.
23. VERMA, G. S. P. and PETERLIN, A. *Kolloid Z.u.Z.Polymere*, **236** (1970) p. 111.
24. DE VRIES, K. L., ROYLANCE, D. K. and WILLIAMS, M. L. *Report UTEC DO 68-056*, College of Eng., Univ. of Utah, Salt Lake City, 1968.
25. KAUSCH, H. H. *J. Macrom. Sci., Revs. Macromal. Chem.*, **C4** (1970) p. 243.
26. DE VRIES, K. L., ROYLANCE, D. K. and Williams, M. L. *J. Polymer Sci. A1*, **8** (1970) p. 237.
27. BACKMAN, D. N. and DE VRIES, K. L. *J. Polymer Sci. A1*, **7** (1969) p. 2125.
28. DE VRIES, K. L., SIMONSON, E. R. and WILLIAMS, M. L. *J. Macrom. Sci., Phys.*, **B4** (1970) p. 671.
29. MEAD, W. T. Molecular Fracture in Mechanically Deformed Polymers, Ph.D. thesis, Univ. of London, 1975.
30. MEAD, W. T. and REED, P. E. *Polymer Eng. Sci.*, **14** (1974) p. 22.
31. MEAD, W. T., PORTER, R. S. and REED, P. E. To be published.
32. KAUSCH, H. H. and BECHT, J. *Rheologica Acta*, **9** (1970) p. 137.
33. NATARAJAN, R. and REED, P. E. *J. Polymer Sci. A2*, **10** (1972) p. 585.
34. ANDREWS, E. H. *The Physics of Glassy Polymers*, ed. R. N. Haward, Applied Science Publishers, London, 1973, p. 394.
35. KAUSCH, H. H. and DE VRIES, K. L. *Int. J. of Fracture*, **11** (1975) p. 727.
36. LLOYD, B. A., DE VRIES, K. L. and WILLIAMS, M. L. *J. Polymer Sci. A2*, **10** (1972) p. 1415.
37. PARK, J. B., STATTON, W. O. and DE VRIES, K. L. *Report UTEC ME 71-122*, Univ. of Utah, Salt Lake City, 1971.

38. ZAKS, B. YU., LEBEDINSKAYA, L. and CHALIDZE, V. N. *Vysokomol. Soyed.*, **A12** (1970) p. 2669.
39. ZAKS, B. YU., LEBENINSKAYA, L. and CHALIDZE, V. N. English translation in *Polymer Sci.*, (*USSR*), **12** (1971) p. 3025.
40. CHIANG, T. C. and SIBILIA, J. P. *J. Polymer Sci.*, **10** (1972) p. 2249.
41. ZHURKOV, S. N. and TOMASHEVSKY, E. E., *Physical Basis of Yield and Fracture*, Institute of Physics, London, 1966, p. 200.
42. CAMPBELL, D. and PETERLIN, A. *Polymer Letters*, **6** (1968) p. 481.
43. ROYLANCE, D. K., DE VRIES, K. L. and WILLIAMS, M. L. *Proc. 2nd. Int. Conf. on Fracture*, Chapman and Hall, London, 1969, p. 551.
44. HASSELL, W. H. MS thesis, Dept. Mech. Eng., Univ. of Utah, Salt Lake City, June 1973.
45. ANDREWS, E. H. *Angew. Chem.*, **86** (1974) p. 151.
46. THOMAS, A. G. *J. Polymer Sci. C, Sympos.*, **48** (1974) p. 145.
47. WILLIAMS, M. L. and DE VRIES, K. L. in *Proc. 5th Int. Congress on Rheology*, ed. S. Onogi, Univ. of Tokyo Press, Tokyo, 1970, Vol. 3, p. 139.
48. REEVE, B. Morphology and Tensile Properties of Polychloroprene, Ph.D. Thesis, Univ. of London, 1972.
49. JOHNSON, U. and KLINKENBERG, D. *Kolloid Z.u.Z.Polymere*, **251** (1973) p. 843.
50. ZHURKOV, S. N. *et al.* Private communication, and *J. Polymer Sci., Polymer Phys. Ed.*, **12** (1974) p. 385.
51. BROWN, N. and IMAI, Y. *J. Materials Sci.*, **11** (1976) p. 425 and references therein.
52. ANDREWS, E. H., LEVY, G. M. and WILLIS, J. *J. Materials Sci.*, **8** (1973) p. 1000.
53. ANDREWS, E. H. and REED, P. E. *Advances in Polymer Science 27*, Springer Verlag, Heidelberg, 1978.

Chapter 3

ENVIRONMENTAL CRACKING OF POLYMERS

EDWARD J. KRAMER
Cornell University, Ithaca, NY, USA

SUMMARY

Environmental cracking of most rigid polymers proceeds by initiation and growth of a craze-like zone of cavitation which breaks down to form a crack. The most important effect of the environment appears to be plasticisation of the craze fibrils, leading to enhanced drawing of these from the craze surface as well as increased fibril creep leading to fibril rupture. The fibril breakdown process is accelerated by decreases in polymer molecular weight; some environments, in addition to being plasticisers, produce chain scission and molecular weight degradation in the fibrils, adversely affecting fibril stability. New experimental methods of investigating craze micromechanics and craze mechanical properties are discussed and it is concluded that these may lead to a greatly improved understanding of the microscopic aspects of environmental cracking.

1 INTRODUCTION

Environmental effects on fracture are well known to materials scientists; these include stress corrosion cracking, hydrogen embrittlement and liquid metal embrittlement of metals and alloys as well as water-vapour-induced 'static fatigue' of oxide glasses. Yet nowhere is the influence of the environment on fracture as important or widespread as it is in the polymeric materials. It is not unusual for normally ductile polymers to fail in an apparently brittle manner at less than one-tenth of the yield stress if placed in contact with certain environments. Indeed, residual stresses from moulding operations are often sufficient to cause environment-assisted cracking, and dipping moulded parts in certain chemical environments is a widely used, but destructive, test for the presence of residual tensile stresses.[1] A dramatic, but not particularly unusual, example of environmental cracking is shown in Fig. 1. A tensile sample of polycarbonate was held under a constant tensile stress of 6·9 MN m^{-2}. It failed in an apparently brittle fashion within 60 s after being placed in contact with a 1 : 1 solution of 6·25N NaOH in ethanol. A similar specimen of polycarbonate held at the same stress in air showed no visible cracks or crazes after one day, the limit of our patience.

There have been a number of previous reviews of environmental cracking of polymers. The most wide ranging of these have been written by Kambour.[2,3] Howard[4,5] has reviewed the phenomenology of environmental stress cracking of polyolefins while Beardmore and Rabinowitz,[6] Gent,[7] and Kambour[8] have reviewed some aspects of environmental cracking of glassy thermoplastics as part of broader reviews of the crazing and fracture of these materials.

In this paper we will refer to specific polymer-environment pairs, where environmental effects on fracture are observed, as environmental fracture systems. Following Kambour's analytical scheme we can attempt to classify these environmental fracture systems depending on whether the environment promotes polymer chain scission, polymer chain cross-linking or neither. For example, in the environmental fracture system ozone-rubber it is well known that ozone attacks double bonds in the main polymer chain resulting in chain cleavage and molecular weight degradation. In the presence of ultra-violet light and oxygen, however, unsaturated rubbers can undergo photolytic oxidation and cross-linking leading ultimately to a drastic loss in ductility.[9] Finally, below its glass transition temperature rubber in contact with cryogenic gases (e.g., liquid N_2) undergoes profuse crazing,[10] followed by crack nucleation and propagation within a craze. In this case, as in most examples of environmental effects on fracture of thermoplastics, there is no evidence that the environment promotes either chain scission or cross-linking.

To further aid classification, one should distinguish between environments which cause embrittlement of a particular polymer and those which cause true stress cracking. In the case of embrittling environments, exposure of the polymers in the absence of stress results in permanent embrittlement so that on subsequent application of stress, even in the absence of that environment, cracks can nucleate and/or propagate at abnormally low stress levels. On the other hand, in true environmental

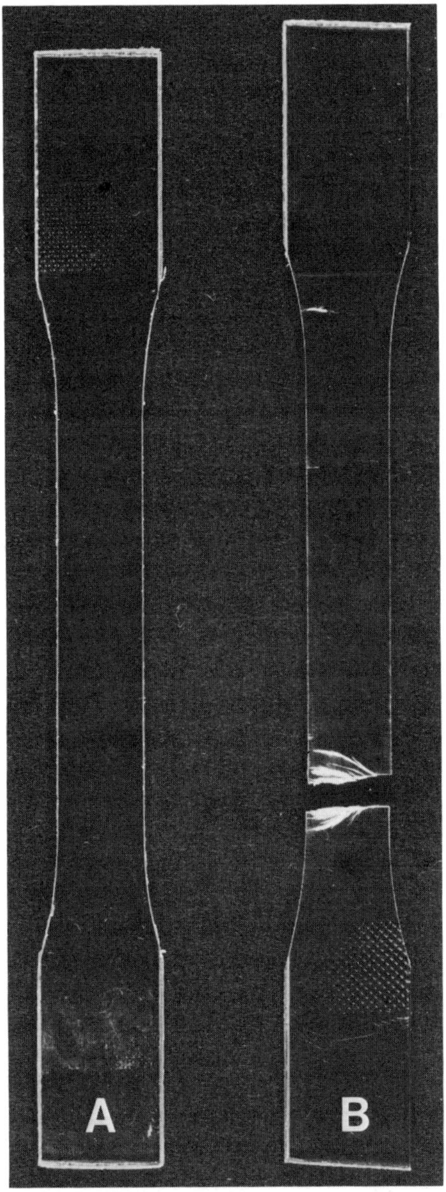

FIG. 1. Specimens of polycarbonate held under a constant tensile stress of 6·9 MN m^{-2}. In air specimen A did not fail after one day, whereas specimen B placed in contact with a solution of one part 6·75 N NaOH and one part ethanol failed in an apparently brittle manner in less than one hour.

stress cracking (ESC), both stress and the environment must be present simultaneously for cracking to occur; exposure to the environment in the absence of stress does not result in subsequent failure at low stress levels out of the environment. In this classification scheme the ozone cracking of rubber is true environmental stress cracking whereas photolytic oxidation and cross-linking of rubber by UV irradiation is embrittlement.

It should not be assumed, however, that all environmental stress cracking agents are necessarily inert to the polymer at zero stress. An example of an environment which is not inert is a liquid that, when absorbed to equilibrium into a glassy thermoplastic, plasticises it so that the T_g of the polymer lies below room temperature. If at the same time the liquid diffuses very slowly into the bulk polymer, exposure of a macroscopic specimen at zero stress will produce no embrittlement and no apparent change in properties since the plasticisation is confined to the polymer surface. On the other hand it will turn out that such liquids are unusually potent ESC agents for reasons that involve both this strong local plasticising ability and the slow bulk diffusion of the liquid in the polymer.

Embrittlement produced by exposure of polymers to sunlight is generally the result of reactions (either cross-linking or chain scission) promoted by absorption in the near-UV. The free radical reactions involved are complex and differ greatly from polymer to polymer, as indeed does the balance between the UV-induced cross-linking rate and the UV-induced chain scission rate. The interested reader is referred to recent reviews[11,12] since detailed coverage of this subject is beyond the scope of this chapter. Presumably the embrittlement by cross-linking occurs because of the decrease in local plastic/anelastic strains that can be achieved near the tip of a crack without local bond rupture in highly cross-linked materials. There has been very little work on the physical mechanisms of this embrittlement, however.

Chain scission is responsible for a much wider variety of both embrittlement and ESC phenomena. Besides the ozone attack of rubber mentioned above, condensation polymers such as silicones, polyesters, polycarbonates, etc. can undergo ESC in basic or acidic conditions. The example shown in Fig. 1 (ESC of polycarbonate by 6·25N NaOH in alcohol) belongs in this category. The overall reaction is hydrolysis of bonds in the main chain and thus falls into the chain scission class. Radiation, such as UV, electron or γ radiation, can result primarily in chain scission embrittlement in certain polymers if the cross-section for chain scission is much larger than that for cross-linking reactions.

While chain scission embrittlement can probably be understood simply as due to a uniform molecular weight reduction of the polymer (see the discussion of molecular weight effects in crazing below), the physical mechanisms of crack advance in chain scission ESC are not obvious. One might postulate a mechanism of sequential attack and rupture of the most highly stressed covalent bonds at the surface of the crack tip, but even in the case of ozone cracking of rubber this simple mechanism apparently does not operate. Gent and co-workers[13–15] have found that the kinetics of ozone-induced subcritical crack growth in both butyl and butadiene-styrene rubbers is controlled by polymer segmental mobility and they concluded (at least in the case of butyl rubber) that diffusion of ozone into the zone in advance of the crack tip was the rate-controlling step. This conclusion implies that the crack propagation mechanism involves a change in the mechanical properties (e.g., the local strain-to-failure and local stress-strain relations) of a finite amount of material in front of the crack produced by chain scission. In the case of butadiene-styrene it is clear that although the ozone-induced chain scission may be very close to the surface of the crack tip, the rate at which energy is dissipated in hysteretic loading and unloading of the rubber in the region surrounding the crack tip actually controls the crack velocity.

There have been very few experiments on hydrolysis-controlled ESC and none which reveal the physical mechanisms of crack advance. In the case of the polycarbonate example shown above crazing may be involved. A very short craze may form at the tip of a flaw so that small fibrils would then be exposed to the basic hydrolytic environment. As the molecular weight of the polymer in these fibrils is degraded by chain scission to below the entanglement molecular weight, the fibrils fail and the crack advances. On the other hand, the cracking might equally well involve a simple stress-induced etching and dissolution of polymer at the crack tip. Which, if either, of these possibilities is correct will not be known until further experiments are done.

The final class of environmental fracture systems is that in which neither chain scission nor cross-linking occurs. This class is by far the largest and most important, encompassing the environmental stress cracking of polyolefins by detergents and alcohols,[4,5] the stress cracking of polyamides by aqueous and alcohol solutions of inorganic metal salts,[16–19] the crazing and cracking of glassy amorphous polymers by organic liquids,[6–8] and the low-temperature crazing/cracking of both partially crystalline and amorphous polymers by gases near their condensation temperatures.[20–22] While there are many differences from one of these systems to another, especially in the concentrations of the environment absorbed into the polymer in the absence of stress, one can form a useful hypothesis that there appears to be a basic similarity in the microscopic mechanism of crack advance. Rather than representing either a true brittle cleavage (which would be ruled out anyway by the absence of chain scission) or chain-by-chain dissolution of material from the crack tip, crack advance appears to occur by a very localised microplastic failure mechanism. A craze-like planar zone of cavitation bridged by polymer fibrils or ligaments develops ahead of the crack and the crack advances into this zone by failure of the highly plastically deformed ligaments at the crack tip. In fact, one can make a plausible argument that all environmental stress cracking of rigid polymers proceeds by such a microscopic mechanism. The role of the environment appears to be to enhance both the localised cavitation and plastic deformation of the ligaments as well as perhaps to alter the failure criterion for fibrils at the crack tip.

In the remainder of this review we will explore both the evidence for, and the consequences of, this hypothesis. In Section 2 we discuss the fundamentals of crazing, including the craze structure and craze micromechanics. Since much more information is often available for air crazes (crazes grown in air), a comparison between air and environmental craze structure and properties will be instructive, especially because air crazing is responsible for the high fracture toughness values observed for many glassy polymers. In Section 3 we discuss the kinetics of environmental stress crazing and cracking, including the application of fracture mechanics concepts to the description of subcritical crack growth and crack initiation, and the prediction of the threshold for ESC in various chemical environments. In each case we will attempt at least to rationalise the phenomenological observations in terms of craze fundamentals and identify areas in which more experimental information is needed.

2 STRUCTURE AND PROPERTIES OF CRAZES

2.1 Craze Morphology and Structure

The microstructure of an air craze in polystyrene (PS) is shown in Fig. 2(a). This is a transmission electron micrograph of a craze in a PS thin film, produced by stretching the film while it is bound to a soft copper grid. The method is an adaptation of one first used by Behan, Bevis and Hull[23–25] for investigating craze microstructures in thin films. Similar microstructures are also observed in microtomed sections of the

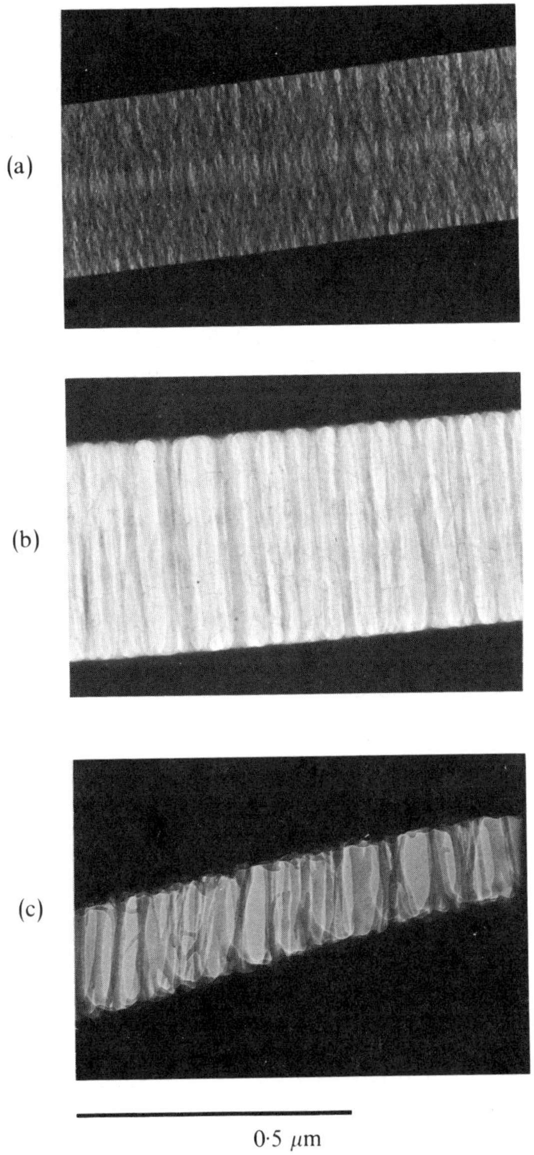

FIG. 2. Transmission electron micrographs of typical crazes in films ≈ 7000 Å thick under the stress at which the crazes originally grew. (a) Microstructure of an air craze in polystyrene; (b) microstructure of a methanol craze in polystyrene; (c) microstructure of an *n*-heptane craze in polystyrene. Bar represents 0·5 μm.

bulk polymer when the crazes have been reinforced by impregnants such as silver nitrate[26] or iodine–sulphur eutectic,[27,28] or when the craze in such a slice has been reopened by stretching on a copper grid.[23,24] If the films used for transmission electron microscopy are sufficiently thick, it appears that the microstructure of crazes formed in these films is characteristic of the microstructure of crazes formed in the bulk.[24]

Air crazes in polystyrene are observed to consist of a dense network of fine fibrils approximately 200 Å in diameter. These crazes have a thickness which tapers to very small dimensions (< 25 Å) at the craze tip and as the craze grows the thickness T at any particular point along the craze increases. A schematic drawing of a craze in cross-section is shown in Fig. 3. The fibrils are formed from a certain original layer of bulk polymer, the primordial craze thickness T_0. The two surfaces of this original layer undergo a displacement w and become the surfaces of the craze. The elongation ratio λ of the craze fibrils is given by

$$\lambda = T/T_0 \quad (1)$$

and the true strain in the fibrils is

$$\varepsilon = \ln \lambda \quad (2)$$

The thickening mechanism is important in determining the microstructure of the craze. Two mechanisms are possible: (1) A surface drawing

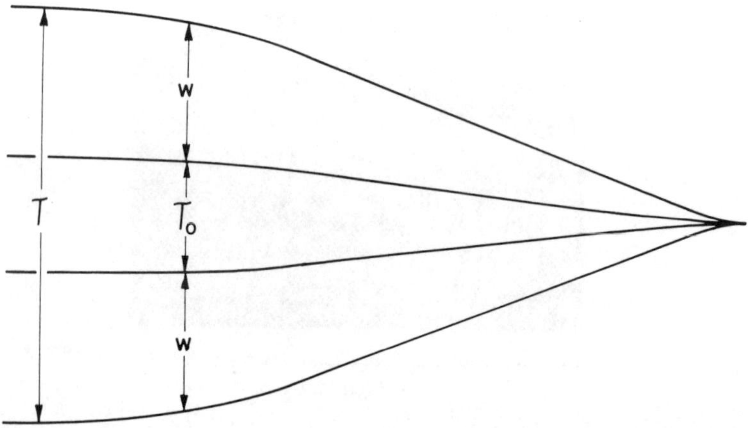

FIG. 3. Schematic drawing of the craze. The final thickness of the craze is T; the thickness of the original or primordial layer of polymer which fibrillated to form the craze is T_0; and w is the displacement of each craze surface.

mechanism in which new fibrillar material is drawn from bulk unoriented polymer at the craze/bulk polymer interface; (2) a fibril creep mechanism in which the thickening occurs by further straining of the existing fibrils. In short air crazes the surface drawing mechanism predominates[29] although for longer crazes there is some evidence that fibril creep may also make a contribution.[30] Fibril creep is an important mechanism of craze failure and of subcritical crack or void growth within the craze.

A very important microstructural parameter of the craze is the volume fraction of fibrils v_f. It is clear that a craze with a low volume fraction of fibrils will be more susceptible to fibril breakdown and craze fracture than a craze with a higher value of v_f if all other conditions are the same. The true stress in the fibrils will be higher by a factor $(v_f)^{-1}$ than the stress S at the craze surface. The value of v_f is determined by the plastic elongation ratio λ of the craze fibrils, i.e.

$$v_f = 1/\lambda \tag{3}$$

since the transformation from bulk unoriented polymer to oriented fibril takes place at approximately constant polymer volume.

Fibril volume fractions can be measured by several different methods. One can measure the refractive index of the craze by measuring the critical angle for reflection of light and convert refractive index to v_f via the Lorentz–Lorenz equation.[26,31–34] One can also measure v_f directly from thin film transmission electron microscopy of the craze by comparing the attenuation of the electron beam due to scattering in the craze, in the solid film and while passing through a hole in the film.[35] In this method values of v_f can be measured in regions smaller than 1000 Å along the craze. In such measurements it is important not to remove the load which caused crazing while making the measurement. Unloading the specimen not only allows retraction of the highly drawn fibrils but actually puts the craze into compression, producing buckling and reverse deformation of the fibrils. The v_f values measured on crazes in unloaded specimens are much larger than the true values.

For air crazes in PS one finds values of fibril volume fraction between 0·35 and 0·15 depending on the conditions of craze growth and fibril drawing.[29,36] Figure 2(a) shows a region ≈ 750 Å thick in the centre of the craze which is lower in v_f than the outer craze layers. This region, which is called the mid-rib,[24,25] is due to the fact that this layer of the fibrils was drawn from the craze surfaces immediately behind the craze tip. There is a region of stress concentration at the craze tip and this high stress does not decrease rapidly with distance behind the tip but remains high for a distance of $\approx 15\ \mu m$.[29] The fibril structure drawn in this region naturally has a somewhat higher draw (elongation) ratio λ, and thus lower v_f, than the outer regions of the fibrils which are produced by drawing under lower stresses in regions further from the craze tip. These results are for an isolated craze grown in a uniform tensile initial stress field, and logically the structure of a craze grown in a spatially inhomogeneous stress field, e.g. at the tip of a crack, will reflect this initial stress inhomogeneity.

The structure of environmental crazes that have been examined to date differs in a number of respects from that of the air crazes outlined above. The most obvious and important additional complication is that the environment can swell and plasticise the craze fibrils and the layer of bulk polymer immediately adjacent to the craze surfaces. The structure of a craze in a glassy amorphous polymer such as polystyrene can depend markedly on the decrease in glass transition temperature of the fibrils produced by this plasticisation.[37] As examples, Figs. 2(b) and 2(c) show the microstructure of a methanol craze in PS and an *n*-heptane craze in PS respectively. Methanol, when absorbed to equilibrium in PS, lowers its T_g by only about 10 °C to a value of about 90 °C. The craze structure in Fig. 2(b) appears to be rather similar to that of the air craze in Fig. 2(a), except that the mid-rib is absent in the solvent craze. *n*-Heptane lowers the glass transition temperature of PS to 6 °C, i.e. to below room temperature. A craze formed by this strong plasticiser is shown in Fig. 2(c). There is a very low fibril volume fraction which decreases markedly from the tip of the craze to its base. The fibrils are coarser than in either the methanol or the air craze. (Whether that is because the finer fibrils rupture early in the craze thickening, or because the average fibril size is large due to the altered rheology of the highly plasticised material immediately ahead of the craze tip and at the craze surfaces, is not known.) Toward the base of the craze one can observe fibrils that appear to have undergone ductile rupture.

The importance of the glass transition temperature of the plasticised fibrils in determining the structure of the environmental craze is underlined by experiments in which the structure of *n*-heptane crazes grown at various temperatures below room temperature were compared.[38,39] As the growth temperature was decreased, the fibril volume fraction of the *n*-heptane craze increased and the fibrils became finer. At a comparable undercooling ($\Delta T \approx 60$ °C) below T_g of the plasticised fibrils, the structure of the *n*-heptane craze was qualitatively similar to that of the methanol craze at room temperature.

The absence of the mid-rib in even the weakly plasticised methanol craze in PS strongly suggests that the thickening mechanism in the environmental crazes differs from that of the air crazes. Indeed holographic interferometry experiments on the highly plasticised ($T_g \approx 40$ °C) methanol crazes in PMMA,[40] and preliminary transmission electron microscopy measurements on *n*-heptane crazes in PS,[41] indicate that thickening of these crazes occurs primarily by fibril creep rather than drawing new material from the craze surfaces.

Although some thickening of these environmental crazes must occur by the surface drawing mechanism, the contribution of this mechanism may be controlled by diffusion of the environment normal to the craze surface. In general, environmental crazing takes place at craze surface stresses which are insufficient to cause air craze thickening. If such crazes are to thicken by surface drawing, diffusion of the environment to plasticise the layer of bulk polymer immediately adjacent to the craze surface must occur first. One can make this hypothesis more quantitative with the aid of Fig. 4 and a simple diffusion model. Suppose that a concentration $c(S)$ of the environmental agent at one fibril radius r_0 from the surface is necessary for drawing from the surface to continue at the imposed craze surface stress S and that the concentration c_0 just beneath the surface is the concentration in local equilibrium with the environment in the craze. The diffusion problem is approximately one of diffusion from an advancing boundary and the result (for Fickian diffusion) is

$$c(S) = c_0 \exp(-vr_0/D) \tag{4}$$

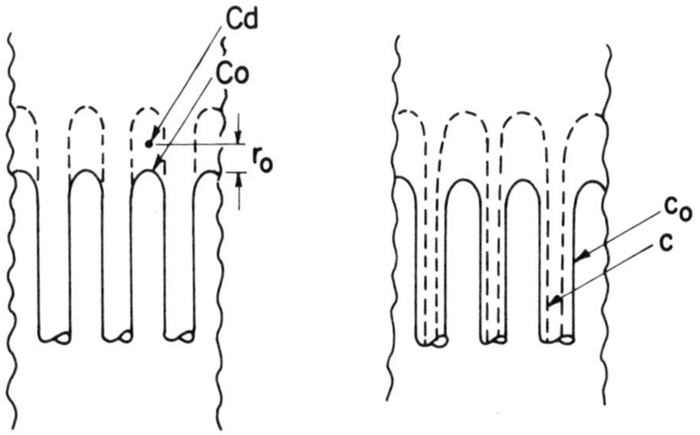

FIG. 4. Schematic drawing of two mechanisms of craze thickening. Left, surface drawing mechanism; $\lambda \approx$ constant. Right, fibril creep mechanism; $T_0 \approx$ constant.

where v is the velocity of the craze surface relative to the bulk polymer and D is the diffusion coefficient of the environment. The thickening velocity of the craze by surface drawing is thus

$$v = (D/r_0)\ \ln(c_0/c_d(S)) \tag{5}$$

whereas the surface displacement rate $\dot{w}$ is

$$\dot{w} = [(\lambda - 1)D/r_0]\ \ln(c_0/c_d(S)) \tag{6}$$

Since $c_d(S)$ decreases approximately linearly with S, the environmental craze thickening by the surface drawing mechanism is insensitive to S and depends primarily on the diffusion coefficient of the environment. Making the approximations that $c_d/c_0 \approx 0{\cdot}5$ and $r_0 \approx 2 \times 10^{-6}$ cm, then $v \approx 3 \times 10^5 D$ if D is expressed in $cm^2\ s^{-1}$. Environmental craze growth rates ordinarily measured are in the range of 10^{-2} cm s^{-1} which imply that v is of the order 10^{-4} cm s^{-1}. Under such conditions the thickening of an environmental craze by drawing from craze surfaces will be diffusion-controlled if $D < 3 \times 10^{-10}\ cm^2\ s^{-1}$. Most organic liquids in PS have $D < 10^{-9}\ cm^2\ s^{-1}$ whereas many gases in PS have $D > 10^{-9}\ cm^2\ s^{-1}$ at room temperature so, as a general rule, one would expect to observe that the surface drawing mechanism of craze thickening will be diffusion-controlled for crazing in liquid environments but not in gaseous environments. A further complication, however, is that one expects diffusion rates to be enhanced by: (1) the local hydrostatic tension stress field near the craze surface;[44] and/or (2) the molecular mobility increase in the glassy polymer which accompanies the local anelastic and plastic deformation. There is very little experimental evidence on either effect that would allow one to extrapolate to the local conditions at the environmental craze surfaces (or at the craze tip).

The other competing mechanism by which an environmental craze can thicken, the creep of the existing craze fibrils, is also illustrated in Fig. 4. The creep rate $\dot{\lambda}$ of the fibrils can be expected to be a function of the true stress σ_t in the fibrils, the concentration of the environment in the fibrils (and thus the T_g depression in the fibrils) and the elongation ratio λ (effect of strain or orientation hardening of the fibrils).

The surface displacement rate for this mechanism is given by

$$\dot{w} = \mathrm{T}\dot{\lambda}/2\lambda = \tfrac{1}{2}\mathrm{T}\dot{\varepsilon}_t \tag{7}$$

where $\dot{\varepsilon}_t$ is the true strain rate in the fibrils. If $c_d(S) < c_0$ an important contribution to $\dot{\lambda}$ in eqn. (7) will occur as c approaches c_0 and the plasticisation of the fibril increases.

A major unknown at present is whether the equilibrium sorption (c_0) as well as the diffusion of the environment into the fibril is markedly enhanced by the stress in the fibril. Gent[7] has proposed that the sorption be computed from the Flory-Huggins-Treloar treatment of polymer-liquid equilibria which relates the volume fraction of polymer v_p in a swollen rubber to the hydrostatic portion of the stress tensor (the dilatant stress) $\sigma_h \approx \sigma_t/3$. This treatment results in the equation

$$\ln(1 - v_p) + v_p + \mu v_p^2 + (\rho\Omega/M_e)\left(v_p^{1/3} - \frac{v_p}{2}\right) = \frac{\sigma_t\Omega}{3RT} \tag{8}$$

where μ is the polymer–liquid interaction parameter, ρ is the polymer density, Ω is the molar volume of the liquid, R is the gas constant, T the temperature and M_e is the molecular weight between entanglements.

While the treatment is appropriate for the rubbery state above T_g, it is uncertain whether it should apply to the glassy state. Equation (8) predicts large increases in swelling of the fibrils for the fibril stresses observed. However, initial measurements of the stress-strain properties of the fibrils of a methanol craze in PMMA suggest that large changes in swelling with stress are not necessary to account for the observed fibril mechanical properties.[40]

Another major unknown in eqn. (8) is the form of the orientation hardening (dependence of $\dot{\lambda}$ on λ) and the effect of the environment on this hardening. There is at present very little data on orientation hardening in the glassy state[47] and none on the effect of environment. Since the degree of such orientation hardening will be an important factor in the damage produced by the fibril creep mechanism, experiments which address these questions are needed.

One final generalisation can be made with regard to the two craze thickening mechanisms. If the diffusion coefficient D of the environment in the bulk polymer is small, the craze which is formed will be more prone to damage (all other factors such as the plasticisation of the fibrils being held constant) than if D is large. A smaller value of D from eqns. (6) and (7) will mean that the fibril creep mechanism will produce a larger relative contribution to the craze surface displacement rate than the surface drawing mechanism. It will be shown in the next section that the craze surface displacement rate determines the stress relaxation rate produced by a craze growing ahead of a crack tip. If D is large the major contribution to the relaxation of stress at the crack tip is the surface drawing mechanism, which does not produce a decrease in fibril volume fraction in the craze. The local crack tip stress can be relaxed to innocuous levels before there is much thinning of the craze due to fibril creep. If D is small, however, the surface drawing mechanism is only a minor contribution to the crack tip stress relaxation, which now must occur primarily by fibril creep producing craze damage in the form of ruptured fibrils and a lower fibril volume fraction v_f. If we assume there is some critical local v_f below which the craze breaks down, it is clear that if D is low this v_f will be reached at earlier times or at smaller initial local stresses than if D is large.

There is some preliminary evidence that these ideas are correct. In particular, from eqns. (1) and (7) one expects the primordial craze thickness T_0 to reflect the contribution of surface drawing to craze thickening; i.e.

$$\mathsf{T}_0 = Dt/r_0 \ln(c_0/c_d(S)) \tag{9}$$

where t is the time of exposure of the craze surface to the environment. Thus one should find that T_0 varies approximately as D from one environment–polymer system to another. For methanol in PS $\mathsf{T}_0 \approx 500$ nm, whereas for n-heptane in PS $\mathsf{T}_0 \approx 30$–60 nm. The diffusion coefficient of methanol in PS is larger than that of n-heptane by approximately a factor of 20. Methanol in PMMA, which diffuses slightly faster than methanol in PS, gives a $\mathsf{T}_0 \approx 1000$ nm.[40]

The above information on crazes has been developed primarily by study of the amorphous glassy polymers PS and PMMA. The reader at this point has a right to ask what relevance this material has to the other environmental stress cracking systems such as the metal salt–polyamide or detergent–polyolefin systems. In the case of the polyamide saturated $ZnCl_2$ the correspondence is clear. Dunn and Sansom[16,17] showed that a complex is formed between the metal salt and the polymer nylon 6 and suggested that it disrupts the inter-chain hydrogen bonding, thus plasticising the amorphous regions of the polyamide. Reimschuessel and Kim[48] examined the fracture surfaces of nylon 6, stress-cracked by aqueous solutions of saturated $ZnCl_2$, and found evidence of highly localised plastic deformation. They also observed that the saturated $ZnCl_2$ solution produced marked swelling of fine powders of any polyamide that it stress-cracked but that thicker sheets $\approx 1{\cdot}5$ mm thick were apparently unswollen in short-term tests. They suggested that this observation could be explained in terms of a very low diffusion coefficient for $ZnCl_2$, and employed much the same argument as used above for PS, namely that the combination of a good plasticising but slowly diffusing environment would be expected to produce a very weak craze.

At Cornell we have produced cracking of thin films of nylon 6–10 bonded to copper grids and have investigated the stress cracking of these films by a saturated aqueous solution of $ZnCl_2$.[49] It was observed that the first step in the cracking was the growth of a craze with a very low volume fraction of fibrils which appears not unlike the *n*-heptane crazes observed in PS. Figure 5 shows an optical micrograph of a craze produced by this environment. These results provide excellent confirmation of the hypothesis of Dunn and Sansom and Reimschuessel and Kim and indicate that environmental stress cracking of the polyamides by metal salts involves the production of a weak craze which breaks down to form a crack.

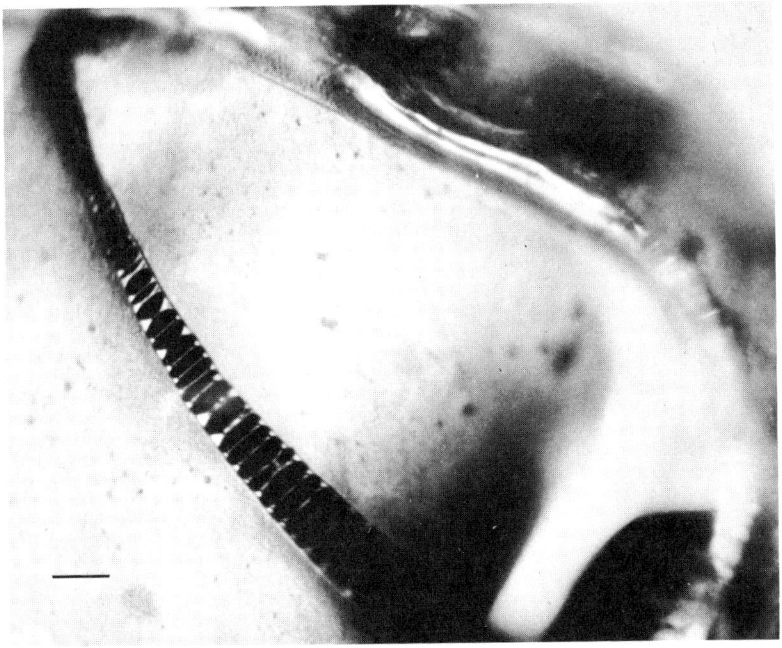

FIG. 5. Optical micrograph of a craze in a film of nylon 6–10 induced by exposure to a saturated aqueous solution of $ZnCl_2$. Bar represents 10 μm.

The mechanism of detergent and alcohol stress cracking of polyolefins appears to be somewhat more complicated. Recent experiments are in agreement that the seemingly brittle plane-strain (thick-section) crack advance actually involves voiding, polymer ligament formation between voids and subsequent plastic failure of the ligaments.[50–53] Most of the evidence comes from examination of the fracture surface resulting from ESC, and there are no direct observations of craze-like cavitation in the bulk material ahead of the crack, so that one has no idea how far ahead of the crack this cavitation zone extends. There are transmission electron microscopy observations of craze-like cavitation zones in both melt-crystallised polyethylene (PE) thin films and in PE single crystals[52,53] drawn on Mylar substrates. Figure 6 shows the craze-like deformation of polyethylene crystals deformed on Mylar substrates in methanol, an ESC agent. No such craze or fibrillation was observed in PE crystals drawn to similar strains in air; only uniform deformation was observed.[51,53]

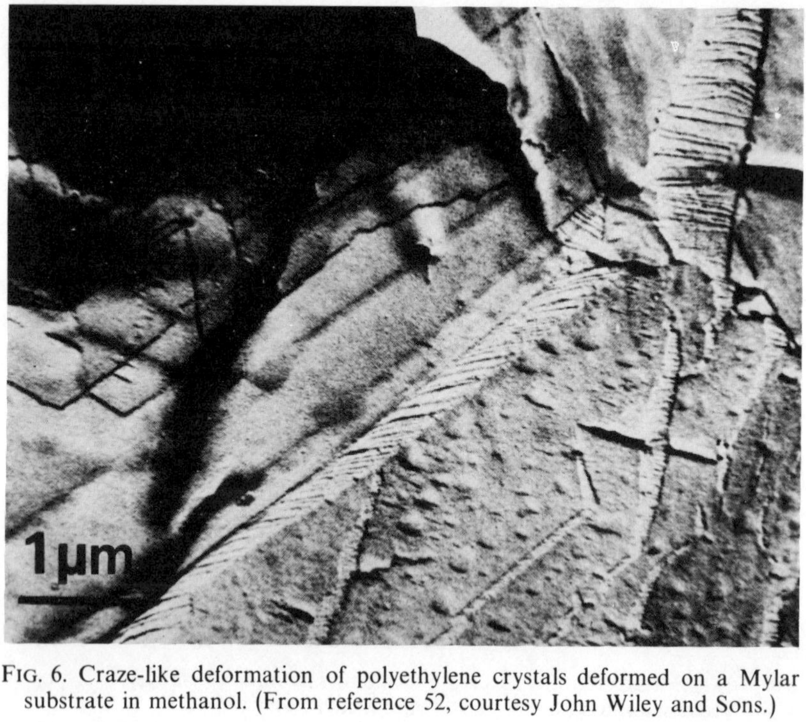

FIG. 6. Craze-like deformation of polyethylene crystals deformed on a Mylar substrate in methanol. (From reference 52, courtesy John Wiley and Sons.)

It thus seems fairly certain that a propagating environmental stress crack in polyethylene or other polyolefins is preceded by a porous cavitated zone much like the craze(s) formed in the amorphous and semi-crystalline glassy polymers and that this zone differs substantially from the plane-strain deformation zone at the advancing crack tip in air. The fundamental cause of this environmental change in the plastic zone is debatable, however. It would seem that the plasticisation arguments given to account for environmental craze formation at low stresses in glassy polymers are not applicable to polyethylene since the T_g of the amorphous regions of PE is generally claimed to be well below room temperature.[54] Conventional wisdom[4,5] also has it that the ESC agents do not dissolve in PE, so a plasticising effect would for this reason, too, appear to be ruled out. Yet Soni and Geil[53] have used the IR spectra of thin (8-μm) films of linear PE soaked in methanol (an ESC agent) to show that more than 2·3% methanol is absorbed in an unstrained film and that approximately double that amount was absorbed in a film stretched 10% in tension. They also found that soaking gold-decorated single crystals of linear PE produced a mobilisation of material on the fold surfaces, similar to that observed with other liquids such as xylene and CCl_4.[55,56]

It has also been suggested[52] that methanol and other ESC agents affect the deformation of the crystal core, largely based on the evidence of the single-crystal 'crazing' referred to above. It is claimed that methanol 'plasticises' mosaic block boundaries, weakening the cohesion at these interfaces.[51,52] While this hypothesis seems superficially attractive, the fact that recent experiments demonstrate that mosaic blocks of small size do not exist in PE single crystals casts considerable doubt on this idea.[57,58] Nevertheless it is possible that the environmental liquid does affect the deformation processes in the crystal core, perhaps by making it easier to nucleate crystal defects such as dislocations and chain defects at the mobilised fold surfaces.

Finally the environmental agent may aid the fibrillation process by reducing the surface energy of the polymer.[51] The whole question of surface energy decrease *versus* plasticisation has a long history in the debate on the mechanism of environmental crazing in glassy polymers and it will be discussed in detail below. For now suffice it to say that if the flow (or drawing) stress of the polymer in the ligaments in the cavitated zone is low enough, and if the ligaments in this zone are small enough, the surface tension of the ligaments can make a large contribution to the stress borne by such ligaments. Thus, decreasing this surface tension with an environmental liquid would enhance cavitation and the breakdown of ligaments. It is difficult to assess this mechanism in the case of environmental stress cracking of polyethylene since there is very little experimental information about the microstructure (e.g. ligament diameter) of the cavitated zone.

Crazes formed in gaseous environments (e.g. N_2, A, O_2, CO_2, etc.) at low temperatures are an interesting exception to the general proposition that environmental crazing normally leads to a decrease in tensile ductility. Crazing can be so profuse in these gaseous environments as to produce the phenomenon of craze yielding, whereby the overall strain rate of the sample due to the crazing actually exceeds the imposed strain rate so that the applied stress drops. Plastic tensile strains-to-fracture of more than 10% can be achieved in gaseous environments at temperatures where the strain-to-failure would be $< 5\%$ in vacuum. Crazing does generally occur at stresses below the fracture stress in vacuum. In a constant load (creep) or static fatigue test, even though the strain-to-failure is usually enhanced by the gaseous atmosphere, the time-to-failure is almost always decreased. One must consider this phenomenon to be environment-assisted failure, although unlike most environmental stress cracking it does not occur in a macroscopically brittle mode.

These observations indicate that crazes produced at low temperatures by gaseous environments are stronger and more resistant to craze breakdown than the usual liquid crazes produced at room temperature. Unfortunately no experiments on the microstructure of such crazes have been conducted to date. Recently at Cornell we have shown that it is possible to grow CO_2 crazes in PS thin films at a temperature (31 °C) above the critical temperature of CO_2 if the CO_2 pressure is increased to approximately 50 atms (5 MPa). This result implies that it should now be possible to begin to investigate the microstructure (and mechanical properties) of gas crazes grown at room temperature and elevated gas pressure (rather than at low temperatures and one atmosphere). Besides convenience, these experiments offer the possibility of studying the effects of gas concentration in the polymer on craze microstructure and strength by changing gas pressure while holding temperature, and thus the mechanical properties of the pure polymer, fixed.

2.2 Craze Micromechanics

2.2.1 Stress–Displacement Relationships

Although the microstructural information in the section above is a valuable qualitative guide to the mechanisms of environmental stress cracking, for many purposes, including treatment of environmental stress cracking by the methods of fracture mechanics, one needs quantitative information on the micromechanics of the craze and the mechanical properties of the fibrils. In this section the available experimental and theoretical methods for obtaining such information are reviewed.

From the mechanics point of view it is difficult to imagine a zone of plastic deformation that is more geometrically simple than an isolated craze in a glassy polymer. The craze has surfaces that are planar, almost parallel, and very close together since craze thicknesses rarely exceed a few microns. The bulk glassy polymer outside the craze can usually be treated as being linear-elastic, a condition that is more easily satisfied at the lower applied stress levels corresponding to environmental crazing. The fibrils inside the craze are usually accurately perpendicular to the craze surfaces (see Fig. 2) and, by virtue of their shape, can only exert normal tensile stresses S on the craze surfaces. For these reasons an isolated craze through the thickness of a film or plate of large extent compared to the craze length can be accurately modelled as a loaded planar slit in an infinite linear-elastic sheet. The problem in linear elasticity is illustrated in Fig. 7. The x-coordinate is taken from the centre of such a craze along the craze length, the tips of the craze are at $x = \pm a$, the y-coordinate is in a direction normal to the plane of the craze and the z-coordinate is orthogonal to x and y. Because of the geometrical simplicity there exist reasonably simple mathematical methods to determine the craze surface stress profile $S(x)$ if the surface displacement profile $w(x)$ is

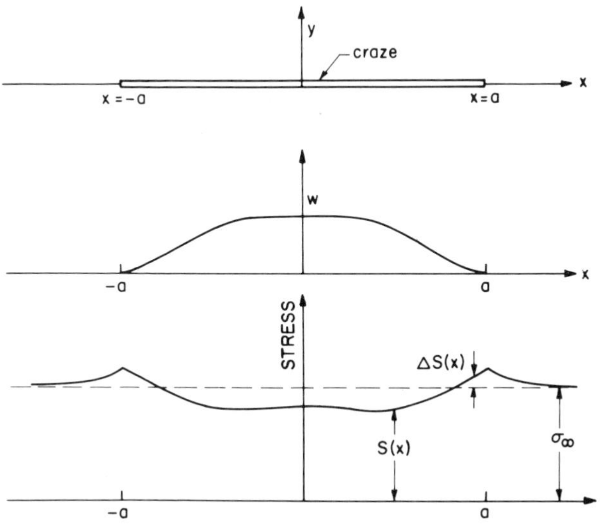

FIG. 7. Craze surface displacement profile $w(x)$ and corresponding craze surface stress profile $S(x)$ (schematic). $\Delta S(x)(= S(x) - \sigma_\infty)$ is the 'self surface stress' profile of the craze, the stress profile that would be necessary if the displacement profile $w(x)$ were maintained in an unloaded sheet ($\sigma_\infty = 0$).

known (see Fig. 3) and *vice versa.* To determine the stresses in the plate one must solve the biharmonic equation

$$\nabla^2\nabla^2\chi = 0 \tag{10}$$

subject to the boundary conditions, where χ is the Airy stress function and where

$$\sigma_{xx} = \frac{\partial^2\chi}{\partial y^2}, \quad \sigma_{yy} = \frac{\partial^2\chi}{\partial x^2} \quad \text{and} \quad \sigma_{xy} = -\frac{\partial^2\chi}{\partial x\,\partial y}$$

To simplify the problem still further it is useful to determine the stresses $\Delta\sigma$ due to $w(x)$ in the absence of the uniform applied stress σ_∞ (assumed to be tensile along the y-direction) and then superpose the uniform stress solution later. The stresses on the surface of the craze are

$$S(x) = \Delta S(x) + \sigma_\infty = \Delta\sigma_{yy}(x, y = 0) + \sigma_\infty \tag{11}$$

Sneddon[59] has developed a Fourier transform method for solving eqn. (10) that allows one to compute $\Delta S(x)$ from $w(x)$, *viz.*

$$\Delta S(x) = -\frac{2}{\pi}\int_0^\infty \bar{p}(\xi)\cos(x\xi)\,\mathrm{d}\xi \tag{12}$$

where $\bar{p}(\xi)$ is given by

$$\bar{p}(\xi) = \frac{\xi E^*}{2}\int_0^a w(x)\cos(x\xi)\,\mathrm{d}x$$

and where

$$E^* = \begin{cases} E/(1 - \nu^2) & \text{for plane strain} \\ E & \text{for plane stress} \end{cases}$$

and where E is Young's modulus of the bulk undeformed polymer. Once $\bar{p}(\xi)$ is known all stresses in the plane at any point outside the craze are determined:

$$\sigma_{yy}(x, y) = -\frac{2}{\pi}\int_0^\infty (1 + \xi|y|)\exp(-\xi|y|)\bar{p}(\xi)\cos(x\xi)\,\mathrm{d}\xi + \sigma_\infty \tag{13a}$$

$$\sigma_{xx}(x, y) = -\frac{2}{\pi}\int_0^\infty (1 - \xi|y|)\exp(-\xi|y|)\bar{p}(\xi)\cos(x\xi)\,\mathrm{d}\xi \tag{13b}$$

$$\sigma_{xy}(x, y) = -\frac{2y}{\pi}\int_0^\infty \xi\exp(-\xi|y|)\bar{p}(\xi)\sin(x\xi)\,\mathrm{d}\xi \tag{13c}$$

The form given here is for the even $w(x)$ profile, $w(x) = w(-x)$, usually found for crazes although profiles of other symmetries can be handled with Sneddon's original formalism. Inverting the problem to find $w(x)$ knowing $S(x)$ between $x = a$ and $x = -a$ is cumbersome using this method and requires the solution of dual integral equations.

As with most elasticity problems there is more than one way to solve this one. A second method that has been used primarily by metallurgists to solve the loaded slit problem is the method of distributed dislocations.[60] One imagines that the surface displacements $w(x)$ are caused by a density $\mathscr{D}(x)$ of dislocations with infinitesimal Burgers vector **b** (in the case of the craze these are edge dislocations lying in the z-direction and **b** is in the y-direction). (It is to be emphasised that the actual structure of the craze is not that of a dislocation array. Rather, the elastic stresses in the region outside the craze are the same whether the surface displacement profile $w(x)$ is produced by an array of dislocations or by a set of craze fibrils. For more discussion of this point see reference 60.) From the definition of the Burgers vector one finds that

$$\mathscr{D}(x) = -\frac{2}{b}\frac{\mathrm{d}w(x)}{\mathrm{d}x} = -\frac{2}{b}w'(x) \tag{14}$$

Since the stress at point x due to the dislocations between x' and $x' + \mathrm{d}x'$ is

$$\sigma_{yy}^{\mathrm{D}}(x, y = 0) = E^* b\mathscr{D}(x')\,\mathrm{d}x'/(4\pi(x - x')) \tag{15}$$

it follows that $\Delta S(x)$ in eqn. (11) is given by

$$\Delta S(x) = -\frac{E^*}{2\pi}\int_{-a}^{a}\frac{w'(x')\,\mathrm{d}x'}{x - x'}. \tag{16}$$

Using the Faltung or convolution theorem one can show that eqns. (16) and (12) are mathematically identical. To compute all components of the stress tensor at an arbitrary point in the x, y plane one can simply generalise eqn. (15) using the expression for the full stress field of an edge dislocation.[61]

Both the above methods are difficult to use to determine $w(x)$ given $\Delta S(x)$ in the range $-a < x < a$. The inversion problem is most conveniently solved using the conformal mapping technique of Muskhelishvili.[62] One first defines the angular variable $\beta = \arccos(x/a)$. The displacements are then given by[63,64]

$$w(\beta) = \frac{a}{\pi E^*}\int_{0}^{\pi/2}\frac{\mathrm{d}\Delta S}{\mathrm{d}\beta'}\mathscr{H}(\beta, \beta')\,\mathrm{d}\beta' \tag{17}$$

where

$$\mathscr{H}(\beta, \beta') = \cos\beta \ln\left[\frac{\sin^2(\beta' - \beta)}{\sin^2(\beta' + \beta)}\right] + \cos\beta' \ln\left[\frac{(\sin\beta' - \sin\beta)^2}{(\sin\beta' + \sin\beta)^2}\right] \quad (18)$$

The possible surface stress profiles are constrained by the condition that no stress singularity appear at the craze tips (at $x = \pm a$), which yields

$$\frac{\pi}{2}\sigma_\infty = \int_0^{\pi/2} S(\beta')\,\mathrm{d}\beta' \quad (19a)$$

or

$$\int_0^{\pi/2} \Delta S(\beta')\,\mathrm{d}\beta' = 0 \quad (19b)$$

This condition is most easily derived from Eshelby's[65] expression for the stress intensity factor K_{I} for a crack loaded only on the crack plane by an even function $S(x)$, i.e.

$$K_{\mathrm{I}} = -\sqrt{\frac{a}{\pi}}\int_{-a}^{a} \frac{\Delta S(x)}{(x^2 - a^2)^{1/2}}\,\mathrm{d}x \quad (20)$$

since the condition of no stress singularity implies that $K_{\mathrm{I}} = 0$.

Although all three methods solve the same elasticity problem, each is more convenient than the others for a particular purpose. For example, if one wishes to compute the full surface stress profile $S(x)$ from experimental displacement profile data, the Fourier transform method is most convenient since fast Fourier transform programs running on modern digital computers are the fastest method of transforming a full set of $w(x)$ data to $S(x)$ data. On the other hand if very local $S(x)$ information is desired at one point along the craze (e.g. the craze tip) the distributed dislocation method offers advantages. Since from eqn. (16) $\Delta S(x)$ is most sensitive to the displacement derivatives in the immediate vicinity of x, very closely spaced $w(x)$ measurements can be made near the desired point and more widely spaced w measurements can be made over the remainder of the craze. Since only one integral is necessary per stress point this method is quite cost-effective if only a few closely spaced points are needed. Furthermore it has an additional advantage over the Fourier transform method if the highest spatial resolution is required. In the Fourier transform method it is necessary to attenuate the high ξ points of $\bar{p}(\xi)$ by multiplying $\bar{p}(\xi)$ by a damping function (typically $\exp(-\alpha\xi^2)$ where α is a constant) to avoid introducing spurious wiggles in $\Delta S(x)$ due to the fact that the numerical evalution of the infinite integral in eqn. (12) must be cut off at an upper limit of $\xi = \xi_{max}$. This procedure leads to a loss of potential spatial resolution (which, however, can be minimised through a proper choice of α) which is avoided completely in the distributed dislocation method.

Finally, the Muskhelishvili method is most useful if one wants to explore the consequences of assuming a certain surface stress profile on the $w(x)$ profile, as well as on the craze length.

2.2.2 *Models of Craze Micromechanics*

These can be divided into models of an isolated craze (with no crack associated with the craze) and models of crazes at crack tips. Chronologically the first model of an isolated craze was published by Knight.[66] He assumed the craze surface displacement was constant over most of the craze and decreased to zero only over a small length near the tip. He claimed that only one tip displacement profile would give finite stresses at the tip and using the Fourier transform method he computed the surface stress profile resulting from this tip displacement profile. However, his claim of a unique $w(x)$ function near the tip is incorrect as was first pointed out by Verhuelpen-Heymans.[67] It is perhaps easiest to show this using the distributed dislocation method. From eqn. (16) it is apparent that any $w(x)$ which has a continuous derivative w' which vanishes at the craze tip (and there are an infinite number of such functions $w(x)$) will give rise only to finite stresses. Unfortunately in correcting Knight's error Verhuelpen-Heymans makes the statement that it is not possible to obtain all components of the stress at any point in the plane using the Fourier transform method. As shown above in eqns. (13a)–(13b) this statement is not true.

Verhuelpen-Heymans and Bauwens[68] have proposed an alternative model of an isolated craze using the Muskhelishvili method. The craze surface is assumed to support a high tensile surface stress S_t over a small distance r behind the craze tip, but behind this craze tip zone the stress drops to a lower value S_c over the rest of the craze. This surface stress distribution is illustrated in Fig. 8(a) along with the predicted displacement profile $w(x)$. Note that eqn. (19) demands that $\beta_t = \arccos(1 - r/a)$ must satisfy the following relationship:

$$(S_t - \sigma_\infty)\beta_t - (\sigma_\infty - S_c)\left(\frac{\pi}{2} - \beta_t\right) = 0$$

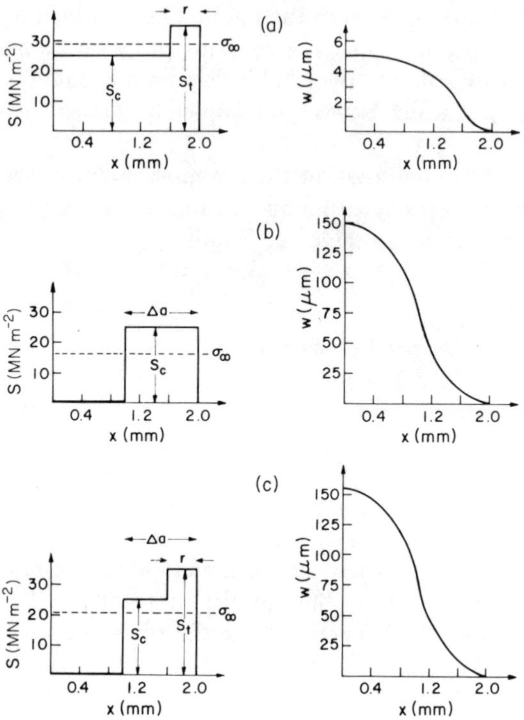

FIG. 8. Surface stress and displacement profile for: (a) the Verhuelpen-Heymans and Bauwens model of an isolated craze (after reference 68); (b) the Dugdale model of a craze at a crack tip (after reference 63); (c) a modified Dugdale model of a craze at a crack tip (after reference 140).

or

$$\beta_t = \frac{\sigma_\infty - S_c}{S_t - S_c}\left(\frac{\pi}{2}\right) \tag{21}$$

The craze displacement profile $w(x)$ is given by eqn. (17):

$$w(\beta) = \frac{a(S_t - S_c)}{\pi E^*}\mathscr{H}(\beta, \beta_t) \tag{22}$$

They make the assumption that the length of the craze tip zone r and craze tip surface stress S_t are constants and thus can predict the length of the craze and the craze growth kinetics from an $S_c(t)$, the time dependence of which they assume to be given by the time-dependent creep of the fibrils. This last assumption ignores the surface drawing contribution to the craze thickening that we have shown to be the major contribution for air crazes in PS and a non-negligible contribution in environments which diffuse rapidly into the bulk polymer. Aside from this assumption the model is the only simple model of an isolated craze which even remotely agrees with experimental measurements of craze mechanics.

Wilczynski, Liu and Hsaio[69] have proposed a different mechanical model for the craze. They represent the craze surfaces as two membranes stretched under tension parallel to the membranes (picture two rectangular drumheads attached at the edges). The space between the membranes is inflated with a uniform pressure representing the applied tensile stress perpendicular to the craze and the tendency for the membranes to be displaced outward is resisted by the fibrils which are modelled as linear elastic springs, the spring constant of which varies with position along the craze. The most charitable comment that can be made about this model is that it is physically unrealistic, both in the assumption of linear elasticity of the fibrils and in the representation of the craze surfaces as membranes under tension. There is no provision to limit the stress in the bulk polymer at the craze tip to finite values. Indeed, eqn. (19b) [$\int_0^{\pi/2} \Delta S(\beta')\,d\beta' = 0$] is not satisfied by their surface stress results; this means that a stress singularity exists at their craze tip, which cannot be allowed on physical grounds. (Another way of putting this point is that their $w'(x = a) \neq 0$.) It is also easily shown that their model will not give the well-known displacement profile for a crack if the spring constant of the fibrils is set equal to zero; surely any model of a craze must do this. Models which ignore the inhomogeneous elastic stress field in the bulk polymer surrounding the craze, as this one does, will not give meaningful results.

Crazes at crack tips may be modelled easily using the Muskhelishvili formalism developed above, setting $S(\beta') = 0$ over the crack. The simplest such model is called the Dugdale model,[70] although the detailed mechanics were first worked out by Goodier and Field.[63] This model, which has been used extensively to represent plane-stress plastic zones at crack tips in metals,[64,71] assumes that the surface stress S_c is constant over the plastic zone or craze in advance of the crack tip as illustrated in Fig. 8(b). To remove the stress singularity at the craze tip the angular variable β_c of the crack tip from eqn. (19) must be

$$\beta_c = \frac{\pi}{2}\frac{\sigma_\infty}{S_c} = \frac{1}{2}\sqrt{\frac{\pi}{a_0}}\frac{K_I}{S_c} \tag{23}$$

and the displacement profile from eqn. (17) is

$$w(\beta) = \frac{aS_c}{\pi E^*} \mathscr{H}(\beta, \beta_c) \tag{24}$$

The displacement at the crack tip, which assumes importance in various fracture criteria, is

$$w_c = w(\beta_c) = \frac{4a_0 S_c}{\pi E^*} \ln(\sec \beta_c) \tag{25}$$

where a_0 is the half-length of the crack. Under conditions where the length of the craze Δa is much less than a_0, the following direct expressions for Δa and w_c can be derived from eqns. (24) and (25):

$$\Delta a = \frac{\pi^2 \sigma_\infty^2 a_0}{8S_c^2} = \frac{\pi}{8} \frac{K_I^2}{S_c^2} \tag{26}$$

and

$$w_c = \frac{\pi \sigma_\infty^2 a_0}{2E^* S_c} = \frac{K_I^2}{2E^* S_c} \tag{27}$$

Here the approximate stress intensity factor for the slit geometry, i.e. $K_I = \sigma_\infty \sqrt{\pi a_0}$, has been substituted to obtain the last equation.

A modification of the simple Dugdale model is prompted by the fact that if an isolated craze (without a crack) truly has a flat $S(x)$ profile it will not grow. (One can satisfy eqn. (19) only when $\sigma_\infty = S_c$.) To remove this difficulty one can add a craze tip zone of length r with a higher surface stress S_t, as in the Verhuelpen-Heymans and Bauwens model of an isolated craze. Models of this type have been proposed by Gerberich,[72] by Williams and Marshall[73] and by Kramer and Bubeck.[43] The surface stress profile for this modified Dugdale model is shown in Fig. 8(c). From eqn. (19) one finds

$$\beta_c = \frac{\pi}{2} \frac{\sigma_\infty}{S_c} - \beta_t \left(\frac{S_t}{S_c} - 1 \right) \tag{28}$$

$$w(\beta) = \frac{a}{\pi E^*} [(S_t - S_c)\mathscr{H}(\beta, \beta_t) + S_c \mathscr{H}(\beta, \beta_c)] \tag{29}$$

$$w_c = \frac{a}{\pi E^*} \left[(S_t - S_c)\mathscr{H}(\beta_c, \beta_t) + \frac{4a_0}{a} S_c \ln(\sec \beta_c) \right] \tag{30}$$

Gerberich[72] assumes the tip zone r is of fixed length and represents the zone of plastic deformation in the bulk polymer ahead of the cavitating craze. Williams and Marshall[73] and Kramer and Bubeck,[43] on the other hand, use their models to describe solvent craze mechanics and r is assumed to be a zone at the craze tip where the fibrils are incompletely plasticised. In one limit this zone can be considered to be a short length of air craze of such a length as is necessary to satisfy eqn. (19) as the craze grows (i.e. as β_c increases).

2.2.3 Experimental Measurements of Craze Mechanics

Perhaps the crudest, but nonetheless effective, method of assessing craze mechanics is to grow a craze from a crack in a given environment and determine the equilibrium or arrest length as a function of applied stress or K_I. One can assume that the Dugdale model holds, then determine β_c from the original crack length and the arrest craze length and then compute S_c from eqn. (23). Graham, Williams and Zichy[74] have made such measurements on polymethylmethacrylate (PMMA) in a variety of environments ranging from air to methanol. As shown in Table 1 there are wide variations in S_c from one environment to another. Also shown in Table 1 are data from Andrews, Levy and Willis[75] for the alcohol environments, including: ϕ_1 (the volume fraction of the liquid absorbed in

TABLE 1

CRAZE SURFACE STRESSES IN PMMA

Environment	S_c (MN m^{-2})	ϕ_1 *at* 293 K	δ_1 (MJ m^{-3})$^{1/2}$	T_g^s °C	K_m MN m$^{-3/2}$
	(a)	(b)	(b)	(b)	(a)
Methanol	7·0	0·23	29·7	36	0·05
Ethanol	5·2	0·24	26·0	30	0·02
n-Propanol	5·3	0·26	24·4	32	0·01
n-Butanol	9·6	0·29	23·3	43	0·07
PMMA–Air	100·0	0·0	19·0	115	0·70
Ethylene glycol	35·3				0·14
Water	86·4				0·40
White spirit	15·2				0·09
Paraffin	40·8				0·22

(a) Reference 74.
(b) Reference 75.

PMMA at equilibrium under zero stress at 293 K); δ_1 (the solubility parameter of the environment—the solubility parameter of PMMA is 19·0); and T_g^s (the glass transition temperature of the PMMA swollen to equilibrium by the environment; it is estimated to be the temperature at which the yield stress of the swollen PMMA extrapolates to zero). The last column gives the minimum value of the stress intensity factor K_I above which craze growth is observed. There appears to be a reasonable correspondence between S_c and T_g^s, in the cases where the latter has been measured, indicating that the major cause of the decrease in S_c from air to alcohol environments is due to plasticisation. Note also that the correlation of S_c with either alcohol absorption or solubility parameter is imperfect, with butanol being absorbed to a larger extent but producing less decrease in T_g than the lower alcohols. On the basis of the Gibbs–Di Marzio theory[76] one would, in fact, expect alcohols of higher molecular weight to be poorer plasticisers (they occupy more sites on the quasi-lattice) than alcohols of lower molecular weight. The S_c value follows the increase in T_g rather than decreasing further with the increase in solute absorption.

The trends observed in the measured values of S_c are satisfying but in fact are not proof that the simple Dugdale model is appropriate, i.e. that $S(\beta') = S_c$ everywhere along the craze. For eqn. (19) any stress profile such that $(2/\pi) \int_0^{\beta_c} S(\beta')\, d\beta' = \sigma_\infty$ will result in a craze of the same length in mechanical equilibrium and the measured value of $S_c = (1/\beta_c) \int_0^{\beta_c} S(\beta') d\beta'$. While one can attempt to vary σ_∞ and observe whether the increase in craze length (and the corresponding increase in β_c) is as predicted for a certain profile, it is also true that one should not necessarily expect a given S profile to be a unique function of β' independent of σ_∞. The only way to check further the validity of the Dugdale model is to develop methods of measuring local values of S.

Bessenov and Kuvshinskii[77] and Kambour[34] first showed that if one views a craze at a crack tip at normal incidence with a reflection microscope one observes interference fringes which can be interpreted in terms of changes in the optical thickness of the craze, $n_c \mathrm{T}$. There have been various attempts[78–83] to use this method quantitatively to determine the $w(x)$ profile and thus check on the validity of the Dugdale model. To determine $w(x)$ rigorously, however, it is necessary to determine the craze refractive index profile $n_c(x)$ as well as the optical thickness profile. (Once $n_c(x)$ is known it can be used to determine both the fibril volume fraction profile $v_f(x)$ and $\mathrm{T}(x)$.[82]) Usually values of the refractive index of a craze not under load are taken from the literature[78,80,81] or determined from critical angle measurements[79] and then the change in the fringe pattern on loading is used to find an average value of n_c, which is assumed to be roughly constant over the craze. Values of $\frac{1}{2}T(x)$ are then computed from the optical thickness and these values are taken as $w(x)$ for comparison with the values predicted for the simple Dugdale model from eqn. (24).

Flawed though these comparisons are by the approximations that are necessary, they indicate that dry crazes at crack tips in PMMA and polyvinylchloride (PVC) follow the predictions of the Dugdale model rather well. It is not possible, however, to determine precisely the displacements (and thus stresses) near the craze tips where there are few fringes (in fact it is difficult to determine exactly the position of the craze tip in this method). Krenz[83] has attempted to apply this method to environmental crazes (ethanol crazes in polycarbonate (PC)) using the shift in the fringe position on drying of the craze under load to determine $n_c(x)$. While this method works, the random error in n_c thus determined is such that one is probably just as well-off assuming a constant value of n_c.

Instead of simply comparing the observed displacement profile with that expected from the Dugdale model, it should be possible to convert the displacement profile to a surface stress profile using either the Fourier transform or the distributed dislocation method. Such a comparison has not been forthcoming for the optical craze thickness data, perhaps because of the difficulties in determining the $n_c(x)$ profile. It is possible to measure $w(x)$ directly from holographic interferometry[39,40,84–86] and to compute $S(x)$ from this profile. Figure 9(a) shows such a determination of $S(x)$ from $w(x)$ for a methanol craze grown from a crack in PMMA.[40] (This craze grew from a crack in the edge of a PMMA strip rather than from a central slit as assumed in the analysis. Nevertheless, good agreement is found if the actual applied stress σ_∞ is used in the computation, rather than the fictitious $\sigma_\infty = \sigma_\infty^1$ needed to match the K_I of the central slit to that of the edge crack.[40]) With the holographic method one also can estimate the $S(x)$ more directly by measuring the change in longitudinal strain $\Delta\varepsilon_{yy} = \Delta w/\Delta y$, due to the craze growth, along a line parallel to the craze but displaced slightly from it. Figure 9(b) shows a determination of $\sigma_{yy}(\approx S(x))$ at x, $y = 0{\cdot}32$ mm for the same craze as that in Fig. 9(a).[40] The two methods are in reasonable agreement. In Fig. 9(a) the results predicted by the simple Dugdale model for a craze of the same length grown at the same σ_∞ are also shown. The stress beyond the tip of the craze is reasonably well predicted by this model but the stress distribution along the craze is nothing like the assumed surface stress distribution of the model. The stress goes through a maximum at or just behind

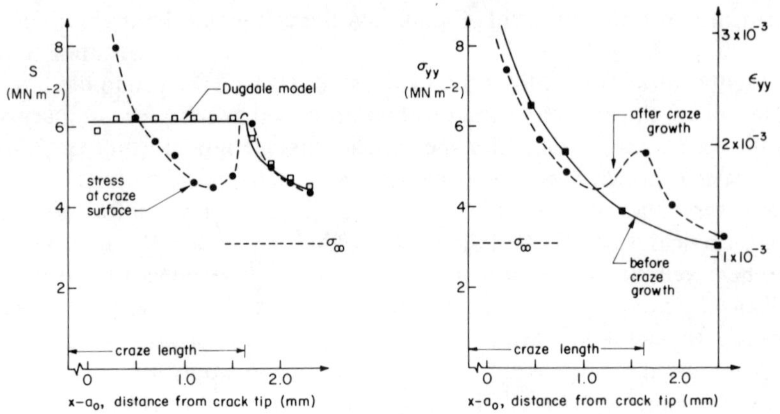

FIG. 9. (a) Surface stress profile for a methanol craze in PMMA from the $w(x)$ profile using the Fourier transform method. (b) Stress profile along a line parallel to the craze in (a) but displaced 0·32 mm, determined from the strain profile measured directly from the fringe pattern.

the craze tip, then decreases and finally increases strongly as the crack tip is approached. The average craze surface stress $\langle S_c \rangle = (1/\beta_c)\int_0^{\beta_c} S(\beta')\,d\beta'$, however, is in good agreement with the value measured in this system by Graham *et al.*[74] from data on craze length *versus* σ_∞. This general shape of the $S(x)$ profile, showing local maxima near the craze and crack tip, seems to be typical of crazes formed in environments which do not plasticise the polymer below the craze growth temperature. Figure 10(a) shows the stress profile from a methanol craze in PS ($T_g^s = 90°C$) which appears to be similar to the methanol craze in PMMA.[84] The measured stress profile for a highly plasticised ($T_g^s = 6°C$) *n*-heptane craze in PS (Fig. 10(b)) is profoundly different. Besides the much lower level of surface stress S borne by this craze, the stress profile is markedly peaked at or just behind the craze tip and decreases continuously toward the crack tip. From the mechanics viewpoint the absence of any hardening of the craze structure in regions of large surface displacement (near the crack tip) is a major factor in the weakness of these *n*-heptane crazes and explains why *n*-heptane is a much more severe stress cracking agent than methanol for PS. From the microstructural point of view, however, it is ultimately the rheology of the fibrils and the resulting craze microstructure that governs this surface stress profile and the mechanical behaviour exhibited is certainly consistent with the differ-

(a)

(b)

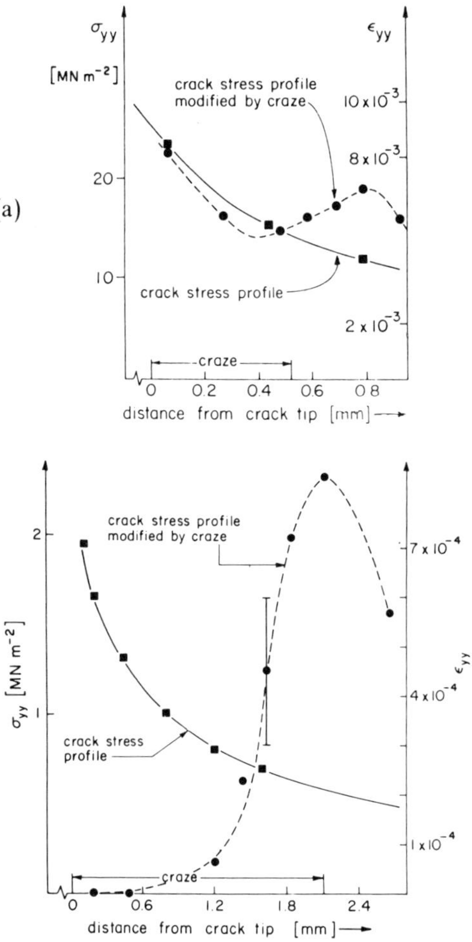

FIG. 10. (a) Surface stress profile for a methanol craze in polystyrene at 363 K. (b) Surface stress profile for an *n*-heptane craze in polystyrene at 279 K.

ence in structure of methanol and *n*-heptane crazes in PS displayed in Fig. 2.

The holographic technique has important limitations. While it can be applied to measure the properties of crazes in systems where long (> 0·2 mm) isolated single crazes can be grown, such as the ones mentioned above, it is not suitable for measuring the properties of short crazes or of resolving an individual craze within a bundle of crazes at a crack tip. The problem is one of both displacement and spatial resolution. The displacement resolution of the technique is about half the wavelength of light used, or $\approx 0{\cdot}3$ μm, which is the displacement increment between fringes on the image. It cannot provide meaningful information on crazes with smaller opening displacements than this resolution limit, and even for crazes with larger displacements, the displacement information is only available at those discrete positions where fringes enter the craze. The relatively small number of displacement points effectively means that the surface stresses computed by either the Fourier transform or the distributed dislocation method are averaged over the distance between fringes. If one wants information on very local stresses, say at intervals of 1 μm behind the craze tip, conventional holographic interferometry (or any optical interference technique such as interference microscopy) cannot provide this information.

Transmission electron microscopy of crazes in polymer films offers a way to improve both the spatial and displacement resolution. As mentioned above it is possible to measure values of the fibril volume fraction v_f in very local regions along the craze by measuring the optical density on an electron image plate. If one also measures the craze thickness T from the plate, one can determine the local craze surface displacement $w(x)$ which is given by $w(x) = \frac{1}{2}T(x)(1 - v_f(x))$. An analysis of the errors involved indicates that $w(x)$ can be determined to within 100 Å.[35] Not only is the displacement resolution much better than with previous methods but one can measure discrete displacement values at intervals as closely spaced as 1000 Å along the craze. Hence one can begin to make statements about stresses on a scale of about 1 μm.

To date only an isolated air craze ≈ 200 μm long in PS has been analysed in detail. The stress profile for this craze shows a small maximum in $S(x)$ (15% above the applied stress) at the craze tip and decreases gradually with distance to a value $\approx 10\%$ below the applied stress which then stays approximately constant over the central portion (≈ 140 μm) of the craze. The best simple model appears to be the two-stage model of Verhuelpen-Heymans and Bauwens although their proposed mechanism of thickening does not operate in this case.

2.2.4 *Mechanical Properties of the Craze Fibrils*

For many purposes one wants more information than simply the surface craze stress, i.e. the overall stress exerted on the craze boundary by the craze fibrils. One would like, for example, the curve of true stress *versus* elongation ratio for the craze fibrils themselves. Kambour and Kopp[87] produced thick single ethanol crazes in PC, dried them under load, removed the load, allowed the craze to relax at zero stress for seven weeks and then measured the subsequent displacements of the dried craze surfaces as a function of applied stress using an optical microscope. Zero true strain in the craze fibrils was difficult to establish (it was difficult to measure the initial strain in the relaxed craze), so strain was measured from the initial relaxed condition and no correction was made to determine the true stress in the fibrils. The stress–strain curve they measured has an initial modulus approximately equal to that of the bulk unoriented polymer, but at moderate stresses 'yielding' of the craze matter occurred with an accompanying large increase in strain, while at still higher stresses the true slope of the curve increases again, perhaps indicating that strain hardening of the fibrils is occurring. Peterson *et al.*[86] observed similar behaviour for dried ethanol crazes in PC that did not traverse the entire specimen and Hoare and Hull[88] observed similar stress–strain curves of profusely air-crazed PS samples. However, none of these measurements of the craze stress–strain curve of pre-crazed specimens is really relevant to the conditions in the fibrils when the craze formed. (In particular, the removal of the ethanol environment must change the properties of the fibrils considerably.)

Recently holographic measurements of the $w(x)$ profile resulting from the growth of methanol crazes in PMMA have been combined with measurements of craze thickness profile $\mathsf{T}(x)$ from scanning electron micrographs of replicas of the specimen surface.[40] From such measurements $S(x)$ is determined using the Fourier transform method discussed above, $\lambda(x)$ is computed $(\lambda(x) = \mathsf{T}(x)/(\mathsf{T}(x) - 2w(x))$ and finally the profile of true stress in the craze fibrils $\sigma_t(x) = S(x)\lambda(x)$ is determined. Since $\lambda(x)$ varies significantly along the craze it is possible to compile a fibril stress–elongation ratio curve from pairs of $\sigma_t(x) - \lambda(x)$ points along the craze. This curve, displayed in Fig. 11, shows that mechanical properties of the fibrils are greatly affected by plasticisation. The extrapolated fibril yield stress of 8 MN m^{-2} is much lower than $\sigma_y \approx 100$ MN m^{-2} for dry PMMA but is very close to the yield stress measured by Andrews *et al.*[75] on bulk specimens of PMMA swollen to equilibrium with methanol. The other feature of note is the remarkable strain hardening (orientation hardening) of the fibrils.

One of the persistent points of controversy in the crazing literature is whether reduction of surface energy plays a role in enhanced environmental crazing. Knowing roughly the fibril diameters in methanol PMMA crazes ($\approx$ 200 Å) one can put an upper limit on the total contribution of

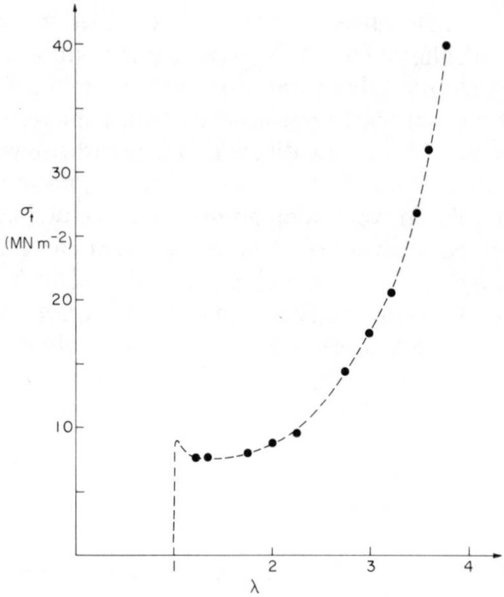

FIG. 11. True stress in the craze fibrils versus fibril elongation ratio for a methanol craze in PMMA.

surface tension to the stress in the fibrils. For PMMA methanol this upper limit is $\approx 3{\cdot}4$ MN m^{-2} and more realistic estimates put the figure at about 0·6 MN m^{-2}. The surface tension contribution will only be a significant contribution to the total stress in the fibrils when these are plasticised to around room temperature and below. One would expect the surface tension contribution to be totally negligible in dry crazing but important in the environmental stress cracking of PE where the amorphous regions at room temperature are above their T_g.[51] Andrews and his co-workers[75,89] have also stressed this point in relation to G_c, the critical value of strain-energy release rate for environmental crack growth, which at high temperatures is independent of T but which at a temperature $T_c \approx T_g$ of the fibrils increases rapidly with decreasing temperature. They propose that the work done at the crack tip above the fibril T_g is done mainly against the force of surface tension, whereas below the fibril T_g the flow stress of the fibrils, which is markedly temperature-dependent, is more important.

2.2.5 *Craze Failure Mechanisms*

Ultimately the environmental stress cracking process involves the advance of a crack into the environmentally affected craze. Fibrils at the crack tip (craze base) must fail by some mechanism. Almost nothing is known experimentally about this process, either in the dry craze or in the environmental craze. It is known from fractography[90–92] on dry crazes that the initial crack advance within a pre-existing craze takes place in the mid-rib, the region of lowest fibril volume fraction and highest λ. The crack front is rounded, with a radius of curvature approximately half the craze thickness. The broken fibrils conform to this interface and therefore, after passage of the crack, point in the direction of crack advance, leading to the familiar brush-mark texture on the fracture surface. In addition to the main crack front advance, secondary fractures can occur ahead of this front.[91] (These often seem to be associated with second-phase particles.)

These observations on dry crazes have led several authors[90,92] to describe the craze failure process as one of viscous rupture of the craze fibrils. This process could be viewed as the ultimate phase of the fibril creep mechanism for craze thickening described above and should therefore depend on the strain hardening capability ($\mathrm{d}\sigma_t/\mathrm{d}\lambda$) of the fibril. A very simple criterion for fibril failure which neglects strain rate effects would hold that fibril failure occurs in a manner much like the plastic failure of ductile metals when $\sigma_t/\lambda > \mathrm{d}\sigma_t/\mathrm{d}\lambda$ for all larger values of λ. (This condition can be derived from Considère's construction using the method outlined by Vincent.[93]) The ultimate properties of the fibril, in particular λ_f, the extension-ratio-to-failure, would depend markedly on the post-yield strain hardening of the fibrils. Similar ideas have been outlined previously by Haward and co-workers.[47,51]

In the case of the environmental stress cracking of PE there is strong circumstantial evidence that changes in the post-yield strain hardening behaviour of PE with molecular weight (MW) are responsible for the increased susceptibility of low-MW PE to environmental stress cracking.[51] A PE with $\overline{M}_n = 1{\cdot}11 \times 10^4$ and $\overline{M}_w = 4{\cdot}70 \times 10^4$ showed little post-yield strain hardening in a normal tensile test and exhibited a stress cracking time-to-failure t_f of 2 h; a PE with $\overline{M}_n = 1{\cdot}55 \times 10^4$ and $\overline{M}_w = 1{\cdot}57 \times 10^5$ showed moderate strain hardening and a stress cracking time of 180 h, whereas a PE with $\overline{M}_n = 7{\cdot}0 \times 10^5$ and $\overline{M}_w = 6{\cdot}4 \times 10^6$ showed strong strain hardening and a $t_f > 4000$ h.

Although no measurements on bulk specimens are available the strain hardening properties of the fibrils in an amorphous polymer should depend markedly on the entanglements between molecules. Under most conditions of interest in environmental stress cracking a single molecule will not span the distance from one craze surface to the other at the crack tip. To do that the molecule in the bulk material must have spanned the primordial craze thickness T_0 before the craze formed. Measurements of the chain dimensions in the glassy polymers are now becoming available from neutron-scattering experiments.[94–96] One would expect that once the root-mean-square end-to-end distance characteristic of the chain, $\langle R^2 \rangle^{1/2}$, is less than T_0 there will be little likelihood that the molecule will span the final craze. Recently, Haward, Daniels and Treloar[97] have treated this problem statistically and have derived the probability that a molecule will span the craze for a given ratio of $T_0/\langle R^2 \rangle^{1/2}$. They find that this probability falls to less than 0·05 for $\langle R^2 \rangle^{1/2} = T_0$. In polystyrene, for example, at a molecular weight $M_n = M_w = 200\,000$, $\langle R^2 \rangle^{1/2} \approx 315$ Å, whereas the typical primordial thickness of an air craze at the crack tip is 4000 Å or larger. Thus one would expect that entanglements between molecules and the stability of these entanglements are very important in both the strain hardening of the fibrils and their ultimate failure.

In amorphous polymers our knowledge of chain entanglements derives primarily from measurements of the rheology and elasticity of melts and concentrated solutions;[98] there are no direct data on chain entanglements or their effects in the solid state. Nevertheless there are striking correspondences between the molecular weight dependence of chain entanglement in the melt and that of craze fracture in the solid.[7] Well above the chain-entanglement molecular weight M_e the critical strain-energy release rate G_{Ic} for first-crack propagation in PMMA and PS in air is generally greater than $10^4 \times 2\gamma$, where γ is the surface energy of the solid; 2γ is the lower limit for G_{Ic} for an ideally brittle solid. Once the molecular weight is decreased close to or below M_e, G_{Ic} drops rather abruptly to values less than $20 \times 2\gamma$.[99,100] The tensile fracture strength of unpre-cracked samples of PS in air also shows a rapid decline with decreasing molecular weight at about M_e,[101,102] probably reflecting both the increasing ease of breakdown of crazes to form cracks and the decrease in G_{Ic}. The microstructure of air crazes in thin films of PS also changes greatly at about M_e.[103–105] Very few fibrils are observed in crazes in PS with $M_w < 30\,000$, whereas a relatively dense fibrillar structure is observed for $M_w > 100\,000$. Molecular weight effects persist, however, for molecular weights well above M_e; for example, the craze optical thickness[82] at the crack tip in PMMA of $M_w = 2{\cdot}2 \times 10^6$ at the onset of rapid crack propagation is ≈ 3 times the craze optical thickness at the crack tip in PMMA of $M_w = 110\,000$, even though both are above the M_e ($\approx 10^4$) for PMMA. The fracture toughness G_{Ic} also increases somewhat with M_w at these molecular weights. These facts, together with the existence of slow (subcritical) crack growth for G below G_{Ic}, lead one to suspect that viscous disentanglement of the chains, rather than chain rupture, is the major mechanism of craze fibril failure in the higher-MW polymers ($M_w > M_e$). The physical principles governing this chain disentanglement in the glassy polymer solid state are essentially unknown and constitute a topic on which fundamental research is urgently needed.

For environmental crazes one expects the molecular weight effects to be accentuated. To start with the chain-entanglement molecular weight of the swollen polymer will be increased approximately as[98]

$$(M_e)_s = M_e/\phi_2 \tag{31}$$

where ϕ_2 is the volume fraction of polymer in the swollen solution and $(M_e)_s$ and M_e are the chain-entanglement molecular weights in the swollen and dry condition respectively. Moreover, it is to be expected that the addition of the small solute molecules will enhance the kinetics of chain-disentanglement processes. The importance of molecular weight on this latter process is vividly illustrated by the results of Rudd.[106] He measured the time-to-failure t_f of PS under creep conditions ($\sigma_\infty = 6{\cdot}9$ MN m^{-2}) in butanol on both a series of narrow molecular weight distribution samples (anionically polymerised) and on a second series of thermally polymerised samples with $\overline{M}_w/\overline{M}_n \approx 2$. As illustrated in Fig. 12, the time-to-failure increases dramatically with molecular weight. From the known data[107] on swelling of PS by other alcohols one can estimate ϕ_2 for butanol swelling to be $\approx 0{\cdot}93$ and the T_g of the fibrils to be ≈ 345 K. Thus M_e is increased by less than 10% and by our previous arguments T_g of the fibrils is far enough above the growth temperature to promote the initial formation of a dense fibril structure, not too different, in fact, from the methanol PS craze structure shown in Fig. 2. A reasonable hypothesis is that the chain-disentanglement process, and thus fibril failure, is highly sensitive to both molecular weight and environment. Further evidence that the craze breakdown is primarily affected is given by Rudd's sketch of the craze morphology in the low and high molecular weight samples. In low molecular weight samples where crazes break down to form cracks rapidly and thus unload their surroundings, a low density of crazes is observed prior to fracture whereas in high molecular weight samples the crazes bear load longer and a very dense system of parallel crazes forms in the sample prior to fracture.

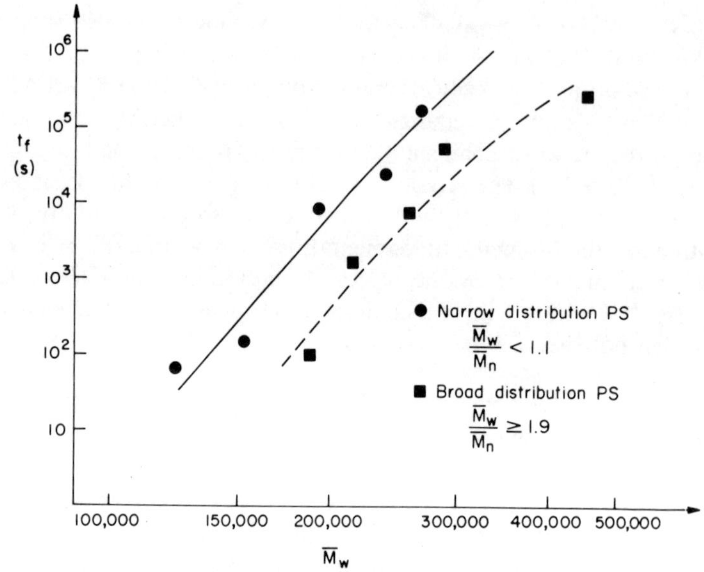

FIG. 12. Time-to-failure of polystyrene in butanol as a function of molecular weight $\bar{M}_w$ under a tensile stress of 6·9 MN m^{-2} (after reference 106).

3 KINETICS OF ENVIRONMENTAL CRAZING AND CRACKING

3.1 Craze Initiation

3.1.1 Phenomenology and Crazing Criteria

Substantial attention has been given in recent years to establishing stress criteria for crazing, the conditions under which crazes will initiate in multi-axial states of stress. Since the experiments and criteria for air crazing have been reviewed extensively[6,8,108] the results will be only briefly summarised. The most complete multi-axial experiments in air are the bi-axial tension and torsion–tension experiments of Sternstein and co-workers[109,110] on PMMA, the bi-axial tension–compression experiments of Oxborough and Bowden[111] on PS and the Argon and Hannoosh[112] torsion–tension experiments on PS. In addition, tensile crazing experiments with a superimposed hydrostatic pressure have been carried out by Matsushige, Radcliffe and Baer on PS samples sealed from high-pressure environments.[113,114] These experiments agree on certain general results, the most important of which is that a superimposed hydrostatic tension aids craze initiation whereas hydrostatic pressure retards it. Sternstein *et al.* describe their results by the following equation in terms of the first stress invariant I_1, and the constants A and B

$$\sigma_b = \sigma_1 - \sigma_2 = A/I_1 + B \tag{32}$$

They call σ_b the 'stress bias' and operationally define it as the difference between the two principal stresses in the plane of the specimen surface. As noted by Oxborough and Bowden this definition is not fundamentally satisfactory as it leads to defining σ_b as the difference between the largest and smallest of the three principal stresses for torsion–tension experiments but as the difference between the largest and the intermediate principal stresses for tension–tension experiments. Oxborough and Bowden propose a maximum principal strain criterion, i.e.

$$\varepsilon_1 = \varepsilon_c = A'/I_1 + B' \tag{33}$$

where

$$\varepsilon_1 = (1/E)(\sigma_1 - \nu\sigma_2 - \nu\sigma_3)$$

Their data are similar to Sternstein's and they show that eqn. (33) fits both sets of data as well as eqn. (32) does. Both equations predict that the stresses to initiate crazing should become infinite at $I_1 = 0$ and both sets of authors treat crazing under hydrostatic pressure ($I_1 < 0$) as being impossible even though both eqns. (32) and (33) formally predict critical negative stress bias or critical strains. Matsushige *et al.*[113,114] find that their data on PS disagree with Oxborough and Bowden. They find the craze initiation occurs at an approximately constant value of the maximum principal stress and more importantly that crazing can occur at negative values of I_1. One problem with such observations is that the local stresses beneath large enough surface imperfections could have a positive I_1 while the global I_1 was negative; however, the authors state that they polish the samples carefully before testing. Our experiences at Cornell with TEM of cast polystyrene films is that air crazes invariably originate from dust particles or other imperfections in the film.

Argon and co-workers[112,115] have recognised the importance of surface imperfections in craze initiations and deliberately introduce a surface of controlled roughness. Although they explored a much smaller region of I_1, their results are generally in accord with Oxborough and Bowden's when the stress concentration of the roughness profile is taken into account. They propose a complex microscopic model of craze initiation, the physical steps of which are nucleation of a microshear region, which gives rise to pore formation, and subsequent plastic expansion of the pores to form a craze nucleus. Like Oxborough and Bowden, Argon considers positive I_1 to be essential for craze initiation. It should also be pointed out that these crazing criteria must ultimately be time-dependent. There is an induction period[116,112] during which no crazes are visible; after this time the craze density and average craze length increase with time. This induction time decreases with increasing applied stress, so any criterion corresponds to a criterion for observation of crazes after a particular time.

Although the experimental situation with regard to air-crazing criteria is unsettled, and there are only a limited number of experiments on environmental craze initiation, one finds a number of differences between environmental crazing and dry crazing. First, there seems to be no induction time for environmental crazing.[118,119] Second, the stress criterion for environmental crazing appears to be simply that the largest principal stress exceed a critical value which does *not* depend on the value of I_1. These results come not only from the tensile experiments on PS in various hydrostatic pressure media (silicone oils and castor oil) of Matsushige *et al.*[113,114] but also from the torsion–tension experiments of Kitagawa[117] on PMMA in kerosene. In both sets of experiments there is no question that craze initiation was observed at negative global values of I_1, in conflict with the predictions of either eqn. (32) or (33). While these results might still be attributed to the effects of surface flaws, the large negative values of I_1 at craze initiation require flaws with much larger stress concentrations than those necessary to explain the air-crazing anomalies. A possible alternative is that the swelling of polymer at the tip of the flaw by the environment creates an additional hydrostatic tensile stress field ahead of the flaw sufficient to offset the global negative value of I_1. That such swelling stresses exist is well known[120] and the stresses developed in the glassy core of polymersheets placed in certain environments are sufficient in themselves to cause crazing.[121]

3.1.2 Chemical and Thermal Effects on Craze Initiation

The effect of changes in both chemical environment and temperature on craze initiation are of great practical and theoretical interest. The usual experiment is to bend a strip of glassy polymer on an elliptical bending jig so that the maximum tensile strain at the surface varies with position along the strip from a large value to a very small one. The jig is then placed in the selected chemical and thermal environment and after a fixed time the position of the boundary between the region where crazes have nucleated and the uncrazed region is measured. From the initial maximum tensile strain at this position a critical crazing strain ε_c is determined. The test has an important flaw in that the stress can be relaxed to zero by homogeneous flow if the environment can rapidly diffuse into the sample and plasticise it to below the testing temperature. Crazes which form initially can heal out by homogeneous viscous flow before the strip is examined. Consequently the test cannot be relied on if the final state of the polymer is a highly swollen glass or rubber.

Kambour and co-workers[2,107,124,150] have used this test to show that the primary determinant in environmental craze initiation is the T_g of the polymer fibrils, which is assumed to be that of the polymer swollen to equilibrium. They measured the T_g of thin films of PS soaked in the liquid environment and showed that ε_c decreased with decreasing T_g of the plasticised PS as shown in Fig. 13(a). Here T_{test} was $\approx 25°C$ and T_g varied. Also shown in this figure (square data points) are results from samples moulded with uniform concentrations of *o*-dichlorobenzene (DCB) which produce a uniformly plasticised PS. T_{test} was again 25°C. The agreement between the critical strain for craze nucleation in external liquid plasticisers with ε_c for the PS plasticised internally with *o*-DCB is good. The final points correspond to PS crazed in air at temperatures of 25°C and above, T_g remaining constant. (These data were taken from critical stress (σ_c) measurements in constant elongation-rate tensile tests,[2] which are effectively higher-rate tests than the bending jig tests. To compensate for the rate effects ε_c was computed as $\alpha_0 \sigma_c / E$, where α_0 was chosen to fit the data at 25°C. No correction was made for the increase in E with T; if such a correction were made, it would improve the fit at low values of $T_g - T_{test}$.) In Fig. 13(b) one observes that the correlation of ε_c with $T_g - T_{test}$ in two other amorphous polymers, a styrene–acrylonitrile (SAN) co-polymer (30% acrylonitrile)[150] and poly(2,6–dimethyl–1,4–phenylene oxide),[124] is just as good. Here T_g varied with the different liquids and T_{test} was 25°C.

Finally it is observed[2] that the critical crazing stress in uni-axial tension undergoes a larger relative decrease with decreases in $T_g - T_{test}$ from $T_g = 100°C$ and $T_{test} = 25°C$ (due to either plasticisation or increase in test temperature) than does the yield stress for homogeneous shear (as measured by the yield stress in compression). Data for unplasticised and plasticised (7·5% *o*-DCB) PS are shown in Fig. 14. Shown also is the fracture stress in tension which runs roughly parallel to the σ_c curve, only higher. This finding is significant in that it indicates that either plasticisation or increases in temperature will bias PS toward a crazing response

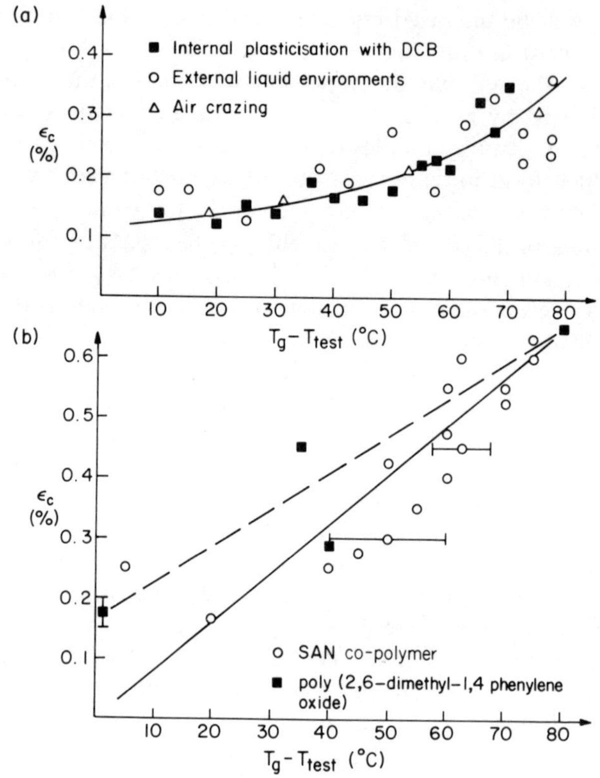

FIG. 13. (a) Critical tensile strain for craze initiation in PS in air and external environments, and in PS internally plasticised with *o*-dichlorobenzene (after references 124 and 2). (b) Critical tensile strain for craze initiation in external environments for a 70% styrene–30% acrylonitrile co-polymer (after reference 150) and for poly(2,6-dimethyl-1,4 phenylene oxide) (after references 124). In each case T_g is the glass transition temperature of a thin film of the polymer soaked to equilibrium in the environment and T_{test} is the testing temperature.

rather than toward a homogeneous yielding response to tensile stress. From a fundamental point of view it would appear to indicate that the controlling microscopic mechanisms for plastic flow are not the same in shear yield and craze initiation, a finding that simply reinforces the results of multi-axial stress experiments presented above.

Although the correlation of craze initiation conditions with the glass transition temperature of the plasticised fibrils appears indisputable, the

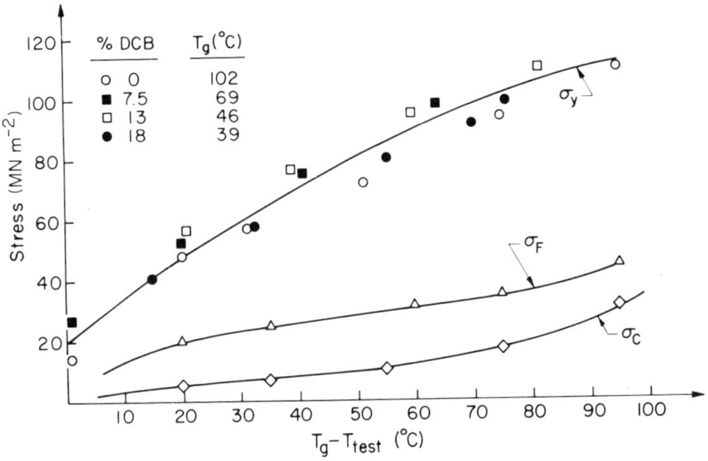

FIG. 14. Crazing stress σ_c and fracture stress σ_f in tension and yield stress σ_y in compression for PS, plasticised and unplasticised, *versus* $T_g - T_{test}$, the difference between the glass transition temperature and the testing temperature.

correlation is not as useful as one might think when one is asked to predict whether a particular environment will lower ε_c to intolerable levels. It is tedious and expensive to have to measure the T_g of a film swollen to equilibrium in the liquid, especially if that liquid diffuses slowly into the polymer. Consequently, various methods of estimating ε_c have been developed,[2,8,107,124,150,152–156] relying primarily on correlations between the difference between the solubility parameters δ_s of the liquid and δ_p of the polymer and the T_g of the plasticised fibrils. (The solubility parameter is the square root of the cohesive energy density of the liquid or polymer.[151]) Such correlations work best for a non-polar polymer (e.g. PS) in non-polar liquids,[107] with critical strains decreasing rapidly with $\delta_s - \delta_p$. For $\delta_s - \delta_p$ less than $2\,\text{cal}^{1/2}\,\text{cm}^{-3/2}$ ($4091\,\text{J}^{1/2}\,\text{m}^{-3/2}$), T_g of the fibrils is close to, or below, room temperature and the critical strain approaches zero. For polycarbonate in aliphatic liquids Jacques and Wyzgoski[156] show that a plot of ε_c *versus* $V_0(\delta_p - \delta_s)^2$ (where V_0 is the molar volume of the liquid) produces a straight line which extrapolates to $\varepsilon_c = 0$ at about the point at which T_g of the plasticised polymer drops below room temperature. The quantity $V_0(\delta_p - \delta_s)^2$ is proportional to the heat of mixing of liquid and polymer according to regular solution theory.[157] For polar liquids it is necessary to separate the solubility parameter into a non-polar and a polar part (including hydrogen bonding), δ_d and δ_a respectively, so that

$$\delta = \sqrt{\delta_d^2 + \delta_a^2}$$

For polar polymers and liquids it is necessary to match both δ_d and δ_a of polymer and liquid to achieve high solubility and large depression of T_g and ε_c.[154–156] Jacques and Wyzgoski[156] plot a contour map of ε_c on a graph of $V_0(\delta_{pd} - \delta_{sd})^2$ *versus* $V_0(\delta_{pa} - \delta_{sa})^2$ and find a very good correlation, with points near the origin representing very low critical strains and points far from the origin high critical strains. While for practical reasons more effort is justified in improving these solubility parameter correlations, it is clear that fundamentally ε_c correlates best with the glass transition temperature of the plasticised polymer.

3.1.3 *Microscopic Craze Initiation Mechanisms*

There is no lack of ideas to describe the microscopic craze initiation step; the difficulty is in attempting to distinguish these mechanisms based on macroscopic experiments. Gent[45] has proposed that the hydrostatic tensile stress at the tip of a surface flaw lowers the local glass transition temperature to below room temperature in the case of air crazes. The decrease in T_g in environments is aided by the additional environment absorption due to the hydrostatic stress (eqn. (8)) and the decrease in T_g due to plasticisation. The same mechanism is proposed as operating at the craze tip, allowing it to advance. However, recent measurements of the local hydrostatic tension $(-P)$ at an air craze tip[41] in PS show that this is relatively small $(-P \approx 15\ \text{MN m}^{-2})$ and the corresponding depression of T_g is much smaller than needed for this mechanism to operate (T_g is depressed only from 373 K to about 370 K). Olf and Peterlin[22] propose much the same physical mechanism (but couch the formal description in terms of free volume theory) to account for low-temperature gas crazing. However, they suggest that it is not necessary to decrease T_g to the craze initiation temperature, but only to such a value as will allow local plastic deformation to occur under the stresses obtaining at the flaw or craze tip.

Argon and co-workers[112,115] have proposed the very detailed microscopic model outlined in the previous section. They differentiate between the processes which take place during craze initiation at the flaw tip and the processes involved in the advance of the craze tip. The initiation model appears well grounded in plasticity theory, employing the same methods used to describe plastic hole growth in ductile metals. Unfortunately there are six parameters in the model that cannot be fixed *ab initio* and although these can be chosen to give a reasonable fit to the extensive measurements of Argon and Hannoosh[112] (they measure the density of crazes as a function of time for various combinations of deviatoric stress and hydrostatic stress, having first carefully characterised the surface flaw distribution) one wonders about the uniqueness of such a description.

None of the microscopic models above can easily explain the degree of strain localisation observed in crazes; i.e. why is the voiding in PS localised in a sheet as small as 25 Å in thickness? This strain localisation appears to require significant initial strain softening of the plastically yielded material. This point has been emphasised by Haward[47] and a finite element computation of the collapse stress and plastic strain distribution has been made by Haward and Owen[122] for both a rigid elastic–plastic material and a strain-softening one. Experimentally the necessity to take local strain softening into account is indicated by the fact that air crazing is not observed in materials in which strain hardening begins at very low strains.[47,123] In the case of environmental crazing, the local plasticisation of the polymeric glass by the environment is a local softening mechanism which has the same effect as strain softening.

A final unanswered question is the importance of surface energy to craze initiation. Although there is a surface energy contribution to the energy of the final craze due to the fine craze fibril sizes, the existence of a surface energy barrier cannot explain craze initiation in the absence of an environment. (Rather the opposite; this barrier actually should make crazing less likely relative to other plastic deformation mechanisms.) However, reduction of fibril surface energy by the environment has been invoked as the most important role of the environment in enhancing crazing.[124,125] Since the scale of the voiding during the initiation processes is unknown, it is impossible to completely rule out such effects, since if the void size is made small enough this contribution must be quite large. Nevertheless, as indicated above, experiments by Kambour and co-workers showed that the critical strain for crazing was governed primarily by the depression of T_g produced by the environment and that approximately the same critical strain for craze initiation at an equivalent T_g depression was observed when a plasticiser (*o*-dichlorobenzene) was dissolved uniformly in the polymer polyphenylene oxide. Since it is unlikely that there is any correlation between the T_g depression produced by the various environments and by various concentrations of *o*-dichlorobenzene and the surface energy of the polymer under these conditions, the role of the surface energy, in these experiments at least, appears to be negligible.

3.2 Craze Growth

3.2.1 Microscopic Mechanisms

The microscopic mechanisms of craze thickening were discussed previously because of their role in the development of the craze microstructure. It remains to discuss the mechanisms by which the craze tip actually advances. Until several years ago it was generally assumed that the advance of the craze tip required the nucleation and expansion of isolated voids in advance of the tip.[127] After some critical value of porosity was reached the void phase was supposed to become interconnected. A difficulty with this mechanism is that it appears to require larger values of stress concentration at the craze tip than the small values actually observed.

It has been recently recognized[130,131] that the Taylor meniscus instability[128,129] can give rise to craze-like structures. This instability is responsible for the break-up of the meniscus between a dense and a less dense viscous fluid, e.g. air, into finger-like protruberances above a certain meniscus velocity. Such instabilities are commonly observed when two flat plates with a layer of liquid between them are forced apart. An example in a liquid layer beneath an advancing roller is shown in Fig. 15(a). The physical mechanism for the instability is the increased hydrostatic tension which develops beneath a convex surface of the meniscus. The conditions necessary for the instability have been worked out for both Newtonian and non-Newtonian fluids[129,130] and Argon and Salama have applied these ideas to the craze tip (Fig. 15(b)). The corrugations in the meniscus eventually give rise to the fibril structure of the craze. The wavelength of the meniscus instability, and thus the craze fibril spacing, is sensitive both to the rheological properties of the polymer ahead of the craze tip and to the surface energy of the polymer. Estimates[131] of this wavelength are of the correct order of magnitude to account for the craze fibril spacing.

The physical evidence for the isolated void mechanism *versus* the meniscus instability mechanism is tenuous at present. Early observations of craze tips by transmission electron microscopy were interpreted as showing isolated voids at the crack tip.[8] However, these observations were made on very thin microtomed sections where the fibril spacing would be expected to be of the order of the film thickness, or on a very thin film

(a)

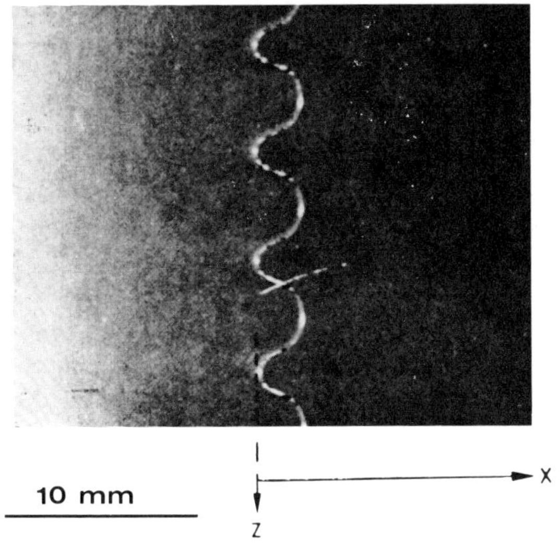

(b)

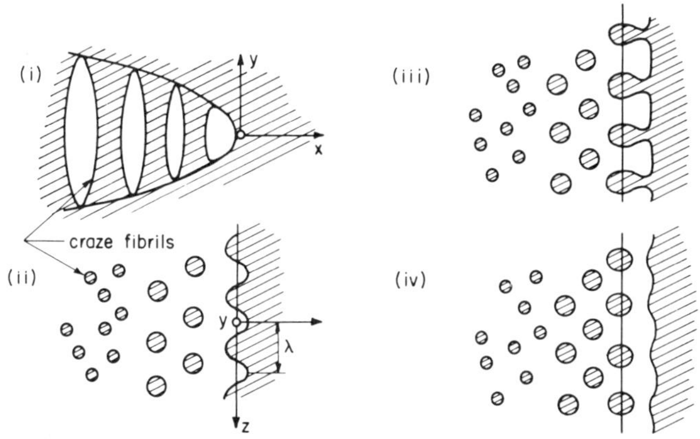

FIG. 15. (a) Onset of macroscopic meniscus instability of a polydimethylsiloxane–air interface. (After reference 129, courtesy of Taylor and Francis, Ltd.) (b) (i) side view of craze tip; (ii)–(iv) sections through the mid-line of the craze, illustrating the advance of the craze tip and fibril formation by the meniscus instability mechanism (after reference 131).

where a meniscus instability mechanism would not be expected to operate. Recent observation on craze tips in thick PS films (≈ 7500 Å) show that the fibril structure of the craze persists to craze thicknesses less than 50 Å. Although it is not possible to rule out isolated voids in these samples, these voids, if present, must be very small.

3.2.2 *Growth Geometry*

Once nucleated, crazes grow both in thickness and length. In isotropic polymers it is well established that crazes grow along planes of maximum principal stress.[108,132] This craze growth trajectory can be predicted from the mechanical principle of maximum plastic resistance[134] (maximum plastic work) together with the models of craze mechanics discussed above, as follows. The work done in a given area of craze growth is

$$W = \int_0^w S(w)\,\mathrm{d}w = \int_0^w \Delta S(w)\,\mathrm{d}w + \sigma_N w \tag{34}$$

where σ_N is the initial stress normal to the craze and $\Delta S(w)$ is the craze surface stress increment due to the displacement w. Since ΔS is independent of the initial stress conditions and the orientation of the craze, due to isotropy, the craze will grow on the plane which maximises σ_N in order to maximise the plastic work. That is, of course, the plane of the maximum principal stress.

(In principle shear forces S_s and displacements u in the x-direction are possible on the craze surface (corresponding to a mode II crack opening). The additional terms which have to be added to eqn. (34) are

$$W_s = \int_0^u (\Delta S_s(w) + \sigma_{yx})\,\mathrm{d}u \tag{34a}$$

where σ_{yx} is the applied shear stress on the plane of the craze in the x-direction. Since for small shear displacements, the total shear stress $S_s = \Delta S_s(u) + \sigma_{xy}$ must be negligible due to the nature of the fibrils, W_s and its variation with angle can be neglected.)

Given that the craze grows along the plane of maximum principal stress in an isotropic polymer one might expect that the craze growth kinetics are only sensitive to the value of the maximum principal stress, unlike the case of craze initiation. That hypothesis has recently been confirmed by Argon and Salama,[131] using torsion–tension experiments for air crazes in PS. They also demonstrated that crazes would grow in a pure shear stress field ($I_1 = 0$), although they were unable to observe craze initiation under such stress conditions, as were Sternstein and Myers.[110]

No similar experiments have been attempted for environmental crazes but it seems safe to predict that when the stress can affect the environmental craze growth kinetics, the relevant stress component will be the maximum principal stress.

In oriented (non-isotropic) polymeric glasses the situation described above changes dramatically. Beardmore and Rabinowitz[135] have investigated the environmental craze growth trajectories in both PMMA and PC in liquid nitrogen at 77 K. They observe that the plane of craze growth deviates from the plane of maximum principal stress toward the direction of molecular orientation, as shown in Fig. 16(a). The angular deviation α, at a given angle θ between σ_∞ and the direction of molecular orientation, is linear in the birefringence (degree of molecular orientation) and is a maximum at $\theta \approx 40°$. Along with this deviation from the plane of maximum principal stress they did find a pronounced decrease in both the crazing stress σ_c and the fracture stress σ_f as the applied stress makes a progressively larger angle to the molecular orientation direction (Fig. 16(b)). Clearly the variation in σ_c and σ_f must reflect an underlying marked variation in the structure and mechanical properties of the craze with direction. At present these variations in either structure or mechanical properties have not been investigated. Beardmore and Rabinowitz hypothesised that the plane of craze follows the plane of maximum principal strain. Predicting the principal (plastic) strain axes from flow stress *versus* θ data in PC they found good agreement with this idea. Kitagawa[117] performed similar experiments at room temperature on rolled PVC and attempted to compare the measured craze planes with the plane of maximum (elastic) principal strain. To obtain agreement he was forced to include a hypothetical residual stress (from the rolling operation) which was a large fraction of the applied stress. It should be possible to apply arguments similar to those which lead to eqn. (34) for isotropic polymers to the oriented polymers. However, the fundamental craze mechanics equations (eqns. 11–16) will have to be re-derived using anisotropic elasticity theory and a complete treatment of the variation of $\Delta S(w)$ with angle will be necessary. Given the importance to the fracture properties of residual molecular orientation due to polymer processing, additional theory and experiments on the structure and mechanical properties of crazes in oriented polymers are needed.

3.2.3 Growth Kinetics

Before discussing the growth kinetics of crazes it is well to point out that a process as complex as craze growth may have several mechanisms which

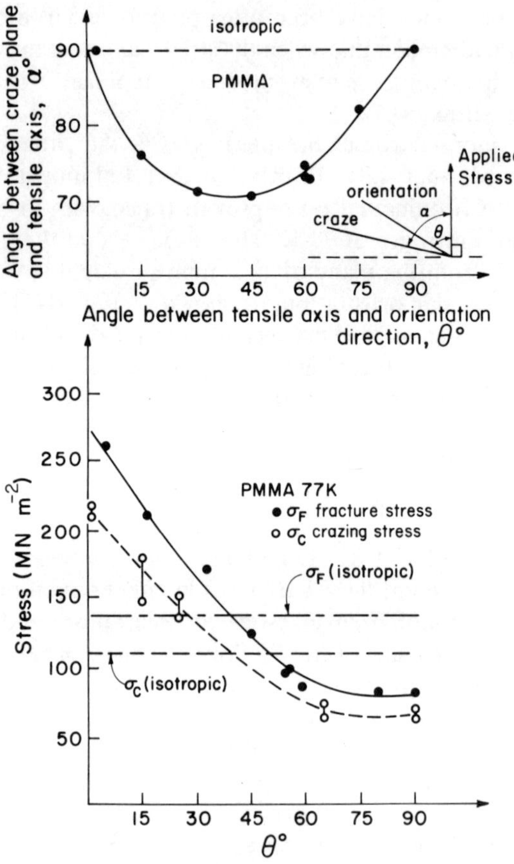

FIG. 16. (a) Deviation of the craze growth plane from the plane of maximum principal stress for oriented PMMA (after reference 135). (b) Anisotropy of the craze initiation and fracture stresses in tension for oriented PMMA at 77 K (after reference 135).

operate in series. Just because it is necessary that a certain mechanism operate for craze growth to occur it does not follow that it is the rate-controlling mechanism. In air crazing, for example, it may be necessary for either the microcavitation or meniscus instability mechanism to operate at a certain rate at the craze tip for the craze tip to advance. The rate of the microscopic mechanism(s) will be determined largely by the local stress conditions at the craze tip ($S(x = a)$), however, and these are determined by the displacement profile $w(x)$ of the craze through eqns. (11) and (12) (or eqn. (16)). If the displacement (and thickness) of the craze behind the tip does not increase, the craze tip will eventually stop at a point where the stress decreases with tip position exactly as $S(x)$ for $x > a$. On the other hand, if the tip remains stationary and $w(x)$ increases, $S(x = a)$ will increase until either the tip advances or plastic flow by some other mechanism occurs. One can make a good case, therefore, that the displacement rate along the craze, due to craze thickening by either surface drawing or fibril creep, controls the advance of the craze tip, and not the specific micromechanism which operates there.

To describe craze growth from crack tips under these conditions (which they call 'relaxation-controlled' craze growth), Williams and Marshall[136] use the simple Dugdale model (eqn. (26)) and assume a power law for the time dependence of craze stress S_c, namely $S_c = S_i t^{-m}$. Substituting this into eqn. (26) they find

$$\Delta a = \frac{\pi}{8} \frac{K_I^2}{S_i^2} t^{2m} \tag{35}$$

They hence determine m from air craze growth experiments to be between 0·11 and 0·04, depending on the polymer, in rough agreement with values of that exponent from separate yield stress *versus* time experiments.

While the agreement with experiment is satisfying, one should realise that the approximation implicit in eqn. (35), that S_c and crack tip displacement w_c are independent, is not correct. One may assume that the craze thickening rate, and thus dw_c/dt, is a power law function of the craze surface stress S_c, i.e.

$$\frac{dw_c}{dt} = \dot{w}_0 \left(\frac{S_c}{S_0}\right)^{(1+m)/m} \tag{36}$$

(This is the equivalent of the power law assumption made by Williams and Marshall.) This displacement rate must match that of the Dugdale model given by differentiating eqn. (27) to yield

$$\frac{dw_c}{dt} = -\frac{K_I^2}{2E^* S_c^2} \frac{dS_c}{dt} \tag{37}$$

The displacement derivative can be eliminated between eqns. (36) and (37) to form a single differential equation for S_c. This can be integrated and the result can be substituted into eqn. (26) to yield

$$\Delta a = \frac{\pi}{8} K_I^{2(1-2\alpha)} \left(\frac{2\alpha \dot{w}_0 E^*}{S_0 \dfrac{1-\alpha}{\alpha}}\right)^{2\alpha} t^{2\alpha} \tag{38}$$

where $\alpha = m/(1 + 2m)$. The quadratic dependence of Δa on K_I at constant t is no longer predicted and eqn. (38) approaches eqn. (35) only in the limit $\alpha \to 0$, i.e. for an ideal elastic-rate-independent plastic solid. Note also that if more than one craze is nucleated and grows from the crack tip both eqns. (36) and (38) will be altered, leading to faster growth. Since the experimental investigations reported in the literature do not distinguish between growth of a single craze and that of a closely spaced bundle of crazes, additional uncertainty is introduced into the interpretation.

The best simple model of isolated air craze growth (no crack present), due to Verhuelpen-Heymans and Bauwens,[68] also uses much the same set of approximations as that of Williams and Marshall.[136] In particular the coupling between the craze displacements and surface stress is ignored in computing the growth law. As demonstrated above, that approximation cannot be rigorously justified.

Craze growth in liquid environments is another example of a growth process where a number of microscopic mechanisms must occur in series. In addition to those mentioned for air crazes, it will be necessary for the environment to diffuse into both the fibrils and the bulk polymer adjacent to the craze surfaces as well as to be transported through the craze structure to a position close behind the craze tip. It turns out that for most liquid crazing systems investigated in detail, the hydrodynamic flow of fluid through the porous craze structure is the rate-controlling step in craze growth.[42,43,74,136–138] The most convincing method by which to demonstrate that hydrodynamic flow controls growth is to alter the flow path through the craze by various surface barriers which restrict the access of liquid to the craze. Consider, for example, a sheet with a craze growing from a crack in its edge. If the liquid is allowed to penetrate the craze from the broad surfaces of the sheet, the distance liquid must be transported (and thus the pressure gradient) is approximately constant as the craze grows in length and a constant growth velocity is observed. If the sheet surfaces are sealed (e.g. with glass plates and silicone grease) so that the liquid can penetrate only along the length of the craze from the craze tip, craze growth is approximately proportional to $\sqrt{t}$, rather than t in the absence of a surface seal, and the growth velocities for long crazes are very much reduced.

The origin of the pressure gradient has been an issue in interpreting these experiments. Marshall, Culver and Williams[42] assumed that the driving force for flow was atmospheric pressure but Kambour[8] pointed out that, due to the small diameters of fibrils and voids in the craze, the capillary (negative) pressure in the craze is much greater than one atmosphere and accordingly that the liquid flow in the craze should be driven by capillarity. Kramer and Bubeck[43] confirmed Kambour's hypothesis by showing that a craze which had been dried under load refills with liquid (driving pressure is capillarity) with the same kinetics as that of the original growth (Fig. 17). The presence of a higher negative capillary pressure P_c in the liquid in the craze (P_c can be as large as 100 atm) also accounts for another puzzling observation, the fact that the presence of dissolved gas in the liquid can lead to as much as a fourfold decrease in craze growth rate.[138] When high dissolved gas contents are present small gas bubbles are observed to nucleate within the craze, partially blocking the flow of liquid through it. Bubble nucleation is enhanced by the high negative capillary pressure of the fluid within the craze.

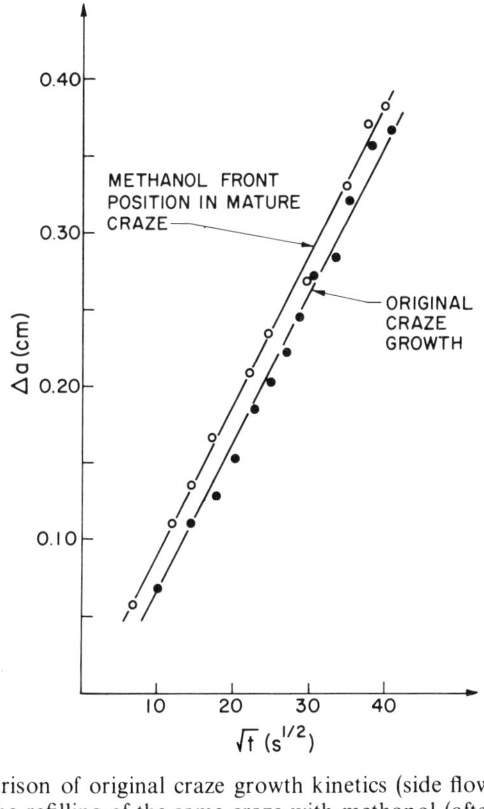

FIG. 17. Comparison of original craze growth kinetics (side flow blocked) with the kinetics of the refilling of the same craze with methanol (after reference 43).

The craze growth law in a sample with sealed sides can be computed theoretically[43] to be

$$\Delta a = \left[\frac{2(P_c + P_a)}{\mu\langle(1 - v_f)/\kappa\rangle}\right]^{1/2} \sqrt{t} \tag{39}$$

where μ, P_a and v_f are the fluid viscosity, ambient pressure and fibril volume fraction. The hydraulic permeability κ for flow perpendicular to an array of cylindrical fibrils can be estimated using Happel's approximate solution[139,140] to be

$$\kappa = (D_f^2/32v_f)[\ln(1/v_f) - (1 - v_f^2)/(1 + v_f^2)] \tag{40}$$

where D_f is the craze fibril diameter. The capillary pressure can be estimated using a similar array of cylindrical fibrils to be

$$P_c = 2\gamma_L\sqrt{v_f}/[(1 - \sqrt{v_f})D_f] \tag{41}$$

where γ_L is the fluid surface tension and where v_f and D_f in eqn. (41) are the values near the craze tip. The brackets $\langle\ \rangle$ in eqn. (39) represent the average of the term $(1 - v_f)/\kappa$ over the craze length; this average is assumed to remain constant as the craze grows.

Equation (39) indicates that increasing viscosity of the liquid should produce a pronounced decrease in the craze growth kinetics if all other parameters (in particular κ and v_f) can be held constant. Graham *et al.*[74] showed that this expectation is verified qualitatively for a variety of liquids and liquid mixtures. Matsushige *et al.*[114] showed that times-to-failure in a high-pressure experiment were much longer for PS in high-viscosity silicone oils than in low-viscosity silicone oils, an observation which is presumably linked to slower craze growth rates in the higher-viscosity silicone oils.

For craze growth from crack tips, the slope of a plot of Δa *versus* $\sqrt{t}$ is observed to increase linearly with increasing K above a threshold value K_m and saturate for higher K.[42] The increase presumably reflects the rise of $\langle\kappa/(1 - v_f)\rangle$ with K for a single craze, while the saturation phenomenon may arise when a threshold is reached for the nucleation of multiple crazes at the crack tip. Experiments to measure the microstructure (D_f, v_f) of the liquid craze as a function of the stress conditions under which it is grown, would be useful in fully verifying the theory of craze growth in liquids.

3.3 Environmental Crack Growth

It is by now conventional to use the language of fracture mechanics to describe the kinetics of crack growth both in air and in liquids. One assumes the existence of a critical stress intensity factor $K_c(\dot{a})$, or equivalently a value of fracture toughness $G_c(\dot{a}) = K_c^2(\dot{a})/E$, that is an increasing function of crack velocity $\dot{a}$. (The dynamic fracture toughness $G_c(\dot{a})$ in this context is often given the symbol R.[141] In the limit $\dot{a} \to 0$, $G_c(\dot{a}) \to 2\mathcal{T}$ where $\mathcal{T}$ is the surface work of fracture used by Andrews and Bevan.[89]) One can then plot the results as K_c or G_c *versus* $\log_{10} \dot{a}$ and check to see if the assumptions are correct, i.e. that a unique curve results regardless of the geometry of the specimen or the initial crack length or loading. Although few experiments on polymers have tested these assumptions in detail, they appear to hold as long as the zone of plastic deformation, i.e. the craze or crazes, is short compared to crack or specimen dimensions. Figure 18(a) shows the K_c *versus* $\log_{10} \dot{a}$ curve for low-density PE of two different melt flow indices (molecular weight) in both methanol and ethanol from Marshall *et al.*,[142] whereas Fig. 18(b) shows similar curves for crack growth in glassy PMMA in air, in CCl_4 and in ethanol environments.[143,144] These curves have several features in common. They are nearly horizontal at low crack growth rates, indicating that for practical purposes there is a threshold K below which measurable slow crack growth does not occur. This feature is common to almost all other environmental cracking systems, including hydrogen environment cracking in steels,[146,147] water vapour induced cracking in silica-based glasses[148,149] and stress corrosion cracking of metals and alloys.[133] In principle it is quite important from an engineering design point of view, since it means one can set an upper limit on the largest pre-existing flaw (by NDT or other methods) and design so that the K of this flaw is less than the threshold K. In polymers, however, the threshold K in many environments is very low, and the flaws or cracks may not exist prior to application of the load but result from breakdown of stress-induced crazes; so the use of such design procedures is not common.

By integrating such curves one can predict the time-to-failure t_f in a constant load (or static fatigue) test.[145] Since most of the time-to-failure is spent at the lower crack growth rates, the portion of the curve just above the threshold K is the critical portion in determining t_f. It is interesting to note that for PMMA in the CCl_4 and ethanol environments larger K is required to produce high $\dot{a}$ than is required in air, even though the threshold K in both environments is much lower than in air. Indeed K_{Ic}, the K above which unstable (and very rapid) crack propagation sets in, is usually larger in various liquid environments than in air. This behaviour is illustrated in Table 2, where G_c and K_c for PMMA at $\dot{a} \approx 0{\cdot}1$ mm sec^{-1} are listed for various environments.[143] The ultimate time-to-failure,

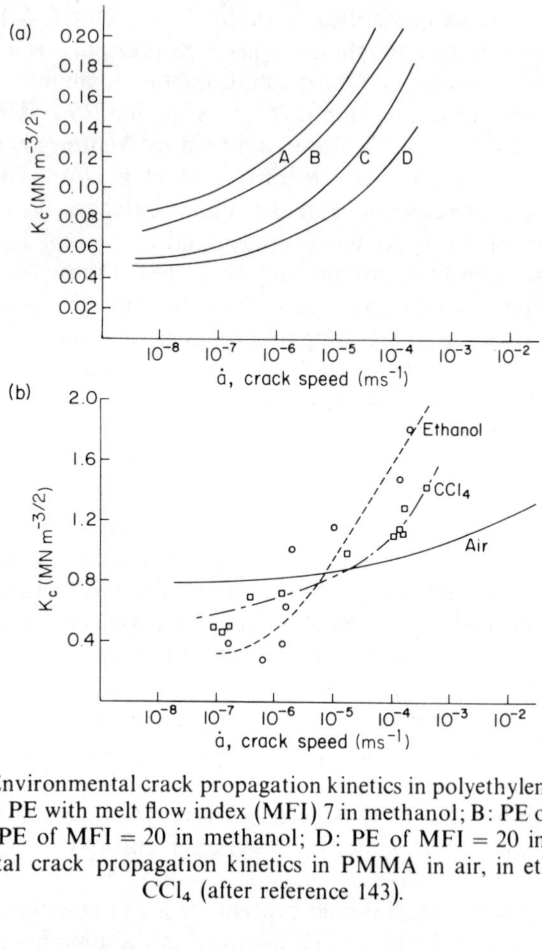

FIG. 18. (a)Environmental crack propagation kinetics in polyethylene (after reference 142). A: PE with melt flow index (MFI) 7 in methanol; B: PE of MFI = 7 in ethanol; C: PE of MFI = 20 in methanol; D: PE of MFI = 20 in ethanol. (b) Environmental crack propagation kinetics in PMMA in air, in ethanol and in CCl_4 (after reference 143).

and whether failure occurs at all, depend on K_c and G_c at smaller $\dot{a}$. Since these K_c and G_c values are much lower in most of these environments than in air, it follows that the results of conventional (high $\dot{a}$) fracture mechanics (K_{Ic}) testing in stress-cracking environments can be quite misleading.

Conceding the engineering significance of the K_c or G_c *versus* $\dot{a}$ curves, the microscopic explanation of these is far from clear for slow crack growth in either air or liquid environments. One can compute G_c for a crack which advances an incremental distance with a craze of steady-state

TABLE 2

FRACTURE TOUGHNESS OF PMMA AT HIGH CRACK VELOCITIES IN VARIOUS ENVIRONMENTS (AFTER REFERENCE 143)

Environment	$G_c(\dot{a} \approx 0{\cdot}1\ \mathrm{mm\ s^{-1}})$ $\mathrm{J\ m^{-2}}$	$K_c(\dot{a} \approx 0{\cdot}1\ \mathrm{mm\ s^{-1}})$ $\mathrm{MN\ m^{-3/2}}$
Air	380	1·02
Lubrication oil	540	1·22
CCl_4	600	1·29
H_2O	680	1·37
Gasoline	1 100	1·74
Ethanol	1 210	1·83
Kerosene	1 400	1·97

length at its tip, i.e.

$$G_c = -2 \int_a^{a+\Delta a} S(x) \frac{\mathrm{d}w}{\mathrm{d}x}(x)\,\mathrm{d}x \tag{42}$$

or

$$G_c = 2 \int_0^{w_c} S(w)\,\mathrm{d}w \tag{43}$$

For the simple Dugdale model these integrals reduce to

$$G_c = 2w_c S_c \tag{44}$$

or

$$K_c = \sqrt{2w_c S_c E^*} \tag{45}$$

where w_c is the displacement at the crack tip (or craze base) and S_c is the craze surface stress (assumed constant along the craze in this model). The usual assumption for cracking in air is that w_c is a constant of the material and that S_c and E^* are time dependent; e.g. $S_c = S_i t^{-m}$ and $E^* = E_i^* t^{-n}$, where t is taken to be $\Delta a/\dot{a}$. If it is further assumed that $m \approx n \approx 0{\cdot}1$ then

$$K_c(\dot{a}) \approx \sqrt{2w_c S_i E_i^*(\dot{a}/\Delta a_i)^m}. \tag{46}$$

where

$$\Delta a_i = \frac{\pi}{8} \frac{K_{\mathrm{threshold}}^2}{S_i^2}$$

J. G. Williams and his students have been the prime advocates of this model, which they call the relaxation-controlled crack growth model, although others have also adopted very similar methods.

While on plots of ln K_c *versus* ln $\dot{a}$ the experimental curves are approximately the straight lines of slope 0·1 implied by eqn. (46), the theoretical derivation, as in the case of craze growth, ignores the coupling between surface stress and surface displacement; namely, the fact that w_c cannot be constant if S_c is spatially constant but time-varying. Moreover, the net effect of eqn. (46) is to replace a macroscopic unknown K_c by a microscopic unknown material parameter w_c which may not have any more fundamental significance. Indeed, for air crazes, craze thickening and craze surface displacement occur primarily due to the surface drawing mechanism, whereas craze failure is primarily attributed to the fibril creep mechanism, which provides only a minor part of the craze surface displacement. In such circumstances it is difficult to see how w_c can *control* microscopic fibril failure, although it may well turn out that craze fibril failure will occur for combinations of initial local crack tip stress conditions (K) and crack velocities ($\dot{a}$) that will result in a $w_c(\dot{a})$ which is roughly constant. An obvious case where w_c should not be constant is when a transition from single-craze nucleation to multiple-craze nucleation at the crack tip occurs.

Additional considerations for liquid environmental crazing are that the craze thickening may be dominated by fibril creep rather than surface drawing and that the steady-state craze length (and thus the stress distribution along the craze) may be governed by hydrodynamic flow of liquid into the craze. Since for a steady-state craze of length Δa, the crack growth rate $\dot{a}$ equals the craze growth rate, one finds from eqn. (39) that

$$\dot{a} = \frac{2(P_c + P_a)}{\mu\langle(1 - v_f)/\kappa\rangle\Delta a} \tag{47}$$

and since for the simple Dugdale model

$$\Delta a = \frac{\pi}{2}\frac{E^{*2}w_c^2}{K^2}$$

(eqns. (26) and (27))

$$K_c = \sqrt{\frac{\pi\mu\langle(1 - v_f)/\kappa\rangle}{4(P_c + P_a)}}\,E^* w_c \dot{a}^{1/2} \tag{48}$$

if $\kappa/(1 - v_f)$ is independent of K. Equation (48) resembles one derived by Williams and Marshall,[136] but their K-dependence essentially comes from their assumption that $\kappa/(1 - v_f)$ increases as K^2. If that assumption is made, our K_c increases as $\dot{a}^{1/4}$ for a constant w_c.

Equations (47) and (48) cannot hold to arbitrarily low $\dot{a}$ since the craze length Δa would then exceed that of a fully plasticised craze; presumably there is a lower limit to S_c, which implies an upper limit for Δa. Under these conditions the craze growth would no longer be transport-controlled and one would expect the slope of the data on a plot of ln K_c *versus* ln $\dot{a}$ to approach the (relaxation-controlled?) slope of $\approx 0{\cdot}1$. Changes in slope on such a plot have been identified[136] but there is as yet no direct proof of a transition from relaxation control to transport control. Similarly it has been difficult[136] to verify the predicted dependence of the $K_c(\dot{a})$ curve on fluid viscosity since the different fluids probably produce different craze structures and result in different values of w_c.

At the same time it is difficult to understand how a fluid which has a $K_{threshold}$ that is much greater than the $K_{threshold}$ in air can have a K_c at higher $\dot{a}$ which exceeds K_{Ic} in air (Fig. 16). While one can explain the lower $K_{threshold}$ *via* eqn. (45) as due to the decrease in S_c by plasticisation, or by a decrease in w_c, or both, a higher K_c at the higher values of $\dot{a}$ appears to require a substantial increase in w_c (there must be some decrease in S_c due to plasticisation, even if it is not complete at these values of $\dot{a}$). It is possible that at the higher values of K multiple crazes (a craze bundle) form at the crack tip in the environments and that these craze bundles do not form to the same extent in air, due to the incubation time for air craze nucleation. Such a craze bundle would produce a much larger crack opening displacement even though the strain-to-fracture in each individual craze was not greatly increased. Unfortunately, there have been no studies of the craze morphology and microstructure at the crack tip in environmental cracking systems simultaneously with the determination of K_c *versus* $\dot{a}$ curves. Such investigations, as well as fundamental studies of the craze failure mechanisms (what determines w_c?), should be encouraged.

4 CONCLUSIONS

(1) Environmental cracking of most rigid polymers proceeds by initiation and growth of a craze-like zone of cavitation which breaks down to form a crack.

(2) The most important role of the environment appears to be plasticisation of the craze fibrils. The depression of T_g in the fibrils can enhance

the drawing of craze fibrils from the craze surface as well as producing increased fibril creep, leading to craze fibril rupture.

(3) The molecular weight of the polymer is of over-riding importance in the breakdown of craze fibrils to form a crack. It is likely that some environments, in addition to being plasticisers, produce chain scission and molecular weight degradation in the fibrils, thus adversely affecting fibril stability. Improving our understanding of the effect of molecular weight and chain entanglements on craze fibril mechanical properties should be a key objective of future research.

(4) Fracture mechanics provides a useful framework for interpreting and systematising the results of environmental crack growth studies. While these studies are of value in themselves for engineering purposes, future progress on the fundamentals of environmental crack growth requires an improved understanding of the microscopic failure processes of the craze at the crack tip.

(5) Important questions remain about the effects of hydrostatic stress and plastic strain rate on sorption and diffusion of the environment into the craze fibrils and the bulk polymer adjacent to the craze surfaces.

(6) New methods of determining craze structure and micromechanics have the potential to greatly increase our knowledge of the microscopic aspects of environmental crazing and cracking. It appears particularly important to begin to apply these methods to systems other than glassy amorphous thermoplastics in organic liquids. Other systems of interest include: semicrystalline thermoplastics, cross-linked resins, oriented thermoplastics and a variety of aqueous and gaseous environments.

ACKNOWLEDGEMENTS

The support of the U.S. Army Research Office, Durham, N.C., is gratefully acknowledged. This work also benefited from the use of the facilities of the Cornell Materials Science Center which is funded by the National Science Foundation. I also appreciate useful correspondence and discussions with Dr. Roger Kambour, Dr. Richard Robertson, Professor J. G. Williams, Professor R. N. Haward and Professor P. H. Geil. I am grateful to my former and present graduate students, Dr. Heider Krenz, Dr. Robert Bubeck, Bruce Lauterwasser, Jay Kita, Andrew Conners and W. C. Wang for their spirited discussions and careful experiments which form an important background for this chapter.

REFERENCES

1. ZIEGLER, E. E. and BROWN, W. E. *Plastics Technology* (August 1955) pp. 341, 409.
2. KAMBOUR, R. P., in *Mechanisms of Environment-Sensitive Cracking of Materials*, ed. P. R. Swann, F. P. Ford and A. R. C. Westwood, Metals Society, London, 1977, p. 213.
3. KAMBOUR, R. P., in *Corrosion Fatigue*, ed. O. F. Deveraux, A. J. McEvily and R. W. Staehle, NACE-2, National Assoc. of Corrosion Eng., Houston, Texas, 1972, p. 681.
4. HOWARD, J. B., in *Engineering Design for Plastics*, ed. E. Baer, Reinhold, New York, 1964, p. 742.
5. HOWARD, J. B., in *Crystalline Olefin Polymers, Part II*, ed. R. A. V. Raff and K. W. Doak, Interscience, New York, 1965, p. 47.
6. RABINOWITZ, S. and BEARDMORE, P. *CRC Critical Reviews in Macromolecular Sci.*, **1** (1972) p. 1.
7. GENT, A. N., in *The Mechanics of Fracture, AMD Vol. 19*, ed. F. Erdogan, American Society of Mechanical Engineers, New York, 1976, p. 55.
8. KAMBOUR, R. P. *J. Polymer Sci., Part D, Macromolecular Reviews*, **7** (1973) p. 1.
9. RANBY, B. and RABEK, J. F. *Photodegradation, Photo-oxidation and Photostabilization of Polymers*, J. Wiley, New York, 1975, p. 206.
10. NATARAJAN, R. and REED, P. E. *J. Polymer Sci. A2*, **10** (1972) p. 585.
11. WINSLOW, F. H. 'Environmental Degradation of Polymers,' in *Treatise on Materials Science and Technology, Vol. 10B*, ed. J. M. Schultz, Academic Press, New York, 1977, p. 741.
12. KAMAL, M. R. *Polymer Sci. and Eng.*, **10** (1970) p. 108.
13. GENT, A. N. and HIRAKAWA, H. *J. Polymer Sci. A2*, **5** (1967) p. 157.
14. GENT, A. N. and MCGRATH, J. E. *J. Polymer Sci. A*, **3** (1965) p. 1475.
15. BRADEN, M. and GENT, A. N. *J. Applied Polymer Sci.*, **3** (1960) p. 90.
16. DUNN, P. and SANSOM, G. F. *J. Applied Polymer Sci.*, **13** (1969) p. 1641.
17. DUNN, P. and SANSOM, G. F. *J. Applied Polymer Sci.*, **13** (1969) p. 1657.
18. DUNN, P. and SANSOM, G. F. *J. Applied Polymer Sci.*, **13** (1969) p. 1673.
19. DUNN, P. and SANSOM, G. F. *J. Applied Polymer Sci.*, **14** (1970) p. 1799.
20. KASTELIC, J. R. and BAER, E. *J. Macromol. Sci.—Phys.*, **B7** (1973) p. 679.
21. BROWN, N. and FISCHER, S. *J. Polymer Sci.—Polymer Phys.*, **13** (1975) p. 1315.
22. OLF, H. G. and PETERLIN, A. *J. Polymer Sci.—Polymer Phys.*, **12** (1974) p. 2209.
23. BEAHAN, P., BEVIS, M. and HULL, D. *Phil. Mag.*, **24** (1971) p. 1267.
24. BEAHAN, P., BEVIS, M. and HULL, D. *J. Materials Sci.*, **8** (1974) p. 162.
25. BEAHAN, P., BEVIS, M. and HULL, D. *Proc. Roy. Soc. (London)*, **A343** (1975) p. 523.
26. KAMBOUR, R. P. *Polymer*, **5** (1964) p. 143.
27. KAMBOUR, R. P. and HOLIK, A. S. *J. Polymer Sci. A2*, **7** (1969) p. 1393.
28. KAMBOUR, R. P. and RUSSELL, R. R. *Polymer*, **12** (1971) p. 237.
29. LAUTERWASSER, B. D. and KRAMER, E. J., to be published.
30. DOYLE, M. J. and WAGNER, J. G., in *ACS Symposium on Toughness and*

Brittleness of Plastics, ed. R. D. Denin and A. D. Crugnola, *Advances in Chemistry Series*, **154** (1976) p. 63.
31. KAMBOUR, R. P. *Nature*, **195** (1962) p. 1299.
32. KAMBOUR, R. P. *J. Polymer Sci. A*, **2** (1964) p. 4159.
33. KAMBOUR, R. P. *J. Polymer Sci. A*, **2** (1964) p. 4165.
34. KAMBOUR, R. P. *J. Polymer Sci. A*, **4** (1966) p. 349.
35. LAUTERWASSER, B. D. and KRAMER, E. J. *Bull. Amer. Physical Soc.*, **23** (1978) p. 370.
36. DOYLE, M. J. *J. Materials Sci.*, **8** (1973) p. 1185.
37. KRENZ, H. G., KRAMER, E. J. and AST, D. G. *J. Materials Sci.*, **11** (1976) p. 2211.
38. LAUTERWASSER, B. D., AST, D. G. and KRAMER, E. J. *The Mechanical Properties of* n-*Heptane Crazes in Polystyrene: The Effect of Growth Temperature, MSC Report No. 2655*, Cornell University, 1976.
39. KRAMER, E. J., in *Mechanisms of Environment-Sensitive Cracking of Materials*, ed. P. R. Swann, F. P. Ford and A. R. C. Westwood, Metals Society, London, 1977, p. 249.
40. KRAMER, E. J., KRENZ, H. G. and AST, D. G. *J. Polymer Sci.—Polymer Phys.*, **16** (1978) p. 349.
41. LAUTERWASSER, B. D., unpublished.
42. MARSHALL, G. P., CULVER, L. E. and WILLIAMS, J. G. *Proc. Roy. Soc. (London)*, **A319** (1970) p. 165.
43. KRAMER, E. J. and BUBECK, R. A. *J. Polymer Sci.—Polymer Phys.*, **16** (1978) p. 1195.
44. PETERLIN, A. *J. Macromol. Sci.—Phys.*, **11** (1975) p. 57.
45. GENT, A. N. *J. Materials Sci.*, **5** (1970) p. 925.
46. TRELOAR, L. R. G. *Physics of Rubber Elasticity*, Oxford University Press, London, 1970, Chap. 7.
47. HAWARD, R. N., in *The Physics of Glassy Polymers*, ed. R. N. Haward, Applied Science, London, 1973, Chap. 6, p. 340.
48. REIMSCHUESSEL, A. C. and KIM, Y. J. *J. Materials Sci.*, **13** (1978) p. 243.
49. BRIBER, R., TANZER, J. and KRAMER, E. J., unpublished.
50. HANNON, M. J. *J. Applied Polymer Sci.*, **18** (1974) p. 3761.
51. HAWARD, R. N. and OWEN, D. R. J. *Proc. Roy. Soc. (London)*, **A352** (1977) p. 505.
52. SINGLETON, C. J., ROCHE, E. and GEIL, P. H. *J. Applied Polymer Sci.*, **21** (1977) p. 2319.
53. SONI, P. and GEIL, P. H. *J. Applied Polymer Sci.*, in press.
54. FAUCHER, J. A. and REDING, F. P., in *Crystalline Olefin Polymers*, ed. R. A. V. Raff and K. W. Doak, Interscience, New York, 1965, p. 706.
55. KELLER, A. and SADLER, D. M. *J. Macromol. Sci.—Phys.*, **B7** (1973) p. 263.
56. UDAGAWA, Y. and KELLER, A. *J. Polymer Sci. A2*, **9** (1971) p. 437.
57. THOMAS, E. L., SASS, S. L. and KRAMER, E. J. *J. Polymer Sci.—Polymer Phys.*, **12** (1974) p. 1015.
58. HARRISON, I. R., KELLER, A., SADLER, D. M. and THOMAS, E. L. *Polymer*, **17** (1976) p. 650.
59. SNEDDON, I. N. *Fourier Transforms*, McGraw-Hill, New York, 1951, pp. 395–430.

60. BILBY, B. A. and ESHELBY, J. D. 'Dislocations and the Theory of Fracture,' in *Fracture, Vol. 1*, ed. H. Liebowitz, Academic Press, New York, 1972, Chap. 2, p. 111.
61. HIRTH, J. P. and LOTHE, J. *Theory of Dislocations*, McGraw-Hill, New York, 1968.
62. MUSKHELISHVILI, N. I. *Some Basic Problems of the Mathematical Theory of Elasticity*, Noordhoff, Groningen, The Netherlands, 1953, p. 340.
63. GOODIER, J. N. and FIELD, F. A. *Proc. of International Conference on Fracture of Solids*, ed. D. C. Drucker and J. J. Gilman, *Met. Soc. Conferences, Vol. 20*, Interscience, New York, 1963, p. 103.
64. HAHN, G. T. and ROSENFIELD, A. R. *Acta Met.*, **13** (1965) p. 293. [Equation (18) is based on a straightforward generalisation of the superposition method of Hahn and Rosenfield; however, their statement immediately after eqn. A(11a) in that paper regarding the method for finding the displacement of an arbitrary loaded slit is incorrect and should be disregarded.]
65. ESHELBY, J. D., in *Fracture Toughness*, The Iron and Steel Institute, *ISI Publication 121*, London, 1968, Chap. 2, p. 55.
66. KNIGHT, A. C. *J. Polymer Sci. A*, **3** (1965) p. 1845.
67. VERHUELPEN-HEYMANS, N. *J. Polymer Sci.—Polymer Phys.*, **14** (1976) p. 93.
68. VERHUELPEN-HEYMANS, N. and BAUWENS, J. C. *J. Materials Sci.*, **11** (1976) p. 7.
69. WILCZYNSKI, A. P., LIU, C. H. and HSIAO, C. C. *J. Appl. Phys.*, **48** (1977) p. 1149.
70. DUGDALE, D. S. *J. Mechanics of Solids*, **8** (1960) p. 100.
71. HAHN, G. T., KANNINEN, M. F. and ROSENFIELD, A. R. *Ann. Revs. of Materials Sci.*, **2** (1972) p. 381.
72. GERBERICH, W. W. *International J. of Fracture*, **13** (1977) p. 535. [There is an error in eqns. (2) and (3) of this paper (eqn. (3) should be replaced with our eqn. (28)). The error has very little effect, however, on his final results.]
73. WILLIAMS, J. G. and MARSHALL, G. P., in *Mechanisms of Environment-Sensitive Cracking of Materials, Proceedings of the International Conference, University of Surrey, April 1977*, ed. P. R. Swann, F. P. Ford and A. R. C. Westwood, The Metals Society, London, 1977, p. 237.
74. GRAHAM, I. D., WILLIAMS, J. G. and ZICHY, E. L. *Polymer*, **17** (1976) p. 439.
75. ANDREWS, E. H., LEVY, G. M. and WILLIS, J. *J. Materials Sci.*, **8** (1973) p. 1000.
76. DI MARZIO, E. A. and GIBBS, J. H. *J. Polymer Sci. A*, **1** (1963) p. 1417.
77. BESSENOV, M. I. and KUVSHINSKII, E. V. *Sov. Phys. (Solid State)*, **3** (1960) p. 1959.
78. BROWN, H. R. and WARD, I. M. *Polymer*, **14** (1973) p. 469.
79. MILLS, N. J. and WALKER, N. *Polymer*, **17** (1976) p. 439.
80. MORGAN, G. P. and WARD, I. M. *Polymer*, **18** (1977) p. 87.
81. FRASER, R. A. W. and WARD, I. M. *Polymer*, **19** (1978) p. 220.
82. WEIDMANN, G. N. and DÖLL, W. *Colloid Polymer Sci.*, **254** (1976) p. 205.
83. KRENZ, H. G. Ph.D. Thesis, Cornell University, 1976.
84. KRENZ, H. G., AST, D. G. and KRAMER, E. J. *J. Materials Sci.*, **11** (1976) p. 2198.

85. KRAMER, E. J., KRENZ, H. G. and AST, D. G. *J. Materials Sci.*, **13** (1978) p. 1093.
86. PETERSON, T. L., AST, D. G. and KRAMER, E. J. *J. Appl. Phys.* **45** (1975) p. 4220.
87. KAMBOUR, R. P. and KOPP, R. W. *J. Polymer Sci. A2*, **7** (1969) p. 183.
88. HOARE, J. and HULL, D. *Phil. Mag.*, **26** (1972) p. 443.
89. ANDREWS, E. H. and BEVAN, L. *Polymer*, **13** (1972) p. 337.
90. MURRAY, J. and HULL, D. *Polymer*, **10** (1969) p. 451.
91. MURRAY, J. and HULL, D. *J. Polymer Sci. A2*, **8** (1970) p. 583.
92. DOYLE, M. J., MARANCI, A., OROWAN, E. and STORK, S. T. *Proc. Roy. Soc. (London)*, **A329** (1972) p. 137.
93. VINCENT, P. I. *Polymer*, **1** (1960) p. 425.
94. KIRSTE, R. G., KRUSE, W. A. and SCHELTEN, J. *Macromol. Chemie*, **162** (1972) p. 199.
95. BENOIT, H., DECKER, D., HIGGINS, J. S., PICOT, C., COTTON, J. P., FARNOUX, B., JANNICK, G. and OBER, R. *Nature—Phys. Sci.*, **245** (1973) p. 13.
96. WIGNALL, G. D., BALLARD, D. G. H. and SCHELTEN, J. *J. Macromol. Sci.—Phys.*, **B12** (1976) p. 75.
97. HAWARD, R. N., DANIELS, H. E. and TRELOAR, L. R. G., 'Molecular Conformation and Craze Fracture,' *J. Polymer Sci.*, in press.
98. PORTER, R. S. and JOHNSON, J. F. *Chem. Reviews*, **66** (1966) p. 1.
99. ROBERTSON, R. E., in *American Chemical Society Symposium on Toughness and Brittleness of Plastics, September 1974*, ed. R. D. Denin and A. D. Crugnola, *Advances in Chemistry Series*, **154** (1976) p. 89.
100. KUSY, R. P. and TURNER, D. T. *Polymer*, **15** (1974) p. 394.
101. GENT, A. N. and THOMAS, A. G. *J. Polymer Sci. A2*, **10** (1972) p. 571.
102. FELLERS, J. F. and KEE, B. F. *J. Applied Polymer Sci.*, **18** (1974) p. 2355.
103. BRADY, T. E. and YEH, G. S. Y. *J. Materials Sci.*, **8** (1973) p. 1083.
104. WELLINGHOFF, S. and BAER, E. *J. Macromol. Sci.—Phys.*, **B11** (1975) p. 367.
105. LAINCHBURY, D. L. G. and BEVIS, M. *J. Materials Sci.*, **11** (1976) p. 2222.
106. RUDD, J. F. *Polymer Letters*, **1** (1963) p. 1.
107. KAMBOUR, R. P., GRUNER, G. L. and ROMAGOSA, E. E. *J. Polymer Sci.—Polymer Phys.*, **11** (1973) p. 1879.
108. STERNSTEIN, S. S. 'Mechanical Properties of Glassy Polymers,' in *Treatise on Materials Science, Vol. 10B*, ed. J. M. Schultz, Academic Press, New York, 1977, p. 541.
109. STERNSTEIN, S. S. *ACS Polymer Preprints*, **10** (1969) p. 117.
110. STERNSTEIN, S. S. and MYERS, F. A. *J. Macromol. Sci.—Phys.*, **B8** (1973) p. 557.
111. OXBOROUGH, R. J. and BOWDEN, P. B. *Phil. Mag.*, **28** (1973) p. 547.
112. ARGON, A. S. and HANNOOSH, J. G. *Phil. Mag.*, **36** (1977) p. 1195.
113. MATSUSHIGE, K., RADCLIFFE, S. V. and BAER, E. *J. Materials Sci.*, **10** (1973) p. 833.
114. MATSUSHIGE, K., BAER, E. and RADCLIFFE, V. *J. Macromol. Sci.—Phys.*, **B11** (1975) p. 565.
115. ARGON, A. S. *Pure and Applied Chemistry*, **43** (1975) p. 247.
116. MAXWELL, B. and RAHM, L. F. *J. Soc. of Plastics Eng.*, **6** (1965) p. 437.
117. KITAGAWA, M. *J. Polymer Sci.—Polymer Phys.*, **14** (1976) p. 2095.

118. Sternstein, S. S. and Sims, K. J. *Polymer Preprints*, **5** (1964) p. 422.
119. Bubeck, R. Ph.D. Thesis, Cornell University, 1976.
120. Alfrey, T., Jr., Gurnee, E. F. and Lloyd, W. G. *J. Polymer Sci. C*, **12** (1966) p. 249.
121. Hoffenberg, H. B., Holley, R. H. and Stannett, V. *Polymer Eng. Sci.*, **9** (1969) p. 242.
122. Haward, R. N. and Owen, D. R. J. *J. Materials Sci.*, **8** (1973) p. 1136.
123. Haward, R. N. and Thackray, G. *Proc. Roy. Soc. (London)*, **A302** (1968) p. 453.
124. Bernier, G. A. and Kambour, R. P. *Macromolecules*, **1** (1968) p. 393.
125. Parrish, M. and Brown, N. *Nature*, **237** (1972) p. 122.
126. Brown, N. *J. Polymer Sci.—Polymer Phys.*, **11** (1973) p. 2099.
127. Argon, A. S. *J. Macromol. Sci.—Phys.*, **B8** (1973) p. 573.
128. Saffman, P. G. and Taylor, G. I. *Proc. Roy. Soc. (London)*, **A245** (1958) p. 312.
129. Fields, R. J. and Ashby, M. F. *Phil. Mag.*, **36** (1976) p. 1217.
130. Argon, A. S. and Salama, M. M. *Materials Sci. and Eng.*, **23** (1976) p. 219.
131. Argon, A. S. and Salama, M. M. *Phil. Mag.*, **36** (1977) p. 1217.
132. Sternstein, S. S., Ongchin, L. and Silverman, A. *Appl. Polymer Symp.*, **7** (1968) p. 175.
133. Speidel, M. O. *Met. Trans.*, **6A** (1975) p. 631.
134. McClintock, F. A. and Argon, A. S. *Mechanical Behavior of Materials*, Addison–Wesley, Reading, Massachusetts, 1966, p. 285.
135. Beardmore, P. and Rabinowitz, S. *J. Materials Sci.*, **10** (1975) p. 1763.
136. Williams, J. G. and Marshall, G. P. *Proc. Roy. Soc. (London)*, **A342** (1975) p. 55.
137. Narisawa, I. and Kondo, T. *J. Polymer Sci.—Polymer Phys.*, **11** (1973) p. 223.
138. Williams, J. G., Marshall, G. P., Graham, I. and Zichy, E. L. *Pure and Applied Chem.*, **39** (1974) p. 275.
139. Happel, J. *A.I.Ch.E. Journal*, **5** (1959) p. 174.
140. Happel, J. and Brenner, H. *Low Reynolds Number Hydrodynamics*, Prentice-Hall International, Hemel-Hempstead, Herts, 1965, p. 395.
141. Gurney, C. and Hunt, J. *Proc. Roy. Soc. (London)*, **A299** (1967) p. 509.
142. Marshall, G. P., Culver, L. E. and Williams, J. G. *Plastics and Polymers*, **37** (1969) p. 95.
143. Mai, Y. W. *J. Materials Sci.*, **10** (1975) p. 943.
144. Marshall, G. P., Culver, L. E. and Williams, J. G. *Plastics and Polymers*, **37** (1969) p. 75.
145. Beaumont, P. W. R. and Young, R. J. *J. Materials Sci.*, **10** (1975) p. 1334.
146. Johnson, H. H. in *Fracture, Vol. 3*, ed. H. Liebowitz, Academic Press, New York, 1971, p. 679.
147. Oriani, R. A. and Josephic, P. H. *Acta Met.*, **22** (1974) p. 1065.
148. Weidmann, G. W. and Holloway, D. G. *Physics and Chemistry of Glasses*, **15** (1974) p. 116.
149. Lawn, B. R. and Wilshaw, T. R. *Fracture of Brittle Solids*, Cambridge University Press, Cambridge, 1975, Chap. 8.

150. Kambour, R. D. and Gruner, C. L. *J. Polymer Sci.—Polymer Phys.*, **16** (1978) p. 703.
151. Barton, A. F. M. *Chem. Reviews*, **75** (1975) p. 731.
152. Kambour, R. P., Romagosa, E. E. and Gruner, C. L. *Macromolecules*, **5** (1972) p. 335.
153. Kambour, R. P., Gruner, G. L. and Romagosa, E. E. *Macromolecules*, **7** (1974) p. 249.
154. Vincent, P. I. and Raha, S. *Polymer*, **13** (1972) p. 283.
155. Henry, L. F. *Polymer Eng. and Sci.*, **14** (1974) p. 167.
156. Jacques, C. H. M. and Wyzgoski, M. G. 'Prediction of Environmental Stress Cracking of Polycarbonate from Solubility Considerations,' *J. Applied Polymer Sci.*, in press.
157. Hildebrand, J. H. and Scott, R. L. *Solubility of Non-Electrolytes*, Dover, New York, 1964.

Chapter 4

IMPACT PERFORMANCE OF POLYMERS

P. E. REED
Queen Mary College, University of London

SUMMARY

Impact performance of a polymeric component is considered to be determined by several interacting complex factors. Impact performance depends not only on the choice of material but also the manner in which the material is deployed within the component. The chapter seeks to explore the development in different avenues of research related to impact performance. Following a review of current test methods used to determine the 'impact strength' of a polymer, consideration is given to the significance of the term 'impact strength' as determined by these tests. Recent studies on the effects of specimen geometry, test temperature and processing variables are reviewed, as well as the analyses of notched-bar tests. Considerations of the response of actual polymeric components to impact loading, as opposed to standardised test pieces of polymeric materials, complete this section.

1 INTRODUCTION

The impact test is one test used to assess the mechanical performance of plastics and other materials. The term 'impact test' suggests a basic uniformity but in 1958 Westover[1] said: 'Notched and un-notched specimens have been made in various shapes and sizes and have been subjected to tensile, compressive, torsion and bending impacts. Materials have been thrown, dropped and subjected to blows from hammers, bullets, falling weights, falling balls, horizontally moving balls, pendulums and projections from flywheels'. The situation has not been simplified in the past 20 years: a bewildering array of test techniques, sometimes producing conflicting results, still exists. Results from different tests cannot be correlated in general.

Impact testing is conducted with several objectives in view. It completes a series of mechanical tests which seek to assess the performance of the material; impact testing is viewed as covering the high strain-rate end of the strain-rate spectrum. However the strain rates in conventional notched-bar impact tests are not necessarily particularly high and orthodox high strain-rate testing may be more appropriate. Newton[2] lists the order-of-magnitude characteristics of various impact tests shown here.

TABLE 1

ORDER-OF-MAGNITUDE CHARACTERISTICS OF VARIOUS IMPACT TESTS

Method	*Order of magnitude of strain rate* (s^{-1})	*Velocity* m s^{-1}
Charpy	(10)	3
Izod	(60)	2·44
Falling ball	10^{-1}–10	1–4
Conventional Instron	10^{-3}–10^{-1}	10^{-5}–10^{-1}
Pneumatic gun	10^{2}–10^{4}	20–240
Hydraulic systems	1–10^{2}	·008–4

Newton did not supply a strain-rate value for the Charpy and Izod tests. The values given are derived from an assumed circular curvature of the specimen in each case, with a central rate of deflection of 3 and 2·44 m s^{-1} respectively. Strain rates at the notch tip are indeterminable but considerably greater than the two values derived. Impact testing is also performed to determine the brittle/ductile transition point of a material under impulsive loading. Since brittle behaviour can lead to catastrophic failure of structures, designers seek to obviate its onset by appropriate selection of materials.

The primary aim of impact studies on plastics materials is to give a basic understanding of the behaviour of such materials under impact loading and hence provide a reliable guide to their impact performance. However, impact tests, although simple to perform, are difficult to analyse in detailed scientific terms, since they involve the dynamic response of structures, stress waves in solids and non-equilibrium situations. Recommendations on the presentation of plastics design data[3] state that: 'Impact behaviour depends on product design, method of manufacture and condition of use, in addition to the basic properties of the material. Because of the many interacting variables, it has not yet been possible to develop formulae for use in the rational design of plastics products with regard to impact behaviour'. Because of the complex nature of impact testing, it is not surprising that many different impact tests have evolved and been used. Each test provides some insight into the impact performance of polymers and polymeric structures.

Problems with impact testing and the application of impact test results to practical situations are not unique to plastics. The same problems face the metallurgist and engineer using metals, and understanding in that area is probably not greatly more advanced than in polymers. However, in contrast to many commonly used metals, polymeric materials are generally highly sensitive to strain rate and temperature in the zone of use. Brittle behaviour can often occur with minor increases in strain rate or modest decreases in temperature. Since the impact test serves to highlight the possibility of brittle behaviour in service, the test takes on more significance in the case of polymeric materials.

Turner[4] classifies impact tests under three headings:

(a) high-speed tensile impact;
(b) flexed-beam impact;
(c) falling-weight impact.

This classification is fairly broad, covering the majority of tests currently used and certainly those covered by British Standard (BS) and American Society for Testing and Materials (ASTM) tests for *material assessment*. However, it possibly excludes certain high strain-rate tests (e.g. split Hopkinson bar) which may be considered as adjuncts to impact testing if not actually classified as impact tests themselves. There are also acceptance impact tests on polymeric *components* covered by the Standards (e.g. references 5–9). It is difficult to set a boundary to the subject of impact testing, in contrast to other mechanical tests. To some the term is synonymous with notched-bar testing, to others it implies high strain-rate testing. Probably all would agree that it involves impact loading of the specimen or component. Turner[4] makes the point that current impact tests are designed specifically to measure the ability of a *specimen* to withstand a sudden blow, drawing a clear distinction between specimen and sample (a specimen is taken from a sample, a sample in turn being taken from a population). With such a broad spectrum of views on impact testing Westover[1] comments: 'There are nearly as many types of testing machine as there are investigators'. We might possibly add that there are as many views as there are investigators.

2 IMPACT TEST TECHNIQUES

2.1 Izod and Charpy Impact Tests

The Izod and Charpy tests for plastics are specified in both BS 2782 and ASTM–D 256.[10,11] Both are swinging pendulum techniques, in which a pendulum is released from a fixed height (h_0), swings down to strike and break the specimen at the bottom of the swing and then continues its swing to a measured maximum height (h_f) at the end of the first half-oscillation. The two tests differ mainly in the manner in which the notched-bar specimens are held. In the Charpy test, the test-piece is a simply supported rectangular beam, struck at mid-span behind the notch as shown in Fig. 1(a). In the Izod test the specimen is a notched cantilever beam which is held in a vice and struck by the pendulum bob just above the notch, as shown in Fig. 1(b). The different Standard specifications state exact dimensions for the specimens (as shown in Fig. 1) but give only broad requirements for the machine. For example:

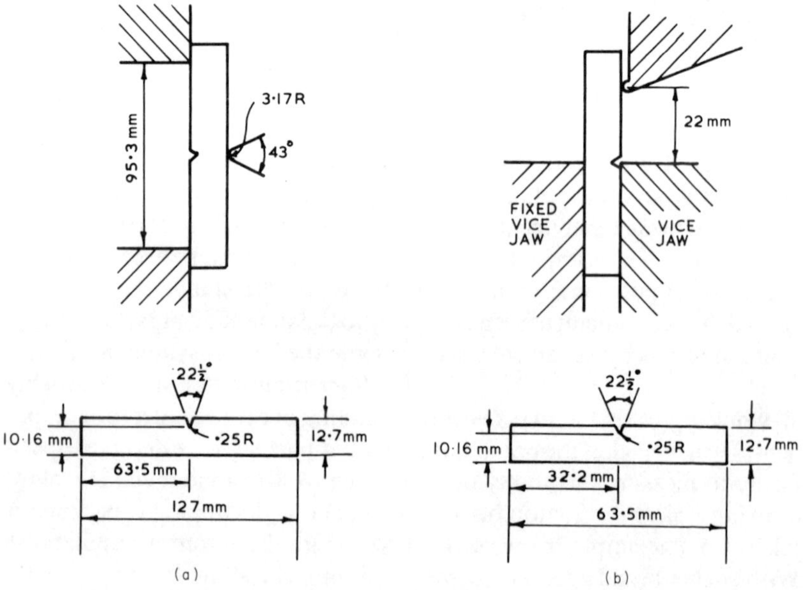

FIG. 1. Geometry of testing and specimen support arrangement for Charpy (Fig. 1a) and Izod (Fig. 1b) tests specified in ASTM-D256.

(a) BS 2782 for the Izod test requires the specimen to be struck with a striker of nose radius 3·2 mm at a velocity of 2·44 m s^{-1};
(b) the centre of percussion of the pendulum is to coincide with the centre of the striking edge;
(c) For the Charpy test, the relationship required between the energy available in the machine and the strike velocity is as follows:

Maximum energy (J)	Strike velocity (m s^{-1})
0·5	2·9
1	2·9
4	2·9
15	3·8

(d) the energy required to fracture the specimen has to be within 10–80% of the blow energy.

The lack of precise specification of the machine to be used has led to the production of many different devices, some commercially available, others privately built. However, the machine variables which are permitted by the various standards lead to a situation where different values of impact strength are obtained for the same material from Izod and Charpy tests and from different machines using the same test.

In both the Izod and Charpy tests, the parameter measured is the energy absorbed from the pendulum during its first half-oscillation. This impact energy, as it is sometimes termed, may be defined as[12]

$$U = \tfrac{1}{2}I(v_0^2 - v_f^2)$$

where I is the moment of inertia of the pendulum, and v_0 and v_f respectively the velocities of the pendulum bob just before and just after striking the specimen. Assuming no windage or friction losses on the pendulum, $v_0^2 = 2gh_0$ and $v_f^2 = 2gh_f$. Then

$$U = gI(h_0 - h_f)$$

or, more realistically, $U = K(h_0 - h_f)$ where K is a machine constant for a given system.

In both the Izod and Charpy test, failure initiates in the vincinity of the notch tip and propagates through the ligament of the specimen in line with the notch. The types of failure that can occur are discussed by Vincent,[13] and are listed as:

(a) the specimen yields without cracking and the specimen deforms sufficiently to allow passage of the pendulum without separation of the specimen into two parts;
(b) the specimen yields and cracks but the entire fracture surface exhibits ductile failure;
(c) the specimen initially cracks in a brittle manner but then changes to a ductile mode of behaviour;
(d) the specimen yields first and then changes to brittle behaviour;
(e) the fracture surfaces appear entirely brittle.

With this variety of behaviour, it has been suggested[3,13] that information on the mode of failure as well as the impact energy or impact strength should be quoted following any test.

The impact strength is defined either in terms of energy per unit width of notch or energy per unit area under the notch. Taking the width (B), depth (D) and notch length (a) of the specimen shown in Fig. 2, the impact strengths are specified as either:

$$U/B(\mathrm{kJ\ m^{-1}})$$

or

$$U/B(D-a)(\mathrm{kJ\ m^{-2}})$$

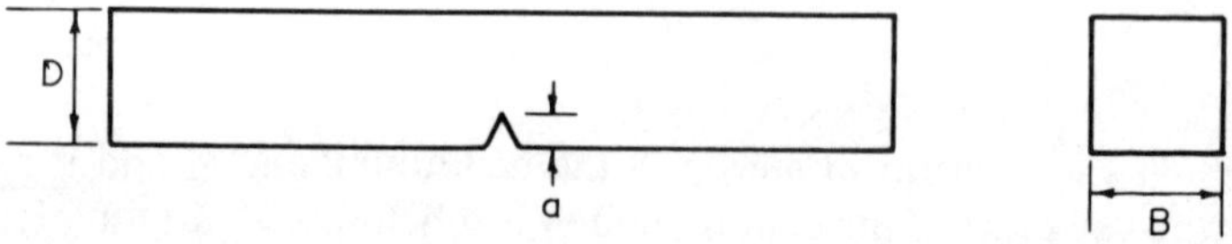

FIG. 2. Characteristic dimensions of notched bar specimen for impact strength determination.

As pointed out by Vincent,[13] the two methods of expressing the impact strength are not convertible unless it is assumed that the energy-to-break is proportional to the ligament length under the notch. Then, assuming $(D-a) = 10{\cdot}16$ mm

$$1\ \mathrm{kJ\ m^{-1}} \equiv 100\ \mathrm{kJ\ m^{-2}}$$

2.2 Falling-weight Impact Testing

Impact testing of sheet or film is difficult by means of conventional notched-bar tests, and falling-weight (or dart-drop) tests are then employed. Test procedures are specified in BS 2782[10] and ASTM–D 1709–75.[14] In both tests, a hemispherically-nosed metal dart of variable weight is released from a fixed height and falls freely to strike the centre of the sheet specimen. The specimen is a circular disc which is supported and clamped at its periphery by one of the prescribed methods. The dart is released so that no rotational movement is imparted to it, often from an electromagnet.

Knowing the weight of the dart and the height through which it falls, it is easy to calculate the energy available for fracturing the specimen. However, in contrast to notched-bar testing, the energy required to fracture the specimen has to be determined from a large number of tests. In principle, the weight of the dart is increased on successive drops until fracture just occurs. The variability of the onset of fracture, particularly in brittle materials, renders the assessment of the limiting fracture energy difficult. One of two alternative statistical methods is employed.

In the probit method a number of specimens are tested at each of several energy levels. A record is made of the percentage of failures at each energy level. Subsequently a graph is drawn (the probit curve) of the percentage failure against the energy of striking. The probit curve is constructed to cover almost the whole range 0 to 100% failure. Finally the impact energy is quoted as that which causes a 50% failure level of the specimens (the F–50 value), if that is an appropriate value for the application considered.

Where there is a rapid change in failure energy from 0 to 100% failure on the probit curve, the alternative step or staircase method can be adopted. In this technique, the dart is always dropped from one height but the dart weight is increased in uniform steps until failure occurs. Testing then continues, the dart weight being decreased by a specified increment if failure occurs or being increased by the same amount if it does not. All the results are tabulated; a typical presentation is shown in Fig. 3. It can be seen from Fig. 3 that the failure rates at 244, 258, 271 and 285 J are 0, 12·5, 60·8 and 100% respectively. The table shows the number of failures and non-failures at each energy level, so that the F–50 energy value can again be determined. Linear interpolation between the 285 and 271 values provides a 50% failure value (F–50) of 268 J.

Falling-weight impact test methods are also adopted to test plastic components.[8] In many cases these are used as a quality assurance test

DROP HEIGHT	JOULES IMPACT	F50 = 268 Joules																																																		TOTAL F	TOTAL N
2.74 m	366																																																				
2.64	352																																																				
2.54	339																																																				
2.44	325																																																				
2.34	312																																																				
2.24	298																																																				
2.13	285		O						O		O										O		O								O		O														O		O			9	0
2.03	271	X		O		O		X		X		O				O		O		X		X		O		O		O		X		X		O		O		O				O		O		X		X		O		14	9
1.93	258				X		X						O		X		X		X						X		X		X						X		X		O		X		X		X						X	2	14
1.83	244													X																										X												0	2
1.73	230																																																				
1.63	217																																																				

FIG. 3. Typical results of a dart-drop test which adopts the 'staircase' method of analysis. (After reference 15, by permission of W. J. Moritz.)

during production of the particular component. One advantage of falling-weight testing is that the impact energy distribution can be made available, so providing information of the probability of failure over a range of impact energy values. An analysis of the statistical approaches used in falling-dart impact tests has been presented by Moritz.[15] Falling-weight tests may also approximate more closely to actual impacts received in service in some instances, but the large number of tests required for one F–50 value is a disadvantage of such methods.

2.3 Tensile Impact Tests

Tensile impact tests may be performed on machines used for Izod and Charpy tests with suitable modifications to the pendulum. Procedures are set down in ASTM–D 1822.[16] In these tests the specimen conforms to 'dumb-bell' or 'dog-bone' shape; one end is clamped to the pendulum bob, while the other is held in clamps attached to a striker plate. When the pendulum is released from its raised position, it swings down between stop brackets, trailing the specimen behind. The striker plate is then arrested by the stop brackets as the pendulum bob swings on, causing the specimen to be strained rapidly in tension to failure. The energy required for fracture is determined in the same manner as for the Charpy and Izod tests. The impact strength of the material under test is defined as

(impact energy)/(minimum cross-section of specimen).

In tensile impact tests, the strain rate averaged over the entire specimen length is approximately constant during deformation.[4] Local strain rates due to stress waves within the solid following impact can account for large deviations from the mean value. In flexural impact and falling weight tests the specimen is deformed at a rate which decreases progressively, and rapidly.

3 FRACTURE PARAMETER

The most commonly quoted fracture parameter is that of impact strength. Definitions of the impact strength as determined by notched-bar and tensile impact tests have been given in the previous section. In each case it is quoted as the energy absorbed from the pendulum bob during impact divided by the minimum cross-section of the specimen where fracture occurs. Defined in this way the 'impact strength' begins to resemble other material properties, such as tensile strength. Turner[4] refers to the impact strength as a 'loose misnomer'. One problem is that the energy measured is an all-inclusive term. Ireland[12] considers that five terms are involved:

$$U = U_{\mathrm{I}} + U_{\mathrm{SD}} + U_{\mathrm{B}} + U_{\mathrm{MV}} + U_{\mathrm{ME}}$$

where U_{I} = energy to accelerate the specimen from rest;
U_{SD} = energy of bending the specimen;
U_{B} = energy consumed by Brinnell deformation under the nose of the hammer;
U_{MV} = energy absorbed by the machine through vibrations after initial contact;
U_{ME} = elastic energy stored by the machine.

The definition of impact strength, suggesting that the energy measured can be assigned solely to the zone under the notch, is misleading. Any attempt to perform an energy balance and determine separately all the terms listed above would be difficult. The only term separated easily is the tossing loss U_{I}, which can be from 29 to 83% of the total value.[17] This can be partially estimated by lightly rejoining the broken pieces of the specimen and repeating the impact test. In some instances, tossing losses so estimated are deducted from the total energy before calculating the impact strength. Attempts to reduce energy losses due to friction and windage on the pendulum are also made by the construction of special high-precision machines,[18] which lead to greater consistency in the energy values determined from any sample.

Beside the problems of determining that portion of the total energy which is to be attributed to failure of the material, the concept of using total energy at all as the measure of the susceptibility of a material to fracture is open to debate.[17] Criteria for failure in other mechanical tests are based on limiting stress, limiting shear-strain energy and, in fracture mechanics, limiting energy release rate. Figure 4 shows a series of linear-elastic load-deformation curves. The areas under the curve are identical, hence the energy absorbed in each deformation is the same. However, the loads- and strains-to-fracture differ greatly.

An alternative method of analysing the data obtained from notched-bar tests, based on linear-elastic fracture mechanics (LEFM) considerations, has been proposed by Williams.[18,19] He considered the linear-elastic deformation of a notched specimen. When this is subjected

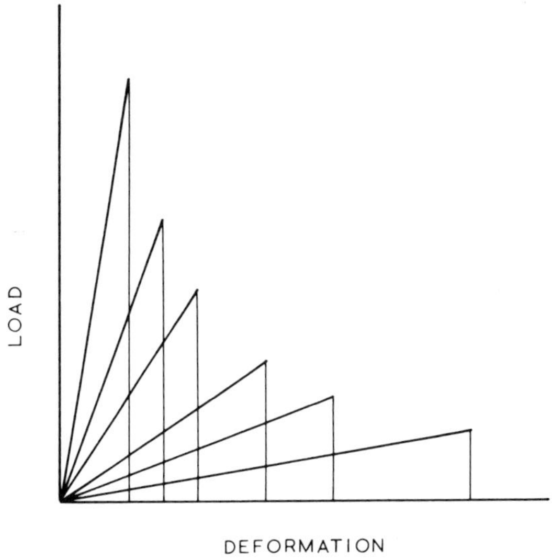

FIG. 4. A series of load-deformation curves for linear elastic materials, for which the energy to rupture is constant.

to an applied load P, resulting in a deflection x under the load, the energy stored is given by

$$U = \tfrac{1}{2}Px \tag{1}$$

Introducing $C = x/P$, the compliance of the beam specimen for a given geometry,

$$U = \tfrac{1}{2}P^2C \tag{2}$$

For a specimen of uniform thickness B, the strain-energy release rate G is

$$G = \frac{1}{B}\frac{\mathrm{d}U}{\mathrm{d}a} = \frac{1}{B}\frac{\mathrm{d}U}{\mathrm{d}C}\frac{\mathrm{d}C}{\mathrm{d}a} \tag{3}$$

where a is the notch length.

When fracture occurs at load P, the critical energy release rate is then

$$G_c = \frac{P^2}{2B}\frac{\mathrm{d}C}{\mathrm{d}a} \tag{4}$$

In terms of energy absorbed, rather than load at fracture,

$$U = G_c B \frac{C}{(\mathrm{d}C/\mathrm{d}a)}$$

Introducing the specimen width D, eqn. (4) may be rewritten

$$G_c = \frac{P^2}{2BD} \frac{dC}{d(a/D)}$$

In terms of energy rather than load at rupture, this becomes

$$G_c = \frac{U}{BDC} \frac{dC}{d(a/D)}$$

By the introduction of the dimensionless factor

$$\phi = \frac{C}{dC/d(a/D)}$$

$$U = G_c BD\phi \tag{5}$$

This analysis predicts that a linear relationship should exist between the total energy absorbed in bending the specimen to fracture and the factor $BD\phi$, the slope of which provides a value for the critical strain energy release rate (G_c) for impact loading conditions. Values for the compliance C, and hence ϕ, are determined from a series of quasi-static loading tests on similar notched bars for a range of a/D ratios.

Williams[18,19] prefers to express the fracture parameter in terms of G_c rather than fracture toughness K_c, since the former can be determined without reference to the strain-rate-dependent Young's modulus E, since

$$K_c^2 = EG_c$$

Plots of G_c against $BD\phi$ are found to be linear for many thermoplastics after certain factors are considered, although considerable scatter in the data exists in some instances, particularly in the low-energy, very brittle materials. Tossing loss energies are normally determined and subtracted from the total energy values. Curves of $U/(BD)$ against ϕ for the more ductile materials are found to exhibit a strong curvature.[19] This is attributed to the formation of a plastic zone at the crack tip and hence the lack of purely elastic deformation. Plati and Williams[18] account for this by the introduction of a plastic zone size r_p such that the crack length is effectively increased to

$$a_f = a + r_p$$

It is found that $U/(BD)$ *versus* ϕ curves for modestly ductile materials can be reduced to a straight line in many instances by the introduction of an arbitrary single value of r_p, and hence a value for G_c is determined.

Strain-energy release rates determined by Plati and Williams[18,57] for a selection of plastics are given in Table 2.

TABLE 2

STRAIN-ENERGY RELEASE RATES FOR VARIOUS PLASTICS

Material	G_c(kJ m^{-2})
PMMA	1·3–1·6
PS	0·3–0·8
PVC	1·3–1·4
PC	2–5
Nylon 66	0·25–4
PE (medium-density)	5–10
PE (low-density)	5–40
HIPS	2–16

This analysis is based on certain basic assumptions:

(a) linear-elastic fracture mechanics is applicable;
(b) the energy U used in eqn. (5), suitably modified for tossing losses, is the energy stored in the specimen which can be released for fracture (i.e. $U = U_{SD}$);
(c) values of the compliance C, and hence ϕ, determined in quasi-static tests are equally applicable to the dynamic situation;
(d) that the quantities U, a and C are explicit functions of each other.

Whether these assumptions are valid in the impact test has not been justified.

Data on stress intensity factors determined under dynamic conditions are also available as a further means of characterising the behaviour of plastics under rapid loading conditions (e.g. references 20, 21). Whether such tests are truly impact tests raises, yet again, the problem of the definition of an impact test.

4 FACTORS AFFECTING THE FRACTURE PARAMETER

Both the impact strength and strain-energy release rate are affected by a number of factors, which include specimen geometry, test temperature and fabrication conditions.

4.1 Effect of Specimen Geometry

The impact strength determined in a conventional notched-bar test is a function of specimen width, specimen length, notch depth and notch tip radius.[13,17] Vincent[13] quotes figures for rigid PVC samples of 3 mm and 6 mm width over a range of temperature. Impact strengths of the narrower specimens are always found to be greater than those for the wider specimens by a factor varying between 1·3 and 3·7. Specimens of larger span, and consequent lower strain rate, give rise to increased values for the impact strength. Decreasing the notch depth, while maintaining the same notch tip geometry, also serves to increase the impact strength. Values given by Vincent[13] for a glass-filled nylon 66 sample show the impact strength increasing from 8 kJ m^{-2} when the notch depth was 2·8 mm to 18·3 kJ m^{-2} for a notch depth of 0·25 mm.

Varying the notch tip radius has a marked effect on the measured impact strength. Data presented by Vincent[13] are shown in Fig. 5.

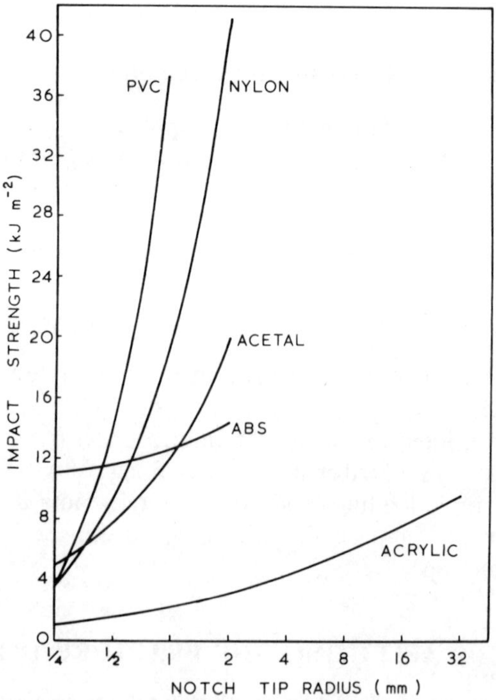

FIG. 5. Impact strength as a function of notch tip radius for samples of five different polymers. (After reference 13, by permission of The Plastics and Rubber Institute.)

Attempts to rationalise these data by the introduction of a stress concentration factor $k = 1 + 2(a/r)^{1/2}$, where r is the notch tip radius, were only partially successful. For some materials it was found that a plot of impact strength against the reciprocal of the stress concentration factor produced a linear relationship, while in other cases it did not (Fig. 6). Cases in

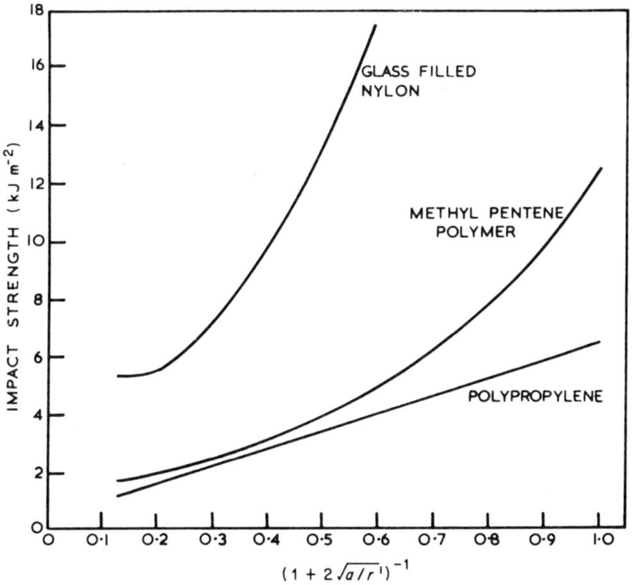

FIG. 6. Variation in the impact strengths of three polymers with approximate stress concentration factor. (After reference 13, by permission of The Plastics and Rubber Institute.)

which the linear relationship was found tended to be those in which the material more nearly obeyed linear-elastic behaviour. Introduction of the stress concentration factor implies that it is the stress at the notch tip which controls fracture. Since the stress concentration factors computed were based on linear-elastic considerations, it was suggested that results from those materials which complied with linear-elastic behaviour would be more readily rationalised by the introduction of the stress concentration factor.

An alternative explanation of the decreasing impact strength with notch tip radius is that the strain rate at the notch tip increases with decrease in tip radius, all other factors remaining constant. This explanation offers more opportunity for unifying the effects of notch tip radius and temperature, particularly with regard to ductile/brittle transitions.[22]

4.2 Effect of Temperature

The test temperature has a marked effect on the impact strength[13] and strain-energy release rates[18] determined. These quantities increase with temperature and often show an abrupt change in value over a short temperature interval, which is associated with a change from a brittle to a ductile mode of failure. The location of this transition is an important aspect of material selection for impact duty. Its position has been found to be a function of rate and notch severity, as shown in Fig. 7, as well as fabrication conditions.[23] Figure 7 shows that the transition in impact strength is a function of notch tip radius, the curves UNIS, IS(2), IS(1) and IS(0·25) corresponding respectively to specimens without notches

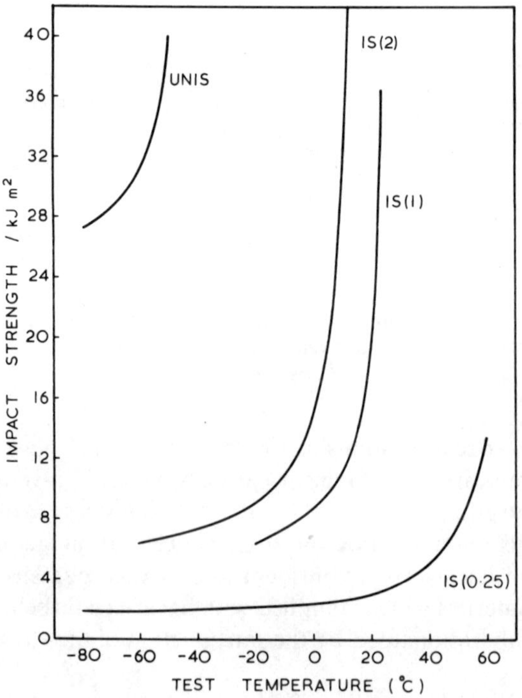

FIG. 7. Effect of test temperature on impact strength of rigid PVC samples containing notches of different severity. (After reference 13, by permission of the Plastics and Rubber Institute.)

and with notches of three different radii, expressed in mm. Vincent found that there was no general relation between the glass transition temperature T_g of the polymer and the brittle/ductile transition temperature.

Similar transitions in the strain-energy release rate with temperature have been reported by Plati and Williams.[18] Once again the transition point is found to move to higher temperatures as the severity of the notch is increased (tip radius decreased). Data for PVC are given by both Vincent[13] and Plati and Williams[18] for identical notch tip radii. The respective transitions in impact strength and G_c for equivalent notch tip radii occur at very similar temperatures. Considering the range of substances covered by the term PVC this agreement is remarkable, although both materials tested were unplasticised PVC. The majority of Plati and Williams' work[18] is concerned with 'sharp' notches, i.e. notches machined with a very sharp cutter. It is interesting to note that, for these sharp-notched specimens, the transition in G_c is often in the region of -10 °C for several polymers which offer widely different general mechanical properties. Their results are interpreted in terms of constraints imposed by the specimen geometry in the vicinity of notch.

An alternative view has been expressed by Ryan.[23] Working with polycarbonate, he found that the position of the transition in impact strength was a function of the notch tip radius and the melt flow rate. It was also found that annealing the specimens prior to test caused a shift of the transition temperature to lower temperatures. Ryan's tests indicate that the position of the transition cannot be explained in terms of geometrical factors alone. It was argued that factors leading to higher rates of strain at the notch tip increase the possibility of a more brittle type of behaviour. Annealing the sample causes a change in the relaxation-time spectrum which was considered to be equivalent to shifting to a higher strain rate.

4.3 Effect of Processing Variables

Many authors have commented on the effects of processing variables on the measured impact strength.[3,4,13,23–28] It has been demonstrated that the impact strength as measured in conventional impact tests is a function of: (i) injection-moulder cylinder temperature; (ii) molecular orientation in injected components; (iii) polymer residence time in the injection-moulder cylinder; (iv) annealing and quenching subsequent to moulding; (vi) melt flow rate in the mould.

Typical effects of varying the injection-moulding cylinder temperature and sample orientation are shown in Table 3 (after Vincent[13]) for polypropylene tested at 22 °C.

TABLE 3

EFFECTS OF PROCESSING VARIABLES ON IMPACT STRENGTH FOR POLYPROPYLENE

Cylinder temperature (°C)	*Impact strength* (kJ m^{-2})		
	Along flow	*Across flow*	*Ratio* (*along*/*across*)
200	17·4	6·1	2·9
220	17·4	6·3	2·8
240	17·9	8·6	2·1
260	14·9	11·8	1·3
280	13·2	10·9	1·2
300	13·0	10·3	1·3
310	8·8	7·8	1·1

Specimens in which the flow is along the length of the specimen and thus normal to the plane of the notch show enhanced strength. In contrast, those specimens cut with the plane of the notch parallel to the flow direction exhibit lower impact strengths. These differences may be directly attributed to molecular orientation in the specimens, the molecules tending to align themselves in the direction of flow. At the lower cylinder temperatures, shear stresses in the melt flow will be greater, thus enhancing molecular orientation and increasing the ratio of the mutually perpendicular impact strengths. Increasing melt temperature leads to a less anisotropic specimen and greater equality of the two impact values. The markedly reduced impact strengths at the highest cylinder temperature may be attributed to degradation of the polymer and consequent reduction in molecular weight. Molecular weight is a further factor affecting the impact strength.[23]

Broutman and Krishnakumar[25,26] have shown that cold rolling of polymeric sheet material can have a marked effect on the impact strength. Small reductions in thickness by cold rolling of poly(vinyl chloride), acrylonitrile–butadiene–styrene (ABS), polyphenylene oxide (PPO), polysulphones and polycarbonate sheets produced large increases in impact strength in these materials. However, similar thickness reductions in sheets of rubber-toughened acrylics resulted in only modest increases in impact strength. Broutman and Krishnakumar argue that the small amount of rolling applied could not induce sufficient orientation in the samples to account for the large increases in toughness. The increase was attributed to residual compressive stresses induced in the surface of the sheet which suppressed fracture. The existence of such stresses was supported by birefringence measurements. Values of the compressive surface stresses were found to be as high as 21 MN m^{-2}.

Annealing and quenching of specimens also modified the impact behaviour.[24,28] The effects produced by thermal treatment were found to be complex, depending on the degree of initial working, material thickness and cooling rate. A transition in impact strength from a high to a low value occurred in polycarbonate as the sheet from which the specimens were taken increased in thickness. Annealing and quenching of samples taken from one sheet thickness produced impact strength values on either side of the transition point, depending on the rate of cooling applied. It was thus argued that the 'thickness effect' observed initially was not due to geometrical factors but could again be attributed to modifications in the residual stresses at the sheet surface caused by the thermal treatments.

Many factors have been shown to affect the impact behaviour of polymeric materials. These factors considerably complicate the interpretation of data from any form of impact test. They also serve to confuse attempts to compare data from different sources, since the test conditions and condition of material under test are rarely identical.

5 INSTRUMENTED IMPACT TESTING

The only parameter determined by the tests described in Section 2 was the impact energy. This is an all-inclusive term, the integral of the load–deformation curve. This curve may take different forms, depending on the nature of the material and the manner in which the specimen fractures. A selection of possible forms is shown in Fig. 8.[29,30] Complex forms of load–deformation curves during impact have also been reported for metals[31] arising from a mixture of brittle and ductile modes of fracture.

It is perhaps optimistic to expect that the single parameter of impact energy can be used to represent and analyse such a wide variety of behaviours, irrespective of the problems discussed in the previous section. A full record of the load–deformation curve during impact can supply much more information, while still supplying the impact energy by integration under the curve. Many types of instrumented impact tests have been described, both for swinging-pendulum and falling-weight systems.

Ireland[12] describes the use of semiconductor strain gauges, bonded to the striker or the specimen supports in pendulum and falling-weight tests, to provide information on the load–time characteristic. Toland[32] sim-

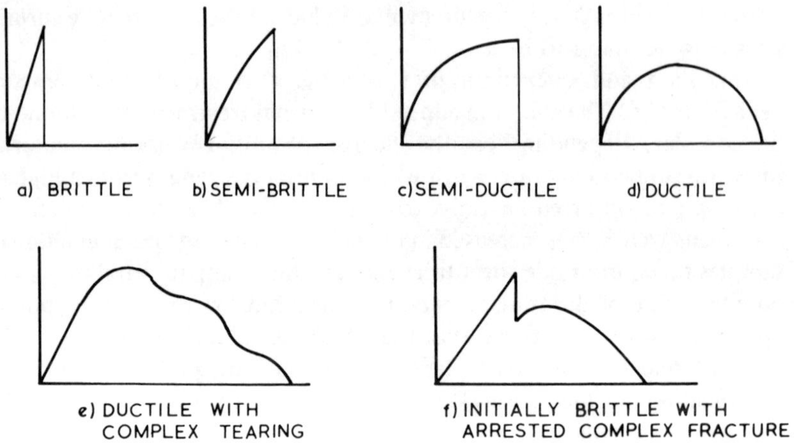

FIG. 8. Typical forms of the load–deformation curve for different types of materials.

ilarly instrumented the Charpy test in order to obtain load–time curves from impact-tested carbon–epoxy composite materials. Lubert *et al.*[33] instrumented an Izod test by attaching a strain gauge to the specimen clamp to measure the force–time characteristic, simultaneously measuring the position of the pendulum arm by means of a semiconductor photocell. Force–displacement curves were hence deduced in this instance. McMillan and Tesh[34] studied the dynamic effects in the impact testing of brittle materials by bonding strain gauges to both the impacter and the specimen. Cessna *et al.*[35] attached an accelerometer to the sledge (dart) of a falling-weight instrument. Knowing the mass of the sledge, the force applied could be calculated from force = mass × acceleration, while the velocity–time and displacement–time characteristics were computed by integrating the acceleration–time profile. Cessna *et al.* also used an alternative form of instrumented plate testing, in which the plate was pulled onto an instrumented stationary rod. Gonzalez and Stowell[30] attached a dynamic load cell to the weight tip in a falling-weight test and hence measured the force–time characteristic.

Instrumentation of the standard impact tests to provide load–deformation data rather than the all-embracing impact energy would seem to be a natural progression. Load–deformation data are recorded in other mechanical tests; it would be most unusual to merely quote the energy-to-rupture of a specimen after a tensile test for example. However, instrumentation of impact testing equipment is not simply a matter of randomly attaching such strain gauges as happen to be available to the pendulum, tip or supports of an impact machine and recording the output. Many artefacts can occur in the outputs obtained, due to inertial loading of the system, inadequacies of the electronic equipment and other factors. Many of these problems are considered in ASTM Special Technical Publication 563, devoted to instrumented impact testing, and ASTM-STP 466, on impact testing of metals.

Instrumented Charpy tests, in which sensors are attached to the pendulum, specimen and anvil show that the form of the load–deformation curve is different in the three locations,[12,36] as shown in Fig. 9. The

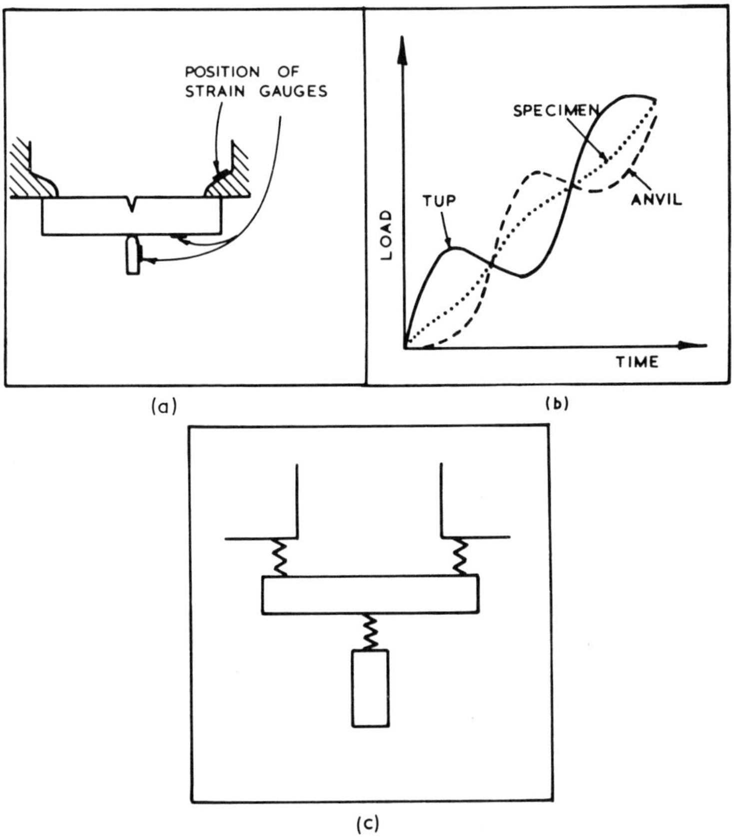

FIG. 9. Instrumented Charpy impact test. (a) Position of strain gauges. (b) Responses measured at the three locations named. (c) Spring-mass representation of the Charpy test. (Reprinted by permission of the American Society for Testing and Materials, 1916 Race Street, Philadelphia, PA 19103; Copyright.)

system considered was essentially linear-elastic. Clearly the load measured at any time, and especially at rupture, will depend on the location of the sensor. Venzi *et al.* analysed the dynamics of the system, by equating it to a series of masses and springs, as shown in Fig. 9(c), assigning mass and spring constants to pendulum, anvil and specimen. The observed load–time profiles for pendulum, anvil and specimen were satisfactorily reproduced by selection of appropriate values for equivalent masses and spring constants. It was found that the natural frequency of the system was given by

$$n = \frac{1}{2\pi}(S_e/M_e)^{1/2}$$

where S_e and M_e are the effective spring constant and equivalent mass for the system. Venzi *et al.* conclude that large errors may result from taking the maximum load in instrumented impact tests as the fracture load in calculations of dynamic fracture-toughness values. They also conclude that there is no simple relationship between the load on the tup and the anvil and the bending stress on the specimen during initial loading.

Similar analyses of the Charpy impact test, considering the dynamics of the system, have been made by Ireland,[12] Turner *et al.*[31] and Victory.[37] Results of Turner *et al.*[31] show that the ratio of pendulum stiffness and specimen stiffness is critical in determining the extent of system oscillations following impact. In certain circumstances it is possible for contact to be lost between the specimen and nose of the pendulum after initial impact, as each member describes its own oscillatory dynamic response.

These analyses serve to emphasise the essentially dynamic nature of any impact test. Application of analyses whose origins are in statics would seem quite inappropriate. It can also be argued that the analysis of standard impact tests in terms of structural dynamic response is more relevant to the dynamic response of structures in practice.

The use of instrumented impact tests is particularly appropriate when testing composites and ductile materials. In these cases the load–deformation curve and/or the fracture pattern can be complex. If the full loading history is available, it is possible to consider the different sections of the load–deformation curves. Lubert *et al.*[33] when testing ABS specimens, divided the curve into three sections: (a) an initial section up to the maximum force measured, which was considered to estimate the energy for failure initiation; (b) a section where the load dropped rapidly without significant oscillation, which was considered to coincide with yielding and crack propagation; and (c) a section where the load decreased only slowly with considerable oscillation, which was taken to cover the period of final plastic drawing and tearing. Hence three different energy terms were calculated which could be assigned to the three parts of the fracture process. Summation of the energy components determined from the instruments was compared with the total energy measured in the orthodox manner by the swing of the pendulum. A close agreement between the two values was found, although the dial energy was slightly greater than that determined by the instrumented test. Close agreement between the total energies determined by orthodox means and integration of load–deformation curves following instrumented tests has also been reported by Cessna *et al.*[35] and Gonzalez and Stowell.[30] Hence instrumentation of the impact test makes further information available without introducing significant errors into the total energy value, if that is all that is required. Using instrumented Charpy tests on un-notched glass-fibre-reinforced polyester and epoxy laminates, Yeung and Broutman[38] examined the effect of the glass–resin interface strength on the impact behaviour. Since the full load–deformation behaviour was available, portions of the total energy could be appropriated to the initiation and propagation stages of fracture. It was found that initiation energy increased with the interlaminar shear strength. The propagation energy for polyester laminates displayed a minimum at a critical value of interlaminar shear stress which was associated with a change in failure characteristic, from that of delamination to that of fibre failure. Hence the availablity of instrumented impact testing in this instance supplied further information and more insight into the nature of the impact failure.

6 IMPACT BEHAVIOUR OF COMPONENTS AND STRUCTURE

The aim of impact testing is to provide reliable data which may be used to assess the impact performance of polymeric components and structures. Interest has so far been concentrated on the impact test itself. However, the impact performance of the finished article cannot be ignored. Some of the drop weight tests for components seek to investigate this aspect of the problem, but usually only as a quality assurance test. A full analysis of the dynamic response of a structure following impact requires a considerable excursion into the subjects of dynamics and mechanics of materials. Such considerations might be considered to extend beyond the realms of mechanical testing of materials. Yet in established texts on the mechanics of materials, 'stress analysis' topics appear beside discussions of mater-

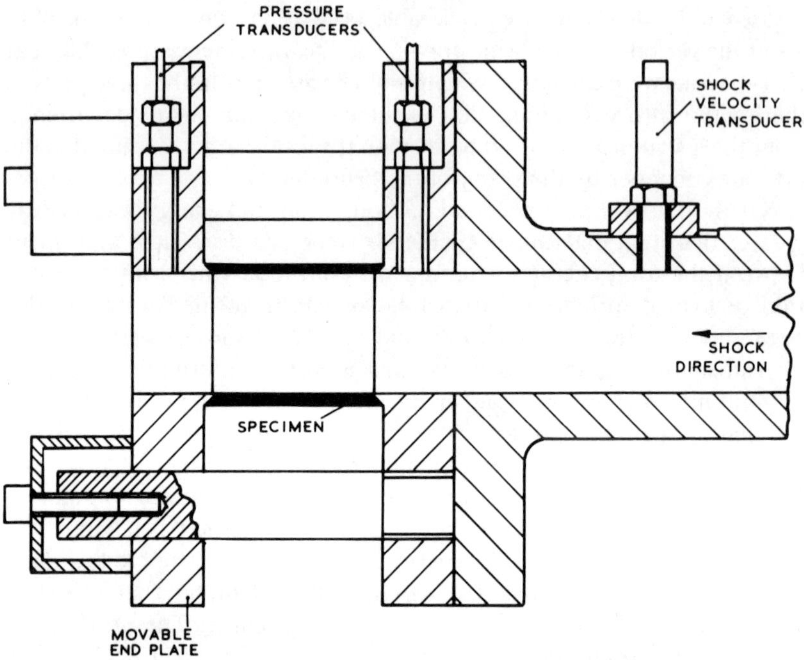

FIG. 10. Method of mounting a thin walled cylindrical specimen in the shock tube and the location of the pressure and shock velocity transducers.

ials testing techniques.[39] Reference was made in the previous Section to analyses of the Charpy test using the dynamics of systems.

In certain situations it is necessary to design structures specifically for known, or limited, impact conditions: e.g. landing of space vehicles on earth and water, blast effects on structures and impact-forming of metals. Analyses for certain prescribed situations may be found in the literature.[40–42] It is important to realise that to contemplate any analysis of an impact situation, the form of the impulse must first be accurately defined. However the situation is often one in which an 'improved' impact performance is required, of imprecise nature. It is also important to remember that the impact performance of a structure depends on both the manner in which the material is distributed (its shape) and the ability of the material to absorb energy. A given volume of steel can be arranged either as a solid cube or as a helically coiled spring: the impact performances of the two structures differ.

The author has sought to derive an impact test in which both the characteristic of the material and the structural response under impact may be considered and inter-related.[43,44] The test uses a conventional shock tube to deliver an internal pressure pulse to a thin-walled cylindrical specimen or a plate across the end of the tube. The arrangement for a cylindrical specimen is shown in Fig. 10. The specimen support system has been designed to minimise interaction between specimen and support system. The pressure pulse may be of either ramp or step form, allowing a wide range of loading rates, and hence strain rates, to be effected on the same equipment. Instrumentation allows the load–time characteristic and resultant strain–time response of the tube to be determined.

Following internal step loading, the specimen describes a damped barrel-mode oscillation, as shown in Fig. 11. This behaviour may be determined theoretically using an analysis of the step loading of a visco-elastic element (Kelvin element). The theoretical prediction is shown

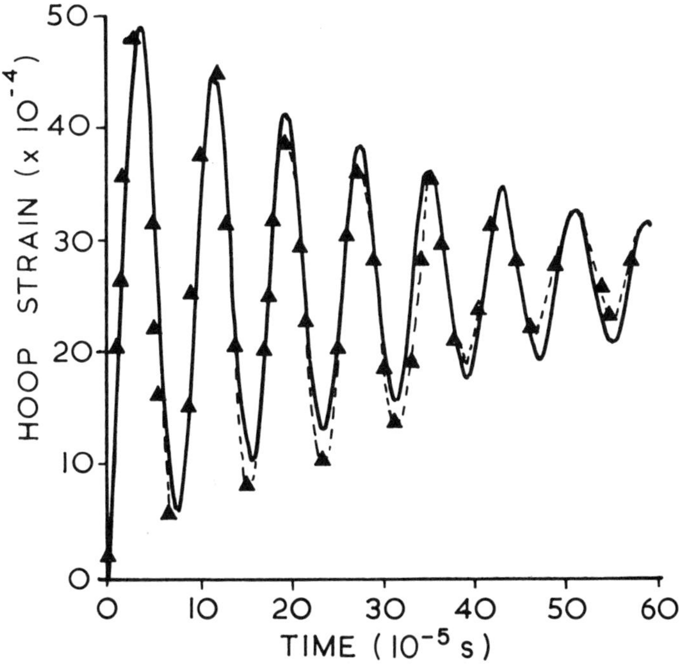

FIG. 11. Comparsion of the theoretical prediction of the hoop strain against time response (shown solid) of a cylindrical PVC tube following step internal pressurisation, with that obtained experimentally.

superimposed in Fig. 11. The cylinder vibrates at a frequency close to its natural frequency, which is given approximately by

$$n = \frac{1}{2\pi}(E/\rho r^2)^{1/2}$$

where E and ρ are respectively the Young's modulus and density of the tube material and r is the radius of the tube. The theory requires only one fitting constant, a relaxation time, to account for the damping in the system. Other factors are dictated by the geometry of the system or material parameters required to fit long-time (quasi-static) characteristics. Values for the relaxation time have been taken as 0·85 μs for a rigid PVC and 0·4 μs for a glass-reinforced epoxy resin system. It is interesting to observe that the frequency of oscillation of the particular system tested is high ($\approx$ 10–20 kHz), as the theory predicts, but that the material still exhibits visco-elastic properties.

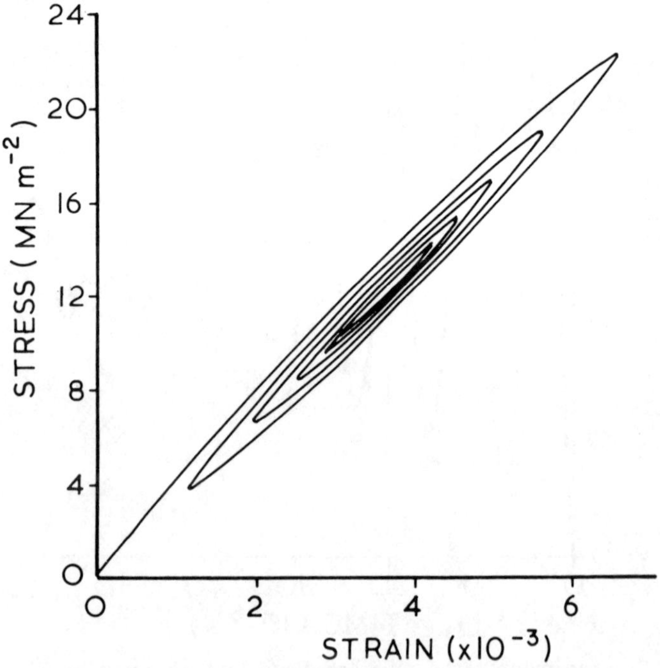

FIG. 12. Typical stress–strain profile sustained by the PVC tube following the step internal pressurisation.

Extension of the analysis permits the stress–strain profile for the material, due to the barrel-mode oscillation, to be determined. Such a curve is shown in Fig. 12, and is seen to take the form of a decaying loop. Peak stress and strain values sustained are found to be nearly twice those predicted from equivalent quasi-static loading. However, these values are not sustained for long. Thus any fracture process initiated under the high stresses or strains must propagate under an oscillating stress field.

The behaviour of structures following impact is complex, involving dynamic oscillations of the structure and stress waves within the solid. Further examination of structural responses may serve to indicate the nature of the impact test data required on materials to predict the impact response of structures generally.

7 MECHANISM OF FRACTURE DUE TO IMPACT

Since much effort has been expended on devising and interpreting impact tests, it is perhaps not surprising to find only fragmentary considerations of the molecular and micromechanics aspects of impact fracture. Considerations of the mechanisms of fracture are more common in discussions on the impact failure of glass-reinforced plastics and rubber-toughened materials, than those dealing with homogeneous thermoplastics.

Wolstenholme *et al.*[45] observed a linear relationship between peak load in an instrumented impact test and the flexural yield stress at corresponding temperatures for several different polymers. These included both homogeneous and rubber-modified plastics. A similar linear relationship has been observed by other workers.[19] Wolstenholme *et al.* have therefore suggested that the primary factors that determine the yield characteristics of thermoplastics are also responsible for their impact behaviour. Increases in the impact energy with temperature were attributed to the increase in energy required for plastic tearing after yielding.

The fracture surfaces of broken impact specimens take many different forms.[13] The surfaces may be those associated with completely brittle failure, or may exhibit shearing at the periphery and brittle fracture in the central region. They can exhibit varying degrees of whitening in the vicinity of the fracture. The whitening is often attributed to crazing, which is one form of ductile behaviour. An interesting observation by Newmann and Williams,[46] concerned with Charpy testing of ductile ABS specimens which exhibited considerable craze whitening in the fracture zone, was that the energy absorbed was directly proportional to the volume of crazed material. A similar observation has been made by the author, working with glass-reinforced epoxy resins. In a falling-weight test, the energy absorbed was found to be approximately proportional to the volume of the whitened zone under the impacter, although whitening in this instance was due to delamination of the fibre from the matrix. While impact-testing notched polycarbonate specimens, Mills[47] observed shear bands in the plastic zone ahead of the notch tip that resembled the slip-line field for yielding from a circular notch. An internal craze then nucleated at the tip of the plastic zone which grew and eventually led to crack formation. It was also found that the stress required for the craze to nucleate was approximately proportional to the yield stress of the material.

Hence there are several references to the close relationship between the impact strength of plastics and the yield characteristics of the material when ductile failure occurs. This close relationship is perhaps to be expected, provided that rate effects do not produce significantly different behaviours in the two tests involved in determining the yield stress and impact strength. A qualitative explanation of the ductile/brittle transition in notched-bar testing has been suggested by Le Grand.[48]

It was proposed that a flaw forms ahead of the notch tip due to the tri-axial tensile field at the centre of the notched specimens. If the material surrounding the flaw can yield more rapidly than the flaw can grow to a critical size, then a ductile fracture mode occurs and impact energies will be consequently relatively high. However, if the material surrounding the void does not yield sufficiently before the crack grows to the critical Griffith size, brittle fracture occurs. This model is similar to others proposed for crazing and fracture under quasi-static loading. It thus helps to unify the approach to the fracture of polymers over a wide range of strain rates.

Peaks in impact data have been reported and attempts have been made to correlate them with molecular motion transitions (loss peaks).[49–51] Such considerations are of interest, since they seek to take the discussion to the molecular level. Polymeric materials exhibit a series of loss peaks when subjected to dynamic testing over ranges of test frequency and/or temperature.[52] Such loss peaks are variously assigned to different types of main-chain or side-group molecular motion. Since both the loss peaks associated with molecular motions and transitions in impact strengths are manifestations of energy absorption changes, a possible correlation of the two quantities has been sought, with varying degrees of success. Plati and Williams[18] found good agreement between impact data peaks at

−60 °C for PMMA and at −20 °C for PVC and β-relaxation peaks for these materials. A similar impact energy peak observed with polycarbonate (PC) at −20 °C could not, however, be related to a loss peak. Sacher[53] has suggested that the high impact strength of PC may be attributed to the co-operative motions associated with many loss peaks rather than one individual peak, which permit the dissipation of impact energy over a large number of degrees of freedom. It has also been suggested[49] that greater impact strength is imparted when the loss peaks are associated with main-chain rather than side-group motions.

The often good agreement between loss peak and impact test data strongly supports such considerations as a basis for understanding the mechanism of energy absorption in impact. However, the common lack of agreement suggests that other factors are important. These other factors may be associated with the microstructure of the polymer. For example, the impact strength of rubber-toughened polystyrene has been found to depend strongly on the particle size of the rubbery block component.[54] Particle sizes of between 2 and 5 μm were found to be necessary for good impact strength. Also the nature of the surface treatment applied to calcium carbonate[55] and fibre glass[32] fillers for polymeric materials can markedly affect their impact strengths.

A considerable body of knowledge exists on the toughening of plastics by rubber modification. Rubber can be added to polymers by blending, mechanical mixing and co-polymerization. The main purpose of adding the rubber is to obviate the brittle characteristics of the basic polymer and thus improve the material's mechanical performance, including impact resistance. The subject of toughening of plastics is extensive and space prevents a proper discussion of it here. It is difficult to separate data on the impact behaviour of these materials from the much broader field of toughening over the full range of mechanical testing. Fuller considerations are given in texts listed in the Bibliography.

Proposed mechanisms to explain the action of rubber in polystyrene fall under four headings:

(a) energy absorption by the rubber particles;
(b) promotion of non-catastrophic crazes, cracks and deformation bands that absorb a large portion of the impact in essentially cold-flow processes;
(c) a tensile strength contribution from the rubber that increases the force required to rupture the specimen and/or prevent a craze from developing into a crack;

(d) interception by rubber particles of cracks that would otherwise grow to catastrophic size.

The four processes are not mutually exclusive. It is apparent that the role of rubber in rubber-extended polymers is complex. It is well established that rubber particles initiate controlled crazing in the glassy polystyrene phase, which leads to high impact strength.

Bucknall and Street[56] have studied the impact strength of a number of rubber-modified polymers as a function of temperature. They find that all notched Izod impact-strength/temperature curves can be divided into three regions, termed I, II and III. The transition from I to II in polybutadiene-toughened PS occurred at approximately $-90\,°C$, which coincides with a peak on the dynamic loss curve of the rubber phase. The transition from region II to region III was found to occur at a temperature which roughly coincided with the β-loss peak for polystyrene. However, Bucknall considered the II–III transition to be associated with the rapid relaxation of the rubber particles, which set up stress concentrations in the surrounding matrix, causing crazing. Thus both transitions were thought to be concerned with the rubber particles.

No clear picture of the mechanism of failure in impact emerges for polymers. Interpretations of fracture are based on a mixture of yield behaviour of the material, stress fields acting in the vicinity of the notch and energy losses associated with different modes of molecular mobility.

8 CONCLUSION

Impact testing of polymers is a complex subject involving many different facets. It is not possible to explore all these aspects in detail in a short review. An attempt has therefore been made to draw the different aspects together in order that the coherence of the subject, or lack of it, can be assessed. Impact tests are inherently difficult to interpret. A simultaneous advantage and disadvantage of all impact tests is their rapidity. They are advantageous in terms of the time taken to perform a test, but the event of interest is so short-lived that much information can only be deduced retrospectively. It is optimistic to expect that one single impact test exists or will emerge that will provide information on all aspects of the problem, including information for the design engineer on one hand and fundamental knowledge of the nature of the mechanics of impact failure in polymers. The existence of many impact tests can provide a variety of information which, if collectively considered, should satisfy the prime aim of assessing and predicting the impact performance of polymers.

REFERENCES

1. R. F. Westover. *Plastics Technology*, **4** (1958) pp. 223 and 348.
2. J. Newton. *Appl. Polymer Symp.*, **17** (1971) p. 95.
3. BS 4618. *Recommendations for the Presentation of Plastic Design Data*, British Standards Institution, 1972.
4. S. Turner. *Mechanical Testing of Plastics*, Iliffe, London, 1973.
5. ASTM–D 2463–74. *Drop-Impact Resistance of Polyethylene Blow-Moulded Containers*, American Society for Testing and Materials.
6. BS 4839. *Blow-Moulded Polyolefin Containers*, British Standards Institution, 1974.
7. ASTM–D 997–50. *Drop Test for Cylindrical Shipping Containers*, American Society for Testing and Materials.
8. ASTM-D 2444–70. *Test for Impact Resistance of Thermoplastic Pipe and Fittings by Means of a Tup (Falling Weight)*, American Society for Testing and Materials.
9. ASTM–D 950–72. *Test for the Impact Strength of an Adhesive Bond*, American Society for Testing and Materials.
10. BS 2782. *Methods of Testing Plastics. Part 3: Mechanical Properties*, British Standards Institution, 1970.
11. ASTM–D 256–73. *Impact Resistance of Plastics and Electrical Insulating Material*, American Society for Testing and Materials.
12. D. R. Ireland. *Procedures and Problems Associated with Reliable Control of the Instrumented Impact Test: Instrumented Impact Testing.* ASTM–STP 563, American Society for Testing and Materials, 1974, pp. 3–29.
13. P. I. Vincent. *Impact Tests and Service Performance of Thermoplastics*, Plastics Institute, London, 1971.
14. ASTM–D 1709–75. *Test for Impact Resistance of Polyethylene Film by the Free-Falling Dart Technique.*
15. W. J. Moritz. Society of Plastics Engineers 33rd. Annual National Technical Conference 1975, p. 540.
16. ASTM–D 1822–68. *Tensile Impact Energy to Break Plastics and Electrical Insulating Materials*, American Society for Testing and Materials.
17. W. Spath and M. Rosner. *Impact Testing of Materials*, Thames and Hudson, London, 1961, pp. 133 and 163.
18. E. Plati and J. G. Williams. *Polymer*, **16** (1975) p. 915.
19. M. W. Birch and J. G. Williams. *Int. J. Frac.*, **14** (1978) p. 69.
20. A. S. Kobayashi. *Eng. Frac. Mech.*, **6** (1974) p. 81.
21. A. K. Green and P. L. Pratt. *Eng. Frac. Mech.*, **6** (1974) p. 71.
22. J. T. Ryan. Society of Plastics Engineers 34th Annual National Technical Conference, **22** (1976) p. 205.

23. J. T. Ryan. *Polymer Eng. Sci.*, **18** (1978) p. 264.
24. L. J. Broutman and S. M. Krishnakumar. *Polymer Eng. Sci.*, **16** (1976) p. 74.
25. L. J. Broutman and S. M. Krishnakumar. *Polymer Eng. Sci.*, **14** (1974) p. 23.
26. L. J. Broutman and S. M. Krishnakumar. *Polymer Eng. Sci.*, **14** (1974) p. 249.
27. M. P. Van der Wielen. *Polymer Eng. Sci.*, **15** (1975) p. 102.
28. L. J. Broutman and S. M. Krishnakumar. Society of Plastics Engineers 33rd. Annual Technical Conference, **21** (1975) p. 413.
29. K. Nikpur and J. G. Williams, in *International Conference on Toughening of Plastics*, Plastics and Rubber Inst., London, 1978, paper 21.
30. H. Gonzalez, Jr., and W. J. Stowell. *J. Appl. Polymer Sci.*, **20** (1976) p. 1389.
31. C. E. Turner, L. E. Culver, J. C. Radon and P. Kennish, *Conf. on Practical Applications of Fracture Mechanics to Pressure Vessel Technology*, I. Mech. Eng., London, 1971, p. 38.
32. R. H. Toland. *Impact Testing of Carbon Epoxy Composite Materials: Instrumented Impact Testing*, ASTM–STP 563, American Society for Testing and Materials, 1973, pp. 133–145.
33. W. Lubert, M. Rink and A. Pavan. *J. Appl. Polymer Sci.*, **20** (1976) p. 1107.
34. P. W. McMillan and J. R. Tesh. *J. Mat. Sci.*, **10** (1975) p. 621.
35. L. C. Cessna, J. P. Lehane, R. H. Ralston and T. Prindle. *Polymer Eng. Sci.*, **16** (1976) p. 419.
36. S. Venzi, A. H. Priest and M. J. May. 'Influence of Inertial Load in Instrumental Impact Tests', in *Impact Testing of Metals*, ASTM–STP 466, American Society for Testing and Materials, 1970, pp. 165–180.
37. E. L. Victory. *Appl. Poly. Symp.*, **1** (1965) p. 163.
38. P. Yeung and L. J. Broutman. *Polymer Eng. Sci.*, **18** (1978) p. 62.
39. S. Timoshenko. *Strength of Materials, Vol. 2.*, D. Van Nostrand, New York, 1956.
40. M. Kornhauser. *Structural Effects of Impact*, Cleaver Hume 1964.
41. W. Johnson. *Impact Strength of Materials*, Arnold, London, 1972.
42. B. Rawlings. *Mechanical Properties of High Rates of Strain: Conference Series No. 21*, Institute of Physics, Oxford, 1974.
43. P. E. Reed, P. J. Nurse and E. H. Andrews. *J. Mat. Sci.*, **9** (1974) p. 1977.
44. P. E. Reed and R. Smith. *Shock Loading of PVC and Glass-Reinforced Plastics.* Final Report for Ministry of Defence under Contract AT/2042/052, London, 1978.
45. W. E. Wolstenholme, S. E. Pregun and C. F. Stark. *J. Appl. Polymer Sci.*, **8** (1964) p. 119.
46. L. V. Newmann and J. G. Williams. *International Conference on Toughening of Plastics*, Plastics and Rubber Institute, London, 1978, paper 18.
47. N. J. Mills. *J. Mat. Sci.*, **11** (1976) p. 363.
48. D. G. LeGrand. *J. Appl. Polymer Sci.*, **13** (1969) p. 2129.
49. J. Heijbour. *J. Polymer Sci. C* **16** (1968) p. 3755.
50. Y. Wada and T. Kasohara. *J. Appl. Polymer Sci.*, **11** (1967) p. 1661.
51. S. G. Turley. *Appl. Polymer Symp.*, **7** (1968) p. 237.
52. N. G. McCrum, B. E. Read and G. Williams. *Anelastic and Dielectric Effects in Polymeric Solids*, Wiley, New York, 1967.

53. E. Sacher. *J. Macromol. Sci.—Phys.*, **B11** (1975) p. 403.
54. R. F. Boyer and H. Keskkula, in *Encyclopaedia of Polymer Science and Technology Vol. 13*, J. Wiley, 1970, p. 392.
55. J. A. Radosta. Society of Plastics Engineers 34th Annual Technical Conference, **22** (1976) p. 465.
56. C. B. Bucknall and D. G. Street. *J. Appl. Polymer Sci.*, **12** (1968) p. 1311.
57. E. Plati and J. G. Williams, *Polymer Eng. Sci.*, **15** (1975) p. 470.

BIBLIOGRAPHY

1. R. F. Boyer (ed.) *Styrene Polymers: Encyclopaedia of Polymer Science and Technology, Vol. 13*, J. Wiley, New York, 1970, pp. 128–447.
2. *International Conference on Toughening of Plastics*, Plastics and Rubber Institute, London, 1978.
3. C. B. Bucknall. *Toughened Plastics*, Applied Science Publishers, London, 1977.

Chapter 5

WEAR AND TEAR OF RUBBER

A. N. GENT and C. T. R. PULFORD
The University of Akron, Ohio, USA

SUMMARY

Elastomers differ from other polymers in having their glass transition temperatures far below room temperature (typically $-50°C$ *to* $-110°C$*). As a result they not only possess a liquid-like structure but also exhibit the molecular mobility of the liquid state. They are solids only in the sense that chemical cross-links between molecules prevent flow. One result of this is that they display large recoverable deformations and non-linear stress-strain behaviour. This has necessitated the development of special fracture mechanics theories to describe their rupture properties.*

The chapter reviews these theories and the many factors that contribute to tear resistance in rubber. It shows how a wide range of failure phenomena, from fatigue fracture to abrasive wear, can be described and unified using the appropriate experimental and analytical techniques.

1 THE RUBBERY STATE

1.1 Molecular structure

A rubbery material is one that is capable of large elastic deformations. This unique property is associated with a particular molecular structure: long, flexible, chain-like molecules that are chemically or physically attached to each other at infrequent intervals and that possess thermal energy exceeding their intermolecular attractive energy. Thus, the molecular chains form a loose, three-dimensional network in which the component parts change shape and slide past each other with ease under

Brownian motion except at the points of attachment. When such attachments are formed by chemically bonding one molecular chain to another the rubber is said to be cross-linked or 'vulcanised'. It then exhibits longer-lasting elasticity than 'raw' or unvulcanised rubber, in which molecular entanglements may act as cross-links temporarily, but eventually slip and allow the material to flow like a liquid.

Elastomers are usually mixed with finely-divided powders, up to about 30% by volume. These stiffen the product considerably and often 'reinforce' the rubber, i.e. increase its toughness, as well.

1.2 Elasticity and Visco-elasticity

Rubberlike elasticity arises because, on deformation, the molecular chains seek to return to their original, more probable, conformations from those less-probable conformations imposed on them by the deformation. Elastic restoring forces are thus entropic in nature, like the pressure exerted by an ideal gas.[1] However, molecular motions take place at a rate that depends on the thermal energy available to overcome various resistive forces which hinder relative motion of chain segments. These resistive forces are due to the finite volume of molecular subunits and internal and external barriers to changing molecular conformations. They have the effect of an internal viscosity and retard the attainment of equilibrium conformations in a similar way.

When the rate of deformation requires that the molecular segments move faster than their normal retarded mobility allows, the resistance to deformation increases and the rubber is stiffer. Conversely, when the deformation rate is slow relative to the rate of molecular motion, the viscous resistance to deformation is small. Indeed, it tends to zero at zero rate of deformation. This rate-sensitive elasticity of rubber is termed *visco-elasticity*.

There are thus two distinct contributions to the elasticity of rubber: equilibrium effects, such as the number and length of molecular chains comprising the network, and non-equilibrium viscous effects. The latter are strongly dependent upon temperature because thermal expansion greatly facilitates the motion of molecular segments. Indeed, below a characteristic temperature, denoted T_g, Brownian motion occurs so slowly, at a rate of less than one movement of a molecular subunit in 10 s, that to all intents and purposes the material is rigid and immobile—a polymeric glass. This glass transition temperature T_g is characteristic of the particular elastomer, ranging from $-120\,°C$ for certain silicone elastomers up to temperatures above ambient for materials, like polystyrene, that only become rubbery when heated to temperatures well above normal.

Because nearly all hydrocarbon polymers have similar intermolecular attractive forces and similar coefficients of thermal expansion, the mobility of molecular subunits increases with temperature in a similar way so that the rate ϕ_T of Brownian motion at a temperature T is given by a 'universal' relation, due to Williams, Landel and Ferry:[2,3]

$$\log_{10}(\phi_T/\phi_{T_g}) = 17{\cdot}6(T - T_g)/(52 + T - T_g) \qquad (1)$$

where ϕ_{T_g} denotes the rate of subunit motion at T_g, about 0·1 jumps/sec.

The magnitude of the viscous contribution to stresses set up in rubber depends upon the rate of deformation $\dot{e}$ relative to the natural mobility ϕ_T of molecular segments at the particular temperature employed, i.e. upon $\dot{e}a_T$, where

$$a_T = \phi_{T_g}/\phi_T \qquad (2)$$

The scaling factor a_T, given by eq. (1), enables measurements at a rate $\dot{e}$ and at any temperature T to be converted into effective rates $\dot{e}a_T$ at a reference temperature, in this case the glass temperature T_g. This rate–temperature equivalence principle has proved to be a powerful aid to the study of viscous processes in simple glass-forming polymers. In particular, it has diagnostic value: if physical measurements are found to depend upon temperature and rate in accordance with this distinctive law, then it is certain that the underlying process is kinetically controlled and depends upon the rate of motion of polymeric subunits. Also, measurements made over a limited range of deformation rates, or times, but over a wide range of temperatures (as is experimentally convenient) can be transposed by means of the shift factors a_T to construct a master relationship expressing the dependence upon deformation rate at T_g, or at any other reference temperature.

2 RESISTANCE TO TEARING

2.1 Fracture Mechanics and Test Methods

The toughness of rubber is conveniently described by its resistance to tearing. In principle, this might be characterised by the magnitude of the stresses at which rupture occurs at the tip of a tear, but in practice the measurement of such localised stresses is extraordinarily difficult. An alternative measure of strength, suggested by Griffith[4] for brittle solids and applied later by Irwin,[5] Orowan[6] and Rivlin and Thomas[7] to imperfectly elastic, non-linear materials, is the amount of work required to advance a tear by unit area. This work can include energy dissipated in a variety of irreversible processes, provided that such energy is dissipated in a region close to the tear tip (so that it is not affected by the way in which forces are applied to the specimen) and the amount of energy expended in this way is proportional to the distance the tear advances. Thus, the critical energy release rate G_c is defined as

$$G_c = -2\left(\frac{\partial \mathscr{E}}{\partial A}\right)_{\text{fracture}} \tag{3}$$

where $\mathscr{E}$ is the total strain energy of the system and A is the area of torn surface.

The quantity G_c is formally equivalent to the fracture energy $2\mathscr{T}$ of the material, and can be calculated for elastic test-pieces of certain simple geometries. Four of these have found wide experimental use:

(*a*) *Split-strip test-piece.* This test-piece is shown in Fig. 1. When the extension in the leg regions I is negligibly small, i.e. when the legs are much wider than they are thick or if they are reinforced with an inextensible fabric layer, then no energy is stored in the legs. Work expended in bending the legs is also zero, since for every new element that becomes bent the same amount is relaxed as the tear advances. Work done by the applied forces F, each acting through the distance Δl, is $2F\Delta l$, where Δl is the distance torn, and this work is applied directly to tearing.[7] Thus $2F\Delta l = G_c\Delta A = G_c t\Delta l$, and hence:

$$G_c = 2F/t \tag{4}$$

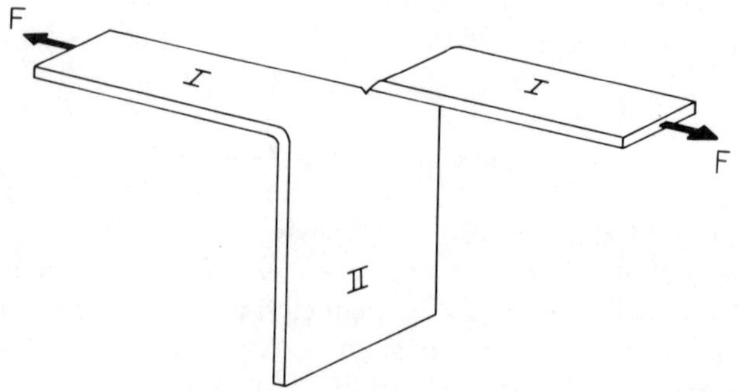

FIG. 1. Split-strip tear test-piece.

where t is the width of the tear path. This is not generally the same as the sheet thickness because the tear tends to run at an angle of about 45° to the thickness direction so that it lies perpendicular to the direction of principal tensile stress (Fig. 2). Thus, whilst failure appears to result from shear stresses, the crack actually follows a path at right angles to the maximum local tensile stress.[8]

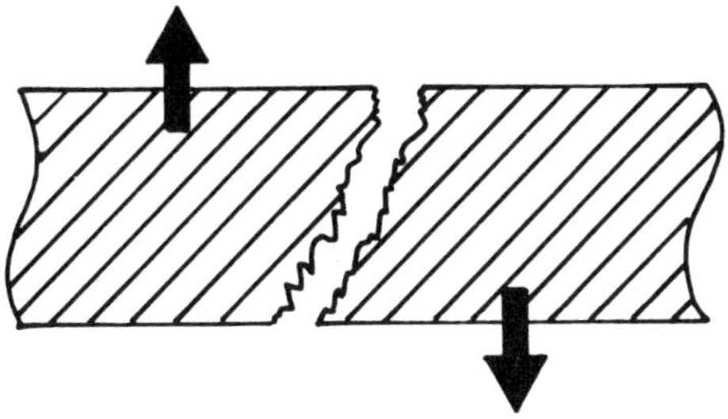

FIG. 2. Cross-section of torn strip.

(*b*) *Cleavage test-piece.* This test-piece is shown in Fig. 3. G_c is related to the tear force F as for the split-strip test-piece:

$$G_c = 2F/t$$

where t is now the width of the narrow web connecting the upper and lower portions of the test-piece.

(*c*) *Tensile test-piece.* A sheet of thickness t having a small cut in one edge, Fig. 4(a), or in the centre, Fig. 4(b), is subjected to a tensile deformation until the cut advances by tearing. At this point, according to Griffith,[4] the decrease in strain energy $\mathscr{E}$ upon growth of the cut becomes exactly sufficient to provide the energy requirements for fracture, i.e.

$$\begin{aligned}(\partial\mathscr{E}/\partial c)_l &= G_c(\partial A/\partial c)\\ &= G_c t\end{aligned} \tag{5}$$

where c is the crack length (or half-length, Fig. 4(b)) and A is the torn area. The suffix 'l' denotes that the grips do not move during propagation of the crack. The strain energy $\mathscr{E}$ is assumed to decrease with c as follows, reflecting the volume of material released from strain around the crack:

$$\mathscr{E} = \mathscr{E}_0 - kc^2tW \tag{6}$$

where $\mathscr{E}_0$ denotes the strain energy for a test-piece containing no cut and W represents the corresponding energy density (energy per unit volume);

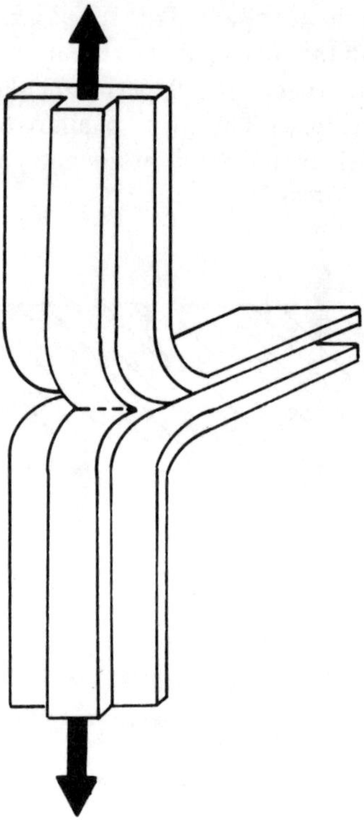

FIG. 3. Cleavage test-piece.

k is a numerical quantity, given by π for infinitesimal strains and by $\pi/(1+e)^{1/2}$, approximately, for materials subjected to a finite tensile strain e.[9]

On substituting from eqn. (6) into eqn. (5), we obtain

$$G_c = 2kcW_b \tag{7}$$

where W_b denotes the mean strain–energy density at the moment of rupture. For linearly elastic materials W_b is given by $\sigma_b^2/2E$, where σ_b denotes the tensile stress at break and E is Young's modulus. Thus, in this simple case the breaking stress for a tensile specimen containing a small cut may be calculated in terms of the fracture energy G_c:

$$\sigma_b = (G_c E/kc)^{1/2} \tag{8}$$

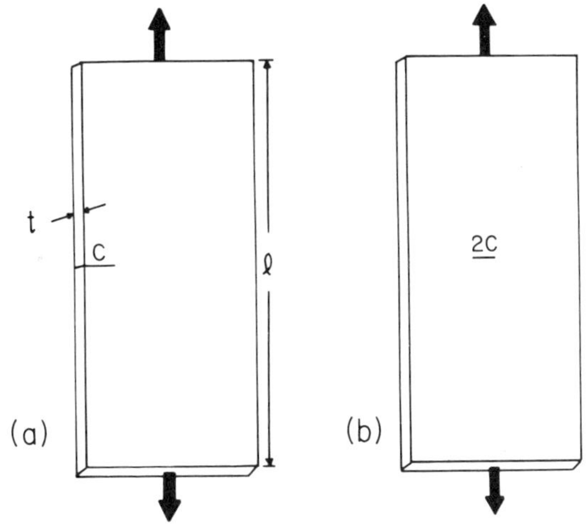

FIG. 4. Tensile test-piece with (a) edge cut and (b) centre cut.

More generally, the energy density at break may be calculated using eqn. (7) if the fracture energy is known, or *vice versa.*

As a means of determining G_c this method has severe disadvantages. The crack grows by tearing at an uncontrollable rate, generally in a catastrophic way, so that the effect of the rate of propagation upon tear strength cannot readily be examined. The start of tearing also depends strongly upon the sharpness of the initial cut, as might be expected. Thus, the value obtained for G_c depends upon the degree of bluntness of the crack. This has been expressed by Thomas[10] in the form

$$G_c = dW_b^* \tag{9}$$

where d denotes the effective diameter of the tear tip and W_b^* represents the local strain-energy density at the point of rupture. Under steady tearing conditions, as in test methods (a) and (b), the tear tip diameter takes up a characteristic size, generally of the order of 0·1–1 mm, and correspondingly G_c takes a value characteristic of the material and conditions of tearing.[11–14] In tensile rupture, on the other hand, the onset of tearing is the critical event and the diameter of the tear at this point is largely governed by the sharpness of the initial cut, i.e. by chance variations in initial conditions.

(*d*) *Pure shear test-piece.* This test-piece, shown in Fig. 5, differs from the previous one in that the initial cut is long in comparison to the initial

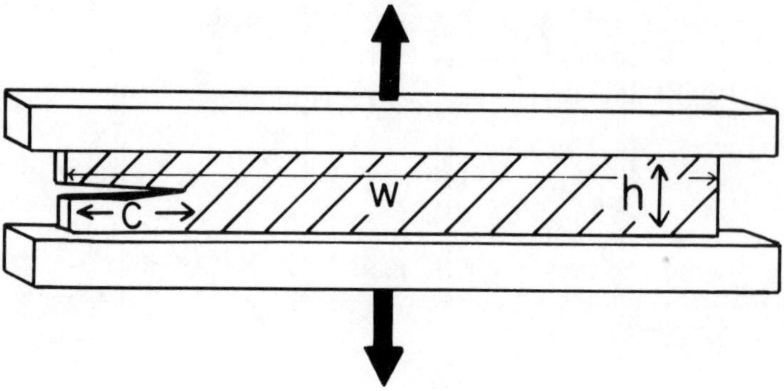

FIG. 5. Pure shear test-piece.

length h_0 of the test-piece (although it must still be short in comparison with the overall width w). The strain energy is therefore given by

$$\mathscr{E} = \mathscr{E}_0 - cth_0 W$$

in place of eqn. (6). On substituting in eqn. (5) G_c is then obtained in the form:[7]

$$G_c = h_0 W_b \tag{10}$$

The reader should note that, in contrast to the previous result, eqn. (7), this relation does not contain the crack length c and thus this tear process is not, in principle, auto-catalytic. Instead, once the conditions for crack propagation are set up, tearing should take place in a steady manner. However, the rate of tearing is not directly controllable and therefore this method is not really as convenient for experimental studies as methods (a) and (b).

The important feature of these test methods is that they all yield the same value for G_c for a particular material under particular test conditions, even though the applied stresses and deformations are quite different.[7] Thus, the basic premise that tearing is governed by a characteristic energy requirement is firmly established for relatively brittle materials, i.e. materials which do not undergo extensive plastic flow. Rubber is brittle in this sense because, although it deforms to a high degree, it recovers almost completely on release.

2.2 Elastic Rubber

(*a*) *Theory*. In the limit, when viscous effects are vanishingly small, many of the physical properties of rubber can be treated theoretically.[1] In particular, the work necessary to propagate a sharp tear may be calculated from the molecular structure without recourse to any adjustable parameters.[15] This remarkable theoretical achievement is outlined here.

Consider a molecular network in which all the strands have the same contour length and end-to-end distance L in the undeformed state, and are composed of n freely jointed subunits. When a force is applied to a given strand, *via* its crosslinkages to other strands, it adjusts its conformation until finally it is fully extended. At this point it resembles a taut string, and in order to break any of the main-chain valence bonds it is necessary to strain every bond to the breaking point (Fig. 6). Thus, if u is

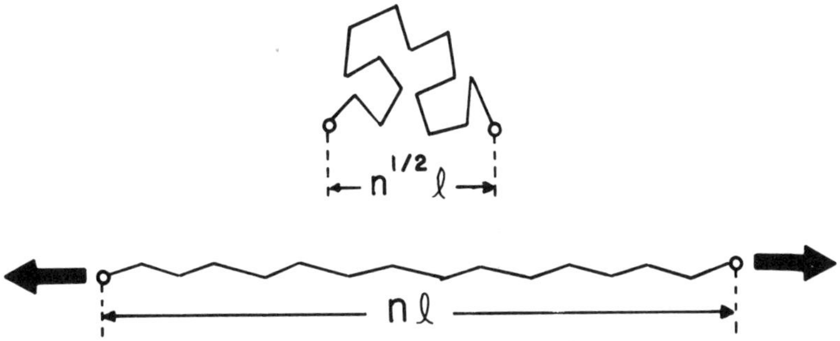

FIG. 6. Network strand composed of a large number n of freely jointed subunits. Length $L = n^{1/2}l$ in unstressed state, where l is the length of a subunit.

the energy required to break (dissociate) a C–C bond and each subunit contains m such bonds, then the energy required to break a single network strand is nmu. If there are N such network strands in unit volume, the number which cross from one side to the other of a randomly chosen plane of unit area in the unstrained state is $(\frac{1}{2})NL$. Thus, in order to break all of the network strands which cross the plane, it is necessary to supply an amount of energy

$$G_{0c} = (\tfrac{1}{2})NLnmu \tag{11}$$

This quantity represents the fracture energy under limiting conditions, when no contribution arises from viscous effects, and thus G_{0c} denotes a threshold strength or minimum tearing energy for an elastomeric material. Equation (11) may be simplified by substituting

$$L = n^{1/2}l$$

for freely jointed chains, where l denotes the length of a subunit, and

$$N = \rho A/nmM_s$$

where M_s is the molecular weight associated with each C–C bond, A is Avogadro's number and ρ is the density of the rubber, yielding

$$G_{0c} = (\tfrac{1}{2})n^{1/2}\, l\rho Au/M_s \tag{12}$$

Numerical substitution leads to a theoretical value for G_{0c} of about 20 J m^{-2} for a hydrocarbon elastomer having a representative strand length of about 100 subunits.[15]

It should be noted that G_{0c} is predicted to increase with strand length n, in proportion to $n^{1/2}$. If the strand length, or more correctly the end-to-end distance L, is taken to denote the minimum possible diameter of the tip of a propagating tear, and the local energy density at break is characterised by $(\frac{1}{2})Nnmu$, then an alternative route to eqn. (12) is obtained by substituting into eqn. (9). This derivation has the merit of relating the maximum sharpness of a tear to the dimensions of the molecular network.

(*b*) *Experiment.* A threshold strength for rubber was first pointed out by Lake and Lindley in studies of crack growth under intermittent stressing.[16] If the energy G available for tearing did not exceed a critical value of about 50 J m^{-2}, then no crack growth occurred. Mueller and Knauss measured a threshold tear energy directly for a swollen polyurethane elastomer, torn slowly at high temperatures so that viscous effects were small, and obtained a value of about 40 J m^{-2} in this way.[17] Ahagon and Gent[14] measured the tear energies of two polybutadiene elastomers, both swollen and unswollen, under threshold conditions and obtained values for G_{0c} ranging from 40 to 80 J m^{-2} as the degree of cross-linking decreased, i.e. as the length of network strands increased. This variation of G_{0c} with cross-linking was in good agreement with the theoretically predicted dependence and the absolute magnitudes obtained for G_{0c} in this and the earlier studies were reasonably close to, but about twice as large as, the theoretical estimates (Fig. 7). Thus, the theory has received rather substantial confirmation.

2.3 Visco-elastic Rubber

Under normal temperatures and rates of tearing the energy required to fracture rubber is many times larger than threshold values, reflecting additional energy expended in viscous processes.[18] Indeed, the tear strength is found to increase with increasing rate of tear propagation R

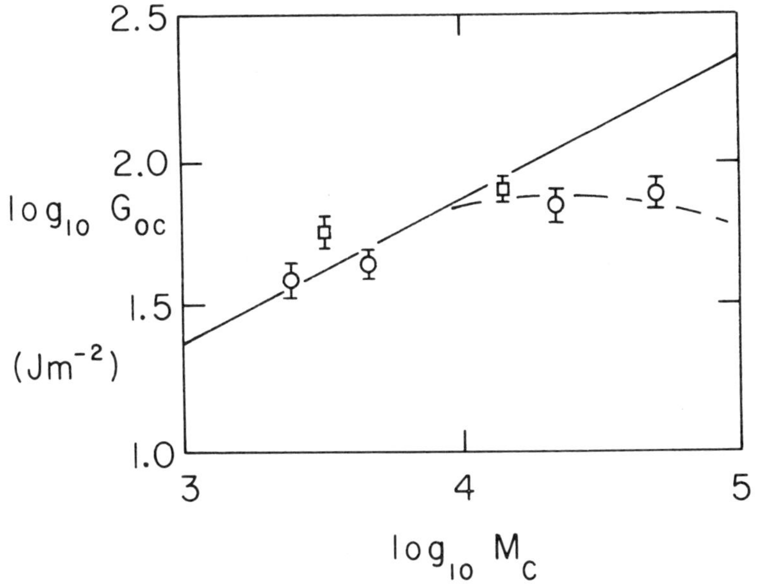

FIG. 7. Threshold tear energy G_{0c} plotted against the molecular weight M_c of network strands. Full line: slope of 1/2 in accordance with eqn. (12). Broken curve: effect of partially unconnected molecules when M_c is comparable to the molecular weight of the original molecules before interlinking. (After reference 14.)

and with decreasing temperature T in the same way that any measure of viscous energy dissipation would do. Moreover, when rates of tearing at different temperatures are converted into equivalent rates at a suitable reference temperature by means of the WLF relationship, eqn. (1), then the corresponding tear energies are found to superpose (Fig. 8), indicating that the tear energy is governed by subunit mobility.[19] In fact, under conditions of equal subunit mobility, i.e. at equal temperature intervals $T - T_g$ above their respective glass temperatures, a wide variety of amorphous elastomers have the same tear energy. Thus, the tear strength of rubber under viscous conditions is apparently independent of the particular molecular structure, except insofar as it determines the degree of internal energy dissipation upon deformation.[19]

Although this general correlation is now well-recognised, it is not an easy task to relate the tear strength of rubber to dissipative parameters in a quantitative way. Andrews[60] has recently given a general theoretical treatment for tear energy in terms of local strain-energy distribution func-

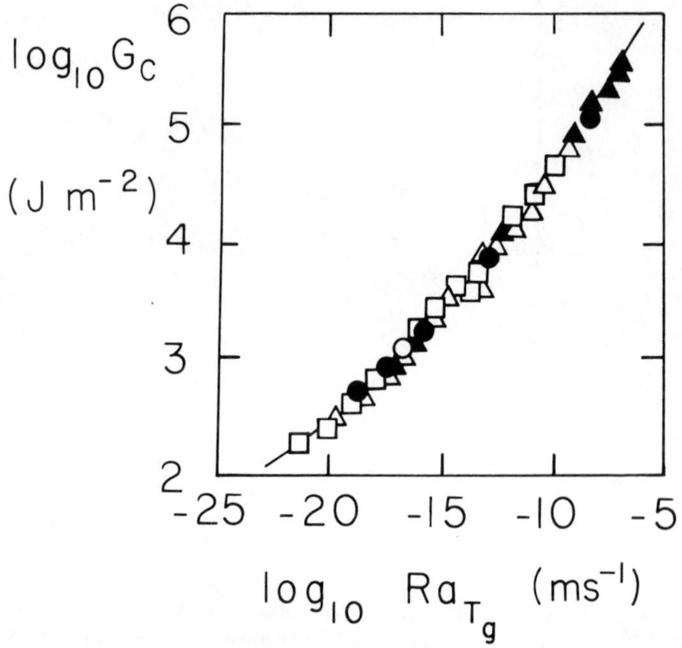

FIG. 8. Tear energy G_c plotted against Ra_{T_g}, the rate of tearing adjusted to T_g by means of eqn. (2). Results are shown for six butadiene–styrene (SBR) and butadiene–acrylonitrile (NBR) elastomers, with T_g ranging from −30 °C to −80 °C. (After reference 19.)

tions, taking into account dissipation, and has shown that experimental results for several rubbery materials are consistent with the theoretical predictions.[61] A quantitative treatment for simple visco-elastic substances thus seems to be close at hand although it has not yet been achieved (to the authors' knowledge).

2.4 Strain-induced Crystallisation

Certain elastomers, notably *cis*-1,4-polyisoprene, crystallise on being stretched by several hundred per cent. The degree of crystallisation attained is small, up to about 30%, but it imparts a high stiffness to the material and a pronounced lack of reversibility due to delays in crystallising and in melting on release. In the highly stressed zone at a crack tip, crystallisation takes place even when the overall strain is relatively small and the energy dissipation accompanying it leads to greatly enhanced tear strength. In effect, the scale of tearing is raised from molecular dimensions to the dimensions of a crystalline domain at the tear tip, perhaps 0·1 mm in length.

Unless the rubber is almost completely released from stress, the crystalline region at the tear tip does not melt on partial retraction. Instead, it remains intact and prevents further crack growth when the stresses are again increased.[20] As a result, strain-crystallising elastomers show an astonishing resistance to crack growth under intermittent stressing conditions when the stress is not completely removed between cycles of deformation.[21]

2.5 Reinforcement by Fillers

Particulate fillers, notably carbon black, increase the tear strength of amorphous elastomers to a remarkable degree. This is only true, however, for a relatively narrow range of tearing rates and temperatures (Fig. 9).[22] The region of high strength corresponds closely to the region in which tearing is discontinuous in a characteristic way, the crack gradually turn-

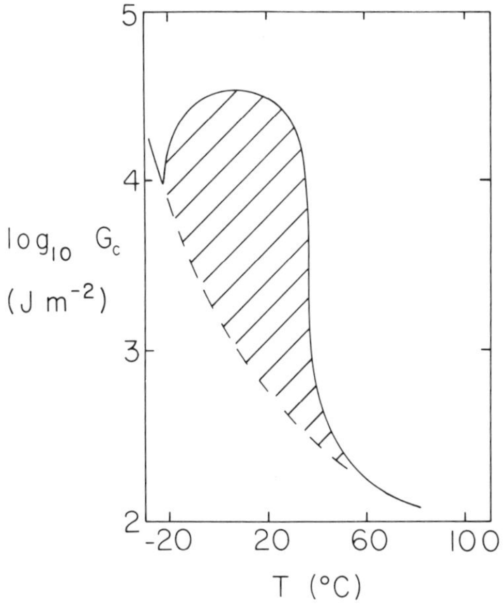

FIG. 9. Tear energy G_c for a butadiene–styrene (SBR) elastomer reinforced with fine thermal (FT) carbon black. The broken curve denotes the relation obtained for a corresponding unreinforced material. Tear rate: 10 μm s^{-1}. (After reference 22.)

ing aside from its initial direction to run parallel to the principal tensile stress and eventually stop. Then a new crack forms and starts off in the original direction before again deviating to run at right angles to it. This intermittent 'knotty' tearing behaviour is the cause of the enhanced strength; when it is prevented by physically forcing the tear to follow a straight line by means of closely spaced metal guides, the tear strength falls to values only two or three times higher than for the unfilled material.[23] Therefore, reinforcement by fillers is due to major deviation of the tear direction augmented by a comparatively minor increase in viscous dissipation.

The reason why the tear tends to follow such a characteristic path is still obscure, however. It appears to reflect a pronounced anisotropy of strength in the highly stretched material at the tear tip[24] so that although the principal tensile stress is applied in one direction and a relatively minor stress is generated at right angles to it, the oriented material is so much more easily split open by the minor stress that the tear veers in this direction.

2.6 Thermoplastic Elastomers

These materials derive their physical characteristics from the fundamental immiscibility of different polymers. They consist of tri-block molecules having the general structure *ABA*, where *A* denotes a glassy polymeric strand, e.g. of polystyrene, and *B* denotes a flexible polymeric strand, e.g. of polybutadiene. The end sequences *A* are immiscible in polymer *B* and hence they tend to aggregate to make up domains of a glassy polymer isolated within an elastomeric matrix. Moreover, because the sequences *A* from one molecule become part of different glassy domains, a network of elastomeric strands is formed, linked together by hard domains (Fig. 10). Materials of this kind behave in a characteristically rubberlike way, at least at temperatures below the glass temperature of polymer *A* and above that of polymer *B*.[25]

The tear strength of a representative thermoplastic elastomer, Kraton 1101 (Shell Chemical Company), is quite comparable to that of a well-reinforced amorphous or strain-crystallising elastomer, about 20 kJ m^{-2} for a tear rate of 1 mm s^{-1} at room temperature.[24] This remarkably high value is attributed to plastic flow of the hard domains under high local stresses, approaching the breaking point.[26] Indeed, such a deformation process seems essential in order for these materials to have the capacity to dissipate strain energy, as any tough material must be able to do.

FIG. 10. Schematic representation of a tri-block elastomer. Spherical domains (A) of the hard end-blocks act as network junction points, linked together by the elastomeric centre blocks (B).

2.7 Tensile Rupture

Breaking of a specimen under a tensile stress σ_b may be regarded as the consequence of catastrophic tearing from some chance nick or flaw, present initially. Indeed, the effective size of accidental flaws may be calculated from eqns. (7) or (8) using the measured tensile strength σ_b or breaking energy density W_b for specimens free from deliberately introduced cuts or flaws. Values obtained for c in this way are about $40 \pm 20\,\mu m$ for typical rubbery materials,[27] a value which is corroborated by extrapolating in a similar way the fatigue life[18] and the critical condition for attack of ozone[28,29] for cut specimens to the corresponding state for specimens having no cuts in them.

Although a variety of fracture processes can be interpreted in terms of the presence of nuclei, equivalent to microscopic cuts, in this way, the exact nature of these fracture precursors is not yet known. They may consist of foreign inclusions—particles of dirt or dust—or they may represent severe structural irregularities of the molecular network. An alternative possibility is that they develop in the course of fracture and thus represent merely a characteristic feature of the fracture process itself rather than physical cracks or cuts in the original material. Research in this area is clearly needed.

Although all rubbery materials appear to break in tension from similar nuclei, they exhibit very great differences in tensile strength[30–35] because of differences in stiffness and in fracture energy G_c (see eqn. (8)). Nevertheless, a striking correlation has been demonstrated between the energy density at break in tension, W_b, and the energy density dissipated in deforming the material almost to break, W_d:

$$W_b = 410 W_d^{2/3}$$

where W_b, W_d are measured in J m^{-3}.[36] This relation, shown in Fig. 11, is obeyed for a wide variety of rubberlike materials, ranging from the weakest to the toughest elastomers, i.e. over the entire range of strength. It reveals the importance of energy dissipation in determining the strength of rubbery materials. The exact mechanism of dissipation appears to be unimportant: whether it is due to strain-induced crystallisation, retarded segmental motion (near T_g), detachment from filler particles, or fracture of weak bonds, for example. We conclude that the tensile strength of elastomers is governed by the extent to which strain energy is dissipated in deforming them, and not by the way in which it is expended.

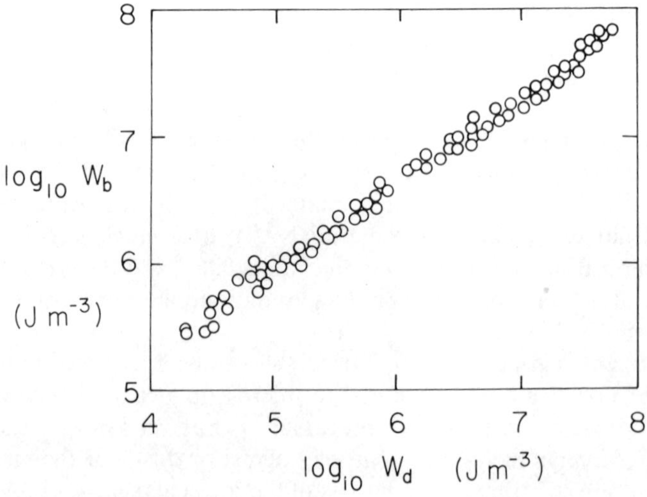

FIG. 11. Energy density W_b required for tensile rupture *versus* energy W_d dissipated in irreversible processes prior to rupture. (After reference 36.)

2.8 Intermittent Stressing

When a rubber component is stressed intermittently, flaws in the material grow at a rate Δc per stress application, where Δc depends upon the rubber employed and upon the severity of stress. A suitable measure of stress severity is afforded by the quantity G, defined by eqn. (3), but without the subscript denoting a critical value. Thus G is the energy *available* for fracture, i.e, the potential energy in the system which would be released by a hypothetical growth of a crack; G_c will continue to be used to denote the corresponding critical value at which a running tear develops.

Even when G is much smaller than G_c, a small growth Δc of the crack takes place each time the stresses are applied.[37] Empirical relations between Δc and G take the form

$$\Delta c = B[(G/G_{0c}) - 1]^2 \tag{13}$$

for natural rubber (strain-crystallising), where B is a crack growth parameter having the approximate value 0·5 Å (0·05 nm), and:

$$\Delta c = D[(G/G_{0c}) - 1]^4 \tag{14}$$

for non-crystallising elastomers, where the crack growth parameter D takes the somewhat similar value 0·2 Å (0·02 nm).[38] As before, G_{0c} denotes the threshold fracture energy, below which no mechanical rupture occurs.

Both of these relations hold reasonably well over a range of values for G between a few hundred J m^{-2} and values approaching G_c, generally around 10000 J m^{-2}. Below about 100 J m^{-2}, crack growth ceases altogether because the threshold condition for any molecular rupture to occur is not met (see Section 2.2). Near G_c, crack growth tends to become catastrophically large. Between these limits, however, there is a wide range of stressing levels where eqns. (13) and (14) give a good representation of the rate of crack growth under intermittent stressing.[21,38]

Crack growth in strain-crystallising materials is not much affected by temperature or rate of stressing, provided that the material at the crack tip is able to crystallise. For amorphous materials, however, the amount of crack growth Δc increases dramatically with increase in temperature because of the lowered dissipation of energy in viscous processes.[18] Indeed, for perfectly elastic rubber the crack growth amount Δc would become infinitely large when $G > G_{0c}$. Thus, a rather general crack-growth law may be formulated for energy-dissipating solids:

$$\Delta c = D[(G/G_{0c}) - 1]^n \tag{15}$$

where the parameter D is of molecular dimensions and the index n depends upon the dissipative character of the material. For completely inelastic dissipative materials (which strain-crystallising elastomers resemble) n takes the value 2. For perfectly elastic, non-dissipative materials, n is infinitely large. And for partially dissipative materials n lies between these two limits (Fig. 12). This variation may be expressed empirically as

$$n = 2/H$$

where H denotes the fraction of input strain energy that is dissipated.

Although eqn. (15) appears to describe the crack-growth behaviour of materials in a semiquantitative way, no satisfactory theoretical derivation of it has been put forward yet, as far as the present authors are aware. Moreover, the growth of cracks under more complex straining conditions than repeated tensile deformations has only been studied to a limited extent so far.[39–41] Thus a unified treatment of crack growth is not yet possible.

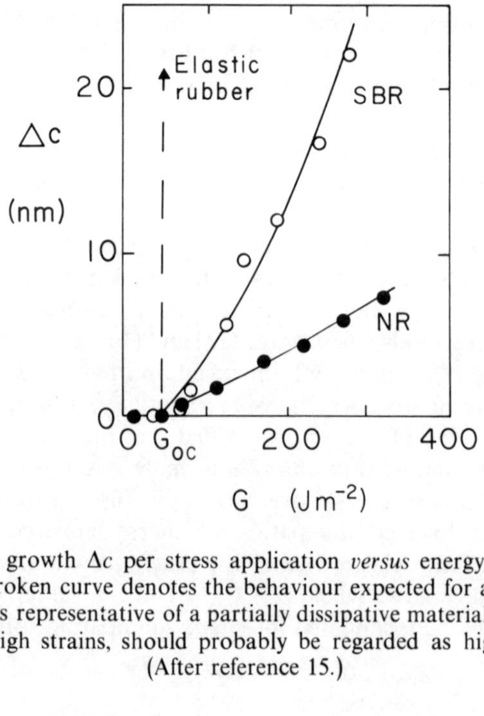

FIG. 12. Crack growth Δc per stress application *versus* energy G available for fracture. The broken curve denotes the behaviour expected for an ideally elastic material, SBR is representative of a partially dissipative material, and NR, being crystalline at high strains, should probably be regarded as highly dissipative. (After reference 15.)

3 ABRASIVE WEAR

3.1 Patterns of Wear.

Under severe conditions, rubber wear takes place in a characteristic way. A series of ridges develops at right angles to the sliding direction with wear occurring mainly at the base of the ridges (Fig. 13). As a result of

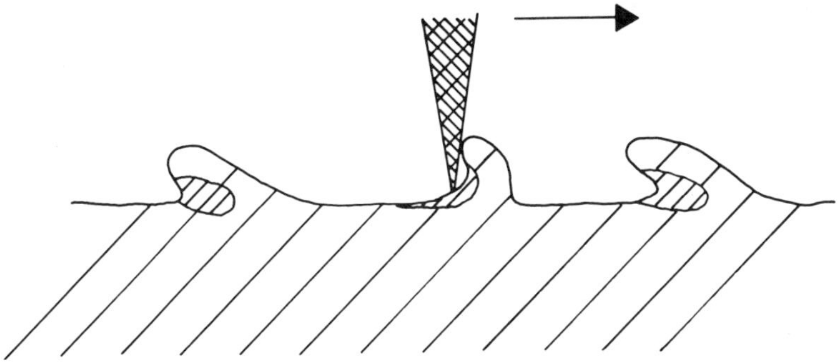

FIG. 13. Sketch of wear pattern, showing regions at the base of the ridges where wear is concentrated. (After reference 42.)

this local wear the ridges become undercut and occasionally the tips become detached to form large pieces of debris. An example is shown in Fig. 14. Fracture events on the scale of a few μm can be seen in the photograph; cumulatively they have brought about the detachment of a particle some 50 μm in size. Thus, two wear processes accompany the formation of abrasion ridge patterns: detachment of relatively large particles from the ridge edges and small-scale 'intrinsic' abrasion from some other fracture process[43].

Both the ridge height and ridge spacing increase with increasing severity of wear, i.e. for sharp abrasive surfaces, high frictional forces and soft rubber. As wear proceeds, the pattern of ridges moves slowly backwards, against the direction of sliding, as a result of the gradual wearing away of the ridge tips[43]. This process may be regarded as the advance into the rubber by tearing of a series of cracks, each crack corresponding to the leading edge of a ridge. A theoretical treatment of pattern abrasion along these lines is described in Section 3.2 below.

If the direction of sliding is changed repeatedly the characteristic abrasion pattern does not form and wear takes place more slowly and on a finer scale—several μm rather than the 50 μm to 5 mm characteristic of

FIG. 14. Photograph of particle of wear debris from an unfilled BR elastomer. The energy available for surface crack growth was $0{\cdot}6\,kJm^{-2}$. (After reference 51.)

pattern spacings. This intrinsic abrasion process, in the absence of ridge formation, appears to consist of the detachment of small particles of rubber plucked out from the surface by frictional forces. It is closely related to tensile rupture, as described later (Section 3.3).

In addition to these two wear processes arising from mechanical rupture of rubber, there is also evidence for chemical deterioration of rubber under milder wear conditions. Some observations on this aspect of abrasion are discussed in the final section of this chapter.

3.2 Pattern Abrasion: Theory and Experiment.

A suitable apparatus for studying wear in uni-directional sliding is shown schematically in Fig. 15. A stationary knife-blade is pressed into the surface of a rubber wheel in a radial direction while the wheel is rotated slowly under it.[44] By measuring the weight loss of the wheel per revolu-

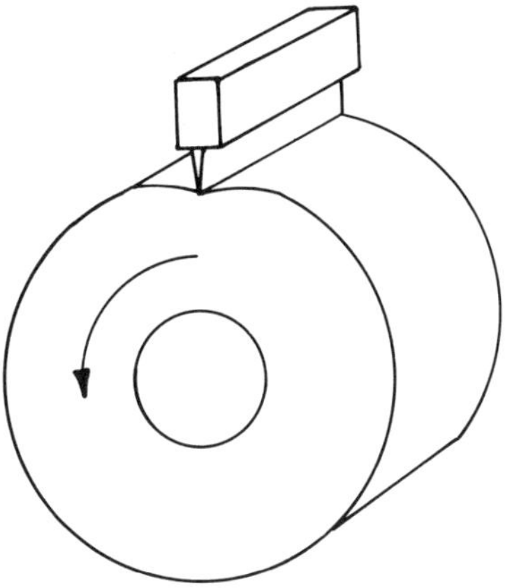

FIG. 15. Knife abrader, used for studying the wear of rubber wheels. (After reference 44.)

tion, the loss in thickness, A, may be calculated. This loss is found to depend strongly upon the frictional force F per unit length of blade edge in contact with the rubber (Fig. 16). Experimental results are shown in Fig. 17 for unfilled rubber.

Southern and Thomas[46] have related the rate of wear, A, to the crack-growth resistance of the rubber by a simple theoretical treatment. A single ridge under the frictional force F is shown in Fig. 16. The energy G available for tear propagation in this configuration is given by $F(1 + \cos\theta)$ where θ is the angle between the direction of crack growth and the surface of the rubber. The crack will therefore advance by a distance Δc, given by eqn. (15). This leads to a loss in thickness of rubber of $\Delta c(\sin\theta)$. Thus,

$$A = DF^n(1 + \cos\theta)^n(\sin\theta)/G_{0c}^n \qquad (16)$$

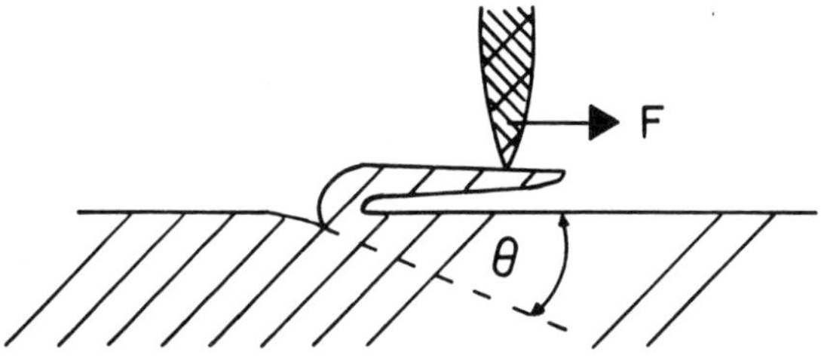

FIG. 16. Sketch of a single ridge subjected to a frictional (tearing) force F per unit width. (After references 44, 46).

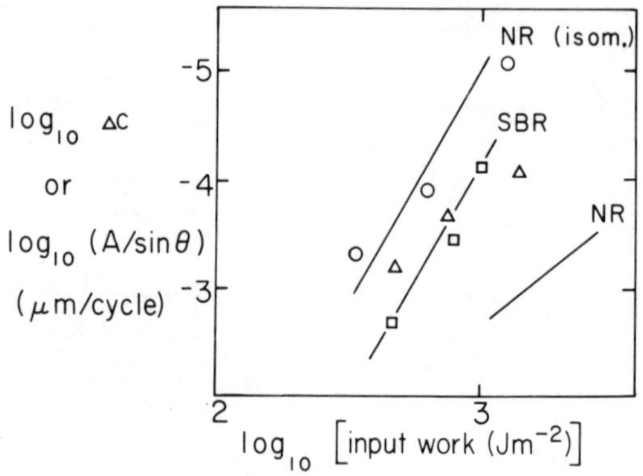

FIG. 17. Rates of wear, A, *versus* frictional work input for NR, Δ; SBR, □; and isomerised (non-crystallizing) NR, ○. The full lines represent Δc, the measured crack-growth properties of the same materials under repeated stressing. (After reference 46.)

The angle θ may be estimated by direct inspection of the way in which abrasion patterns move over the surface during wear. All other terms in eqn. (16) can be determined from tear-growth measurements. Thus, the theory does not involve any fitting constants. In these circumstances, it is remarkably successful in accounting for the rate of wear of several unfilled elastomers under severe conditions of pattern abrasion (Fig. 17).

Two difficulties must be mentioned, however. The agreement is unsatisfactory for unfilled natural rubber which wears away much more rapidly than crack-growth measurements would predict (Fig. 17). Possibly, under abrasive conditions, this material does not undergo strain-induced crystallisation and the development of cracks is therefore unexpectedly rapid. Also, the wear of reinforced rubber is much slower than would be predicted on the basis of crack-growth measurements. Further work is needed to clarify this point, which is of great practical importance.

3.3 Intrinsic Abrasion.

When the direction of sliding is changed frequently, so that the characteristic pattern of ridges does not form, values obtained for the volume A abraded per unit sliding distance are much smaller than in uni-directional wear, perhaps only one-tenth as large[47]. Also they do not seem to be so strongly dependent upon the frictional force F. Grosch and Schallamach[45] employed the quantity A/F as a suitable measure of intrinsic abrasion, implying that the amount of wear is approximately proportional to the energy expended in sliding in this case.

Intrinsic abrasion is affected strongly by the speed of sliding and the temperature. For simple non-crystallising materials a master relation can be constructed from measurements at various temperatures, reducing them to a reference temperature by means of the WLF relation, eqn. (1). As shown in Fig. 18, A/F is found to pass through a minimum as the

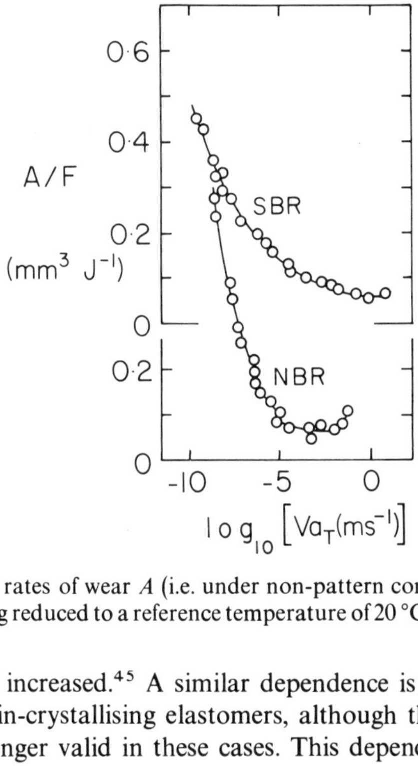

FIG. 18. Intrinsic rates of wear A (i.e. under non-pattern conditions) *versus* Va_T, the speed of sliding reduced to a reference temperature of 20 °C. (After reference 45.)

sliding speed is increased.[45] A similar dependence is obtained for reinforced and strain-crystallising elastomers, although the WLF reduction scheme is no longer valid in these cases. This dependence upon sliding speed resembles the variation of the reciprocal of the energy density at break W_b with rate of extension (a reciprocal relation because A is larger for weaker materials). Grosch and Schallamach[45] concluded that the basic mechanism of intrinsic abrasion is tensile rupture, small pieces of rubber being plucked from the surface. On this basis they surmised that

$$A/F = C/W_b$$

where C is a constant, having the approximate value of 1×10^{-3} for a number of different elastomers. In order for this relation to hold over a range of temperatures, it was necessary to employ values of W_b at rates of elongation $\dot{e}$ of 100 s^{-1}, to give agreement with wear rates measured at a sliding speed of 10 mm s^{-1}. Thus, the scale of the deformation process was relatively small, about 0·1 mm, a value close to the size and spacing of asperities in the abrasive surface used in their experiments.

The empirically determined constant C represents the ratio of the volume of rubber abraded away to that volume which would be brought to rupture if an amount of strain energy equal to the work expended in sliding friction were applied to stretch the rubber uniformly up to the breaking point. Because C is found to be quite small, it is clear that only a small fraction of the frictional work causes rupture. The major part must be dissipated in other ways.

3.4 Chemical Effects.

Under mild abrasion conditions, chemical changes within the elastomer become important in wear.[48,49] The scale of wearing remains small but the particles of debris are often sticky and agglutinate to form larger particles, several mm in size. Indeed, *cis*-polyisoprene and poly(ethylene–copropylene), for which molecular rupture under shearing conditions is particularly pronounced, both develop a tarry liquid surface during abrasion. In contrast, *cis*–polybutadiene shows no signs of structural deterioration. The debris appears to be unchanged chemically and non-adhering. Evidently, different chemical changes are undergone by different elastomers and are responsible for the different types of wear.

Two reactions can occur during wear: oxidative degradation due to frictional heating in the contact zone and mechano-chemical degradation initiated by shear-induced rupture of chemical bonds. Present evidence favours the latter process. For example, in the absence of oxygen the wear of *cis*-polyisoprene changes to resemble that of *cis*-polybutadiene[50] whereas the wear of poly(ethylene–copropylene) is unaltered.[51] These results are in accord with the response of these materials to free-radical reactions.

Surprisingly, the products of chemical changes within the elastomer are capable of causing rapid wear of metals used as abraders.[52–54] This is evident with the knife-blade abrader described in Section 3.2; the blade becomes blunted during wear experiments and the volume of metal removed from the blade is substantially greater when the elastomer undergoes chemical deterioration. This wear of the metal is attributed to direct chemical attack by reactive polymeric species, probably free radicals, during frictional contact.[54] Similar wear has been observed with polymers in other physical states[55]; it appears to be a quite general phenomenon when molecular rupture takes place during sliding.

ACKNOWLEDGEMENTS

This review was prepared in the course of a programme of research on fracture supported by a Research Grant from the Engineering Division of the National Science Foundation. It is based on recent research carried out in this programme, together with some earlier reviews.[56–59] The authors are indebted to E. H. Andrews, T. L. Smith, A. G. Thomas and other authors cited in the text for many helpful discussions, and to C. L. Slezak for preparing the diagrams.

REFERENCES

1. TRELOAR, L. R. G. *Physics of Rubber Elasticity*, 2nd ed., Oxford University Press, Oxford, 1958.
2. FERRY, J. D. *Viscoelastic Properties of Polymers*, 2nd ed., John Wiley and Sons, New York, 1970.
3. BUECHE, F. *Physical Properties of Polymers*, Interscience, New York, 1962.
4. GRIFFITH, A. A. *Phil. Trans. Roy. Soc.* (*London*), **A221** (1921) p. 163.
5. IRWIN, G. R. *J. Appl. Mech.*, **24** (1957) p. 361.
6. OROWAN, E. *Rep. Progr. Phys.*, **12**, (1949) p. 185.
7. RIVLIN, R. S., and THOMAS, A. G. *J. Polymer Sci.*, **10** (1953) p. 291.
8. AHAGON, A., GENT, A. N., KIM, H. J., and KUMAGAI, Y. *Rubb. Chem. Technol.*, **48**, (1975) p. 896.
9. GREENSMITH, H. W. *J. Appl. Polymer Sci.*, **7** (1963) p. 993; Lake, G. J., private communication.
10. THOMAS, A. G. *J. Polymer Sci.*, **18** (1955) p. 177.
11. GREENSMITH, H. W., and THOMAS, A. G. *J. Polymer Sci.*, **18** (1955) p. 189.
12. GREENSMITH, H. W. *J. Polymer Sci.*, **21** (1956) p. 175.
13. VEITH, A. G. *Rubb. Chem. Technol.*, **38** (1965) p. 700.
14. AHAGON, A., and GENT, A. N. *J. Polymer Sci., Polymer Phys. Ed.*, **13** (1975) p. 1903.
15. LAKE, G. J., and THOMAS, A. G. *Proc. Roy. Soc.* (*London*), **A300** (1967) p. 108.
16. LAKE, G. J., and LINDLEY, P. B. *J. Appl. Polymer Sci.*, **9** (1965) p. 1233.
17. MUELLER, H. K., and KNAUSS, W. G. *Trans. Soc. Rheol.*, **15** (1971) p. 217.
18. GREENSMITH, H. W., MULLINS, L., and THOMAS, A. G., in *The Chemistry and Physics of Rubberlike Substances*, ed. L. Bateman, John Wiley and Sons, New York 1963, chap. 10.
19. MULLINS, L. *Trans. Instn. Rubb. Ind.*, **35** (1959) p. 213.
20. ANDREWS, E. H. *J. Appl. Phys.*, **32** (1961) p. 542.
21. LAKE, G. J., and LINDLEY, P. B. *Rubb. J. Internatl. Plastics*, **146** (1964) 10, p. 24; 11, p. 30.

22. GREENSMITH, H. W. *J. Polymer Sci.*, **21** (1956) p. 175.
23. GENT, A. N., and HENRY, A. W. *Proc. Internatl. Rubb. Conf., 1967*, Maclaren and Sons, London, 1968, pp. 193–204.
24. GENT, A. N., and KIM, H. J. *Rubb. Chem. Technol.*, **51** (1978) p. 35.
25. FETTERS, L. J., and MORTON, M. *Macromolecules*, **2** (1969) p. 453.
26. SMITH, T. L. *J. Polymer Sci., Polymer Phys. Ed.*, **12** (1974) p. 1825.
27. THOMAS, A. G. (1970), private communication.
28. BRADEN, M., and GENT, A. N. *J. Appl. Polymer Sci.*, **3** (1960) p. 100.
29. BRADEN, M., and GENT, A. N. *Kautschuk u. Gummi.*, **14** (1961) p. WT157.
30. EPSTEIN, L. M., and SMITH, R. P. *Trans. Soc. Rheol.*, **2** (1958) p. 219.
31. BUECHE, F., and DUDEK, T. J. *Rubb. Chem. Technol.*, **36** (1963) p. 1.
32. SMITH, T. L. *J. Polymer Sci. A*, **1** p. 3597.
33. LANDEL, R. F. and FEDORS, R. F. *Proc. 1st Internatl. Conf. on Fracture, Sendai, Vol. 2*, Japan Soc. Strength and Fracture of Materials, 1966, p. 1247.
34. SMITH, T. L., in *Rheology, Vol. 5*, ed. F. R. Eirich, Academic Press, New York, 1969, chap. 4.
35. THOMAS, A. G., and WHITTLE, J. M. *Rubb. Chem. Technol.*, **43** (1970) p. 222.
36. GROSCH, K. A., HARWOOD, J. A. C., and PAYNE, A. R. *Nature*, **212** (1966) p. 497.
37. THOMAS, A. G. *J. Polymer Sci.*, **31** (1958) p. 467.
38. LAKE, G. J., and LINDLEY, P. B. *J. Appl. Polymer Sci.*, **10** (1966) p. 343.
39. DICKIE, R. A., and SMITH, T. L. *J. Polymer Sci. A2*, **7** (1969) p. 687.
40. LAKE, G. J., and CLAPSON, B. E. *Rubb. J.*, **152** (1970) pp. 12, 36.
41. ROBERTS, B. J., and BENZIES, J. B. *Proc. Internatl. Rubb. Conf., Brighton, England, May 1977*, The Plastics and Rubber Institute, London, 1977.
42. SCHALLAMACH, A. *Trans. Inst. Rubb. Ind.*, **28** (1952) p. 256.
43. SCHALLAMACH, A. *Rubb. Chem. Technol.*, **41** (1968) p. 209.
44. CHAMP, D. H., SOUTHERN, E., and THOMAS, A. G. *ACS Coatings and Plastics Division Preprints*, **34** (1974) p. 237; THOMAS, A. G., *J. Polymer Sci: Symposium No. 48*, (1974) p. 145.
45. GROSCH, K. A., and SCHALLAMACH, A. *Trans. Inst. Rubb. Ind.*, **41** (1965) p. T80.
46. SOUTHERN, E., and THOMAS, A. G. *Proc. Internatl. Rubb. Conf., Brighton, England, May 1977*, The Plastics and Rubber Institute, London, 1977.
47. SCHALLAMACH, A., in *The Chemistry and Physics of Rubberlike Substances*, ed. L. Bateman, John Wiley and Sons, New York, 1963, chap. 13.
48. SCHALLAMACH, A. *J. Appl. Polymer Sci.*, **12** (1968) p. 281.
49. BRODSKII, G. I., SAKHNOVSKII, N. L., REZNIKOVSKII, M. M., and EVSTRATOV, V. F. *Soviet Rubb. Technol.*, **19**(8) (1960) p. 22.
50. GENT, A. N., HENRY, R. L., PADEN, R. A., and PULFORD, C. T. R. *Technical Report No. 2*, NASA–Ames Research Grant NSG–2061, Chemical Research Projects Office, NASA–Ames Research Center, Moffet Field, California, 1976.
51. GENT, A. N., and PULFORD, C. T. R. (1978), unpublished.
52. GOROKHOVSKII, G. A., CHERNENKO, P. A., and SMIRNOV, V. A. *Soviet Material Science*, **8** (1972) p. 557.
53. EVDOKIMOV, Y. A., SANCHES, S. S., and SUKHORUKOV, N. A. *Polymer Mechanics*, **9** (1973) p. 460.
54. GENT, A. N., and PULFORD, C. T. R. *Wear*, **49** (1978) p. 135.

55. VINOGRADOV, G. V., MUSTAFAEV, V. A., and PODOLSKY, Y. Y. *Wear*, **8** (1965) p. 358.
56. ANDREWS, E. H. *Fracture in Polymers*, American Elsevier, New York, 1968.
57. GENT, A. N., in *Fracture: An Advanced Treatise, Vol. 7: Fracture of Non-Metals and Composites*, ed. H. Liebowitz, Academic Press, New York, 1972. chap. 6.
58. EIRICH, F. R., and SMITH, T. L., in *Fracture: An Advanced Treatise. Vol. 7: Fracture of Non-Metals and Composites*, ed. H. Liebowitz, Academic Press, New York, 1972, chap. 7.
59. GENT, A. N., in *Rubber Science and Technology*, ed. F. R. Eirich, Academic Press, New York, 1978, chap. 10.
60. ANDREWS, E. H. *J. Materials Sci.*, **9** (1974) p. 887.
61. ANDREWS, E. H., and FUKAHORI, Y. *J. Materials Sci.*, **12** (1977) p. 1307.

Chapter 6

FRACTURE OF THERMOSETTING RESINS

ROBERT J. YOUNG
Queen Mary College, University of London, UK

SUMMARY

Recent developments in the fracture of thermosetting resins have been reviewed. They are generally brittle materials and have mainly been studied by the use of linear-elastic fracture mechanics (LEFM). The development of LEFM from the Griffith theory of fracture has been discussed and the types of specimen geometries that are used and the application of LEFM to the study of crack propagation in thermosetting polymers have been reviewed. In particular the stability of crack propagation and the effect of crack velocity are considered. Finally, the possible mechanisms of failure, which are at the moment virtually unknown, have been discussed.

1 INTRODUCTION

Thermosetting polymers are being used increasingly in structural engineering applications. They are now widely used as matrix materials for both particulate and fibre composites, and epoxy resins in particular are commonly used as adhesives in aerospace technology and as insulators in many electrical applications. As structural materials they must be able to withstand high stresses, which may be static, dynamic or cyclic, and also they are often exposed to potentially aggressive environments. Since thermosetting polymers are invariably brittle materials, a knowledge of their fracture properties and hence their durability under extreme service conditions is vitally important in aiding their efficient utilisation. This type of knowledge is normally obtained through the use of fracture mechanics. It is only within the last 10 years that fracture mechanics has been applied to any extent to the fracture of brittle polymers. The only fracture data which had been obtained previously were from simple tensile or bend tests, and recent developments in this area have been almost exclusively concerned with the application of linear-elastic fracture mechanics (LEFM) to the initiation and propagation of cracks.

The term 'thermosetting resins' covers a wide range of cross-linked polymers, which are principally epoxy resins, unsaturated polyesters, phenol–formaldehyde and amino resins. Although the properties of these materials differ in detail there are many characteristics which are common to all of them. Most of the detailed investigations into the fracture of thermosets have been concerned with the fracture of epoxy resins, which are generally the strongest thermosetting materials, and there has been rather less emphasis upon the other materials. An important variable which affects the fracture behaviour of all thermosetting polymers is chemical structure, which is in part controlled by such factors as the type and amount of curing agent or hardener and hence the degree of cross-linking. The chemistry of thermosets and of cross-linking reactions will not be covered in any detail here and the reader is directed to general textbooks concerned with such matters.[1] Chemical structure is only covered in this chapter when work is discussed in which it has been shown to have a direct effect upon the fracture behaviour.

An important means of improving the physical characteristics of thermosetting polymers, such as their mechanical strength, has been the addition of second-phase particles such as brittle fillers[2] or rubber particles.[3] The effect of these additives upon the fracture of thermosets will not be considered. An important use of thermosetting polymers and especially epoxy resins is as high-strength adhesives. In this case only developments in the fracture of adhesive joints which are relevant to fracture in the cohesive situation will be considered. Another important use of thermosets is as matrix materials for fibre-reinforced composites. The fracture of these composite materials has been considered in another chapter in this book[4] and has been reviewed by other authors elsewhere.[2]

2 GRIFFITH FRACTURE

It can be shown that the theoretical stress to cause cleavage fracture in a brittle solid is of the order of $E/10$; that is, one-tenth of its Young's modulus.[5] The modulus of a brittle polymer is typically 3 GPa and so the theoretical strength of such a material should be 300 MPa. The fracture strengths of glassy polymers are well below the theoretical value (typically 10–100 MPa), as is the case for many other materials. This shortfall in strength was recognised many years ago by Griffith,[6] who showed that the relatively low strength of a brittle solid could be explained by the presence of flaws. Using classical elasticity theory he showed that the tensile fracture stress, σ_f, of a specimen containing an elliptical centre crack of length $2a$ and subjected to plane stress loading is given by an equation of the form

$$\sigma_f = \sqrt{2E\gamma/\pi a} \qquad (1)$$

where E is the Young's modulus of the material and γ is the energy required to create unit area of the fracture surface. Over the years the approach of Griffith has undergone many modifications and developments but it is still useful since it clearly relates the fracture stress of any specimen to material constants E and γ and the size of any pre-existing flaw in the specimen. The Griffith approach has been applied to many brittle materials such as some metals and ceramics. In the early 1960s Berry applied it to the fracture of amorphous glassy polymers such as poly(methylmethacrylate)[7] and polystyrene[8] by determining the dependence of the tensile strength of such materials upon the size of flaws that had been deliberately introduced into specimens. With this type of experiment there is always the problem that the results might be influenced by the nature of the artificial flaws and so care was taken by Berry to produce 'natural' cracks by carefully controlling crack length and quality.[9] The value of γ can be calculated from fitting eqn. (1) to a plot of σ_f against a or directly from the slope of a plot of σ_f against $a^{-1/2}$ if E is determined from a separate experiment. Berry showed that the Griffith equation was obeyed by both polymethylmethacrylate and polystyrene.[9]

Although, because of their brittleness, thermosetting polymers would appear to be ideal candidates for the study of Griffith fracture, it is only recently that a particular thermosetting system has been investigated systematically in this manner. Nelson and Turner[10] examined the variation of fracture stress σ_f with the size of artificially induced flaws, a, for a phenol–formaldehyde resin. Their results are given in Fig. 1, where σ_f is plotted against $a^{-1/2}$. Using eqn. (1) they found from the slope of the line that γ was about 50 J m^{-2} for this resin. They used a modulus of 9·2 GPa, which would appear to be somewhat high for a glassy polymer. A lower modulus would give a higher value of γ using the same measured slope. They compared their value of γ with the value of 0·7 J m^{-2} which

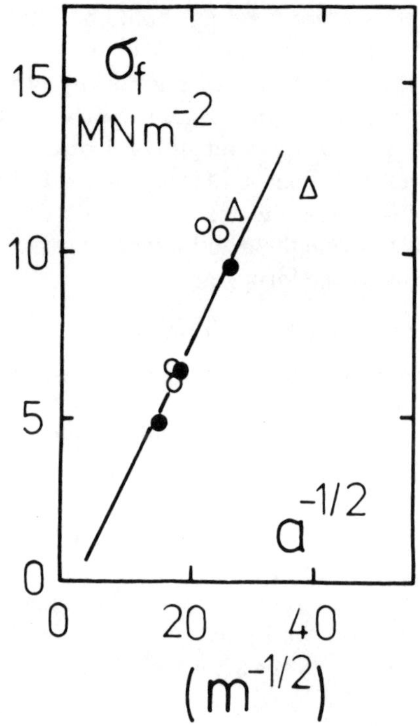

FIG. 1. Plot of fracture stress σ_f against $a^{-1/2}$ for a phenol–formaldehyde resin. (After reference 10.)

they calculated theoretically from the energy required to create new surfaces simply by breaking covalent bonds.

The Griffith theory as originally devised considers γ to be a surface energy. However it has long been recognised from measurements of γ for many materials that the measured value contains contributions from other sources as well as the energy required to break bonds and create new surface. In fact it is thought that even for apparently brittle materials there is an appreciable amount of plastic deformation in areas close to the fracture plane and the energy absorbed by plastic deformation swamps the surface-energy term. This has led to a modified version of eqn. (1):

$$\sigma_f = \sqrt{2E\gamma_p/\pi a} \tag{2}$$

where γ_p is the energy absorbed due to plastic deformation in the creation of unit area of fracture surface. Berry has obtained values of γ_p for poly(methylmethacrylate)[7] and polystyrene[8] and they are given in Table 1 with γ_p for the phenol–formaldehyde resin of Nelson and Turner.[10] For the two thermoplastics the values of γ_p are well in excess of any calculated values and in thermoplastics of this kind the energy absorption process is thought to be due to plastic deformation by means of crazing[9] (see Section 4.2 below).

TABLE 1

QUOTED VALUES OF γ_p (eqn. (2)) FOR VARIOUS LINEAR AND CROSS-LINKED POLYMERS (AFTER REFERENCE 12).

Polymer	γ_p(J m^{-2})	*Reference*
Polystyrene	1700	8
Poly(methylmethacrylate)	300	7
Poly(methylmethacrylate) cross-linked with ethylene glycol dimethacrylate	46	9
Poly(methylmethacrylate) (lightly cross-linked)	80	11
Poly(methylmethacrylate) (heavily cross-linked)	23	11
Phenol–formaldehyde resin	50	10

A link between thermoplastics and thermosets has been established by Berry[9] and Broutman and McGarry[11] by investigating the effect of various degrees of cross-linking upon the value of γ_p for the fracture of poly(methylmethacrylate). Although the degree of cross-linking was not precisely defined in either case they were able to show that cross-linking reduced the value of γ_p from about 300 J m^{-2} for the linear polymer to as low as 23 J m^{-2} for the most highly cross-linked material. In fact this is normally the case; thermosets generally have lower values of γ_p than thermoplastics (Table 1).

Another aspect of the Griffith approach that is particularly interesting is the observation that unnotched tensile specimens of a brittle polymer with the best possible surface preparation (i.e. highly polished) tend to have fracture stresses within a narrow range which are well below the theoretical fracture strengths. This has lead to the concept of a material having a characteristic 'inherent flaw size' (a_0). This can be readily calculated from eqn. (2) by substituting measured values of σ_f, E and γ_p to yield a value of a_0. There have been few reports of values of a_0 in the literature for thermosets. However they can often be calculated from quoted values of fracture stress and fracture energy. Broutman and McGarry[11] have measured the inherent flaw size of poly(methylmethacrylate) and showed that it tended to decrease as the degree of cross-linking increased, in a similar way to the variation of γ_p with cross-linking.

3 FRACTURE MECHANICS

Although the Griffith approach is valuable in its simple approach and yields useful information, there are many phenomena that it is unable to account for, even in relatively brittle materials such as thermosetting polymers. For example, it cannot explain time-dependent failure or fatigue. Also γ_p can vary with testing rate or crack velocity. These and other problems have led to the development of linear-elastic fracture mechanics (LEFM), which has recently been applied extensively to the fracture of thermosetting polymers. The first step in this direction was the replacement of γ in eqn. (1) by γ_p, as described in the previous section. Detailed calculations of the stresses around a crack led to further developments.[13] Equation (2) can be recast into the form

$$\sigma_f^2 a = \text{constant} \times 2E\gamma_p \quad (3)$$

A parameter K, known as the stress intensity factor, can be defined such that

$$K^2 = Y\sigma^2 a \quad (4)$$

where Y is a geometrical factor. Fracture will occur when K reaches a critical value K_c and so

$$K_c^2 = Y\sigma_f^2 a \quad (5)$$

and

$$K_c^2 = EG_c \quad (6)$$

where G_c is equivalent to $2\gamma_p$. The stress intensity factor K_c is a more useful concept for the design engineer than G_c, which has several different names but is most commonly called the 'strain-energy release rate'. However, K_c has rather unusual units ($\text{Nm}^{-3/2}$) and G_c, which has units of energy (J m^{-2}), can be related more directly to the microscopic failure processes.

An important point that is often overlooked when applying LEFM to polymers is that the theory was developed for materials which have linear-elastic stress–strain curves up to the point of fracture. With thermosetting polymers well below their glass-transition temperatures this is nearly always a good approximation and so LEFM can normally be safely applied to these materials. However, there can be difficulties with other polymeric solids which show non-linear deformation. With these materials it is more appropriate to use the *J*-integral of Rice[13] or the generalised theory of fracture developed by Andrews[14] and described elsewhere in this book (see Chapter 1).

3.1 Fracture Mechanics Testing

At the same time that the theory of LEFM was being developed new specimen geometries were being devised to facilitate easier and more reliable measurements of K_c and G_c. Many of the recent advances in the application of LEFM to thermosetting polymers have centered upon the use of test-pieces with particular geometries which give more control of crack propagation and so lead to useful and reproducible results. The particular geometries that have been used to study the fracture of brittle polymers are shown in Fig. 2. All of these specimens measure K_c and G_c for the tensile opening mode of fracture (mode I). These values of K_{Ic} and G_{Ic} are the ones of most interest and hence are the values normally quoted.

K_{Ic} is normally obtained from an expression which has been obtained either analytically or numerically for the particular specimen geometry. It is normally a function of the applied load, crack length and specimen dimensions. There are many sources of these expressions in the literature.[13] Rooke and Cartwright[16] have recently produced a compendium of stress intensity factors for a wide range of specimen geometries. The expressions for K_I for the six geometries illustrated in Fig. 2 are given in Table 2. These expressions for K_I are often only strictly applicable to certain specimen dimensions and crack lengths and so care must be taken to ensure that particular specimens comply with these requirements before a valid K_I value can be quoted. This should be done by checking the original references.

G_{Ic} can be obtained if K_{Ic} and the Young's modulus E are known by using eqn. (6). For many materials this can be readily done but care is needed when this equation is used for brittle polymers. The modulus is a strong function of testing rate for most polymers and so it is essential to use a value of E obtained at an appropriate strain-rate. In many fracture-mechanics test-pieces the strain-rate is difficult to define and so the use of

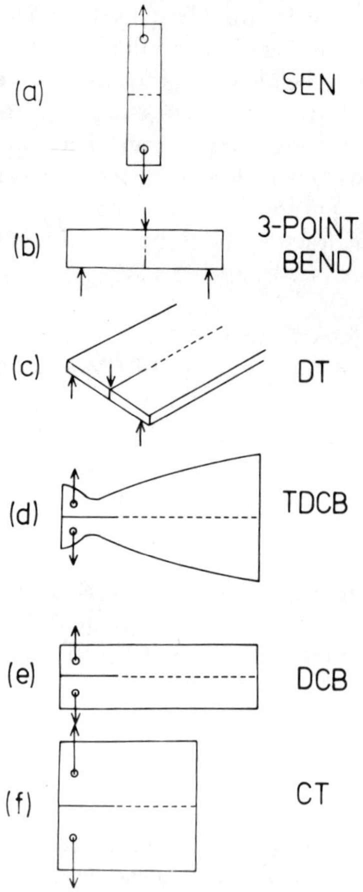

FIG. 2. Schematic diagrams of the various fracture mechanics specimens that have been used with thermosetting resins. (a) Single-edge notched (SEN); (b) three-point bend; (c) double torsion (DT); (d) tapered double cantilever beam (TDCB); (e) double cantilever beam (DCB); (f) compact tension (CT).

eqn. (6) for polymers is restricted. An alternative way of determining G_{Ic} is through the use of compliance methods which measure directly the amount of energy expended within a material when a crack is caused to grow.[13] It can be shown[13] that G_{Ic} is given by the equation

$$G_{Ic} = \frac{P_c^2}{2} \cdot \frac{\partial C}{\partial A} \tag{7}$$

where P_c is the applied load for crack propagation, C is the specimen compliance and A is the area of the crack. The value, $\partial C/\partial A$ is obtained from the slope of a compliance calibration curve, which is simply a plot of compliance against crack area (or length for specimens of uniform thickness). It must be noted that since the compliance of a specimen depends upon the modulus of the material, compliance calibration curves should be obtained for all the strain-rates employed with polymeric specimens.

If the single-edge notched (SEN) and three-point bend specimens (Fig. 2) are tested at constant cross-head speed in a hard testing machine, a crack, once it starts to propagate at a particular load, will become unstable and accelerate until it reaches the end of the specimen. This means that with the SEN or three-point bend specimens it is difficult to correlate K_{Ic} or G_{Ic} with crack velocity.

The double torsion (DT) and certain geometries of tapered double cantilever beam (TDCB) specimens are much more stable in this respect. In this case $\partial C/\partial A$ is constant,[2] and such specimens are often called linear compliance variation specimens since their compliance varies linearly with crack area (or length). It can also be seen from Table 2 that for both specimens K_I is independent of crack length, a. With these two specimens it is much easier to relate K_{Ic} or G_{Ic} to crack velocity.

The final category of specimen is characterised by the double cantilever beam (DCB) and compact tension (CT) specimens. For these specimens, when a constant cross-head speed is used, cracks tend to decelerate as they propagate and so these specimens give very good control of crack propagation.

3.2 Crack Propagation

Most of the recent investigations into the fracture of thermosetting polymers have been concerned with the propagation of cracks. Griffith fracture essentially employs the SEN geometry and so there is very little control of crack propagation. Using the more stable types of test piece it has been possible to gain a good deal of knowledge concerning the factors which control propagation. A variety of different resins, curing agents and formulations have been employed and this can make comparisons between the results of different groups of workers rather difficult. The principal materials studied have been epoxy resins. The main resins that have been used for fracture studies are DER 332 (Dow), Epon 828 or Epikote 828 (Shell) and MY750 or CT200 (Ciba-Geigy). These are all based upon diglycidyl ethers of bisphenol A (DGEBA) and differ only in properties

TABLE 2.

EXPRESSIONS FOR K_I FOR THE VARIOUS FRACTURE MECHANICS TEST-PIECES USED FOR THE STUDY OF FRACTURE IN BRITTLE POLYMERS (SEE FIG. 2)

Test-piece geometry	K_I *stress intensity factor*	*Restrictions and notes*	*Ref.*
Single-edge notch (SEN)	$\frac{P}{bt}a^{1/2}\left[1{\cdot}99 - 0{\cdot}41\left(\frac{a}{b}\right) + 18{\cdot}7\left(\frac{a}{b}\right)^2 - 38{\cdot}48\left(\frac{a}{b}\right)^3 + 53{\cdot}85\left(\frac{a}{b}\right)^4\right]$		17
Three-point bend	$\frac{P}{tw^{1/2}}\left[1{\cdot}93\left(\frac{a}{w}\right)^{1/2} - 3{\cdot}07\left(\frac{a}{w}\right)^{3/2} + 14{\cdot}53\left(\frac{a}{w}\right)^{5/2} - 25{\cdot}11\left(\frac{a}{w}\right)^{7/2} + 25{\cdot}80\left(\frac{a}{w}\right)^{9/2}\right]$	$4w$ = distance between supports	13
Double torsion (DT)	$PW_m\left[\frac{3(1+\nu)}{Wt^3t_n}\right]^{1/2}$ $(W/2 \gg t)$	W_m = length of moment arm W = specimen width ν = Poisson's ratio	18

Tapered double cantilever beam (TDCM)	$2P\left(\frac{m}{t_n t}\right)^{1/2}$	$m = \left(\frac{3a^2}{h^3} + \frac{1}{h}\right)$ specimen contoured such that m constant	19
Double cantilever beam (DCB)	$\frac{2P}{t}\left(\frac{3a^2}{h^3} + \frac{1}{h}\right)^{1/2}$	$t = \sqrt{tt_n}$ if grooved	19
Compact tension (CT)	$\frac{P}{tb^{1/2}}\left[29{\cdot}6\left(\frac{a}{b}\right)^{1/2} - 185{\cdot}5\left(\frac{a}{b}\right)^{3/2} + 655{\cdot}7\left(\frac{a}{b}\right)^{5/2} - 1017\left(\frac{a}{b}\right)^{7/2} + 638{\cdot}9\left(\frac{a}{b}\right)^{9/2}\right]$	$b = 2t$ $h = 0{\cdot}6b$	13

P = applied load
b = specimen length
t_n = sheet thickness in place of crack
t = sheet thickness
h = distance of specimen edge from fracture plane
a = crack length

such as molecular weight. Occasionally other types of resin have been used. Epoxy resins are very versatile thermosetting polymers and can be hardened by many different curing agents. For the materials used in fracture studies these have been principally a variety of primary, secondary and tertiary amines and several different acid anhydrides. On some occasions stoichiometric amounts of these curing agents have been used but often non-stoichiometric formulations have been employed. Heat is usually needed to cure the systems fully but a whole variety of different curing temperatures and times have been used even with the same curing agents. With all of these complications it would seem that any detailed comparisons between different work would be impossible. However it is possible to find general points of agreement and features which are common to the crack propagation behaviour of most systems.

The earliest investigations into crack propagation in epoxy resins were by Broutman and McGarry[11] and Mostovoy and Ripling.[20] A feature that became immediately apparent from these investigations was the observation that crack propagation tended to take place in epoxy resins by means of a discontinuous 'stick-slip' mechanism. Broutman and McGarry[11] found that crack propagation was usually of a continuous type in brittle thermoplastics but tended to be discontinuous in thermosets. Mostovoy and Ripling[20] examined the behaviour of epoxy resins so that they could compare it with crack propagation in adhesive joints. They pioneered the use of TDCB specimens for this type of investigation. The advantage of TDCB specimens in the study of crack stability can be appreciated from the schematic load-displacement curves in Fig. 3. When propagation takes place in a continuous manner, as for example with polymethylmethacrylate, cracks grow at a constant load, as in Fig. 3(a). However, if the cracks propagate in a stick-slip manner the load-displacement curves take on a saw-tooth appearance, as in Fig. 3(b). The load at which the cracks jump is P_i and the load at which they arrest is P_a. Since the TDCB (and DT) specimen are of the linear compliance-variation type, the values of P_i and P_a are approximately constant at all crack lengths. The phenomenon of crack stability and the factors which control it are discussed in detail in the next section.

Since these early studies there have been several more recent investigations which have looked at the effect of different curing agents and conditions upon the fracture of DGEBA epoxy resins.[19–26] The large range of different resins, hardeners and curing conditions makes any detailed comparisons difficult. However, most workers have measured fracture energies or G_{Ic} for the different systems and they are listed in Table 3. The

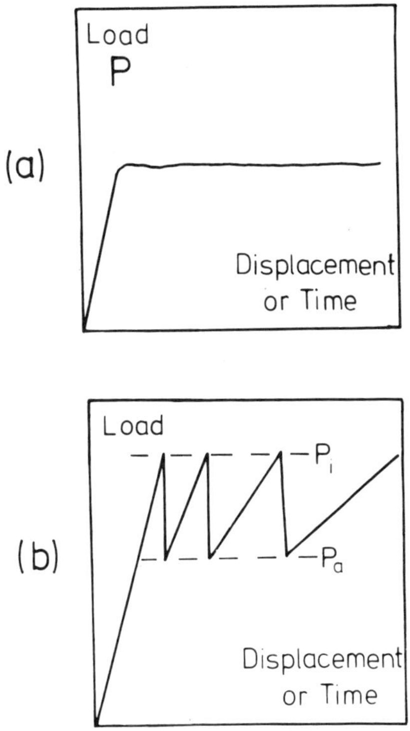

FIG. 3. Schematic load–displacement or load–time curves for crack propagation in thermosetting resins. (a) Stable; continuous propagation; (b) unstable, stick-slip propagation.

values quoted are for crack initiation, using P_i in eqn. (7). It can be seen that there is a large variation in the measured values of G_{Ic}, ranging from quite low ones appropriate to brittle solids up to about 600 J m^{-2}, similar to those of thermoplastics. It has been shown recently that for the DT geometry P_i depends upon cross-head speed.[27] It had been shown earlier that there was some variation of G_{Ic} with cross-head speed using other testing geometries.[23] Since different groups of workers have employed a variety of specimen geometries, sizes and cross-head speeds, absolute comparisons are difficult, even using identical materials.

The effect of the amount of hardener upon the value of G_{Ic} for different systems has been investigated on several occasions.[23,24] For a particular resin/hardener system there is usually one particular hardener content which gives a maximum value of G_{Ic}. For example, Selby and Miller[23]

TABLE 3

VALUES OF G_{Ic} MEASURED FOR CRACK PROPAGATION IN DGEBA EPOXY RESINS WITH VARIOUS HARDENERS

Resin	*phr Hardener*	G_{Ic} (J m^{-2})	*Ref.*
Epikote 828	10 DETA	86	11
	95 MNA + 0·5 BDMA	154	21
	27 DDM	340	22
	14·6 MPD	110	22
	4 DMP	180	22
	27 DDM	312	23
CT 200	13 PA	220	24
	?	115	25
DER 332	5 piperidine	121	19
MY 750	8·3 EDA	329	26
	12·2 TDA	489	26
	16·1 HDA	575	26
	11·5 DETA	130	26
	11·0 TETA	141	26
	15·0 TEPA	136	26

Hardener code:

DETA = diethylene triamine
TETA = triethylene tetramine
TEPA = tetraethylene pentamine
MNA = methyl nadic anhydride
BDMA = benzyl dimethylamine
DDM = diphenyl diaminomethane
MPD = *m*-phenylene diamine
DMP = tris(dimethylaminomethyl) phenol
PA = phthalic anhydride
EDA = ethylene diamine
TDA = tetramethylene diamine
HDA = hexamethylene diamine

obtained a maximum value of G_{Ic} of about 1000 J m^{-2} for 35 parts per hundred parts of resin (phr) of DDM in a DGEBA (Epikote 828). Using a different DGEBA resin (CT200) Griffith and Holloway[24] showed that there was a peak value of G_{Ic} of just over 200 J m^{-2} with 13 phr of phthalic anhydride. Figure 4 shows the plots of G_{Ic} as a function of hardener content for the two systems. The variation of Young's modulus, E, with hardener content was also measured for the two systems[23,24] and

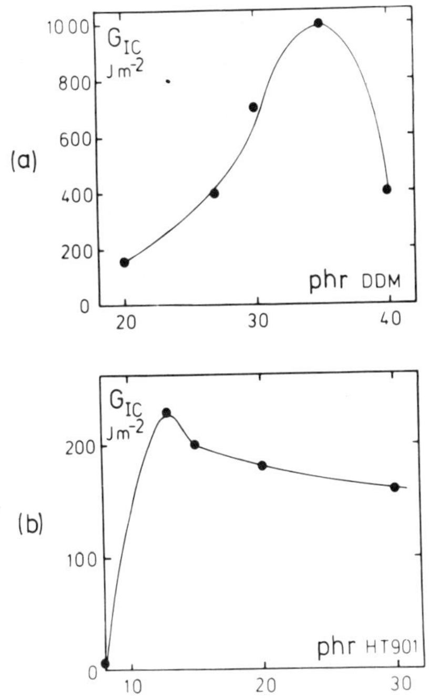

FIG. 4. G_{Ic} as a function of hardener content for two different DGEBA systems. (a) Epikote 828 cured with DDM (after reference 23); (b) CT 200 cured with HT 901 (after reference 24); phr = parts by weight of hardener per hundred parts of resin.

in both cases it was found that there was a different relationship between E and hardener content than between G_{Ic} and hardener content. This means that although one particular hardener content will give a maximum value of G_{Ic} there may not be a corresponding maximum value of K_{Ic} at this composition (eqn. (6)).

The only other thermosetting resins in which crack propagation has been studied in any detail are polyesters.[11,12,21,28–30] Any attempts to compare the results of different groups of workers suffer from the same problems as with epoxy resins since several different materials and formulations have been used. In general the polyester thermosetting resins are found to be rather less tough than epoxy resins and this is reflected in lower values of K_{Ic} and G_{Ic} being measured. When stable test-pieces are used[11,21,29] crack propagation is normally found to be of the unstable stick-slip type as with epoxy resins. There have been several attempts to change the fracture behaviour of polyester resins by the addition of flexibilisers[28,29] or by modifying the curing schedule[29] but no dramatic changes in properties have been found. A considerable change in the critical stress intensity factor K_{Ic} has been recently reported[30] on the exposure of a highly unsaturated isophthalic polyester resin to water at elevated temperatures. K_{Ic} was found to fall from a value of 0·84 MN m$^{-3/2}$ for the unexposed resin to a value of about 0·24 MN m$^{-3/2}$ after immersion in water at 80 °C for up to 90 days. This large reduction was attributed to the leaching of non-bound extractable matter from the resin by the hot water.[30] The reduction is most noticeable after exposure at elevated temperature and for long periods of time and this probably explains why other investigators have found only small changes in fracture behaviour by the exposure of polyester resins to water at ambient temperature and for only relatively short periods of time.[21,29]

There have been only relatively few crack propagation studies in thermosets other than epoxy or polyester resins. Brown and Ward looked at the effect of styrene cross-linking agent concentration upon the fracture of polyurethanes[31] and found that there was a minimum of K_{Ic} at a concentration of about 30% styrene. There have been few measurements of crack propagation in phenol–formaldehyde resins, probably because of the difficulty of preparing sheet specimens in the cast form and also because these materials are invariably used commercially in the reinforced state.[12]

3.3 Stability of Cracking

It has been known for many years that in thermosetting polymers such as epoxy resins and polyester resins, cracks can propagate in an unstable stick-slip manner, even in fracture mechanics test-pieces shaped to promote stable propagation (Section 3.1). However it is only relatively recently that the factors which control the stability of propagation have been isolated. Young and Beaumont,[27] using the DT testing geometry, showed that in a DGEBA epoxy resin (CT200) cured with phthalic anhydride (HT 901) the stability of crack propagation was controlled principally by the rate of testing (i.e. the cross-head speed). They found that at high rates of testing cracks propagated in a stable continuous manner (Fig. 3(a)) and at low rates they moved by an unstable stick-slip mode (Fig. 3(b)). Yamini and Young[32] and Phillips *et al.*[26] extended this work and have shown that the effect of cross-head speed upon the stability of propagation in epoxy resins is quite general and can be found in DGEBA resins (Epikote 828) cured with a variety of primary, secondary, tertiary and mixed amines. Typical load-displacement curves have been given in

Fig. 3. During stick-slip propagation (Fig. 3(b)) there is a critical load at which the cracks propagate, P_i, and this leads to a critical stress intensity factor K_{Ici} or a critical strain energy release rate G_{Ici} for crack initiation. Similarly the value of load at crack arrest P_a corresponds to K_{Ica} or G_{Ica}. The magnitude of the jump is characterised by the size of the difference between K_{Ici} and K_{Ica} or between G_{Ici} and G_{Ica}. When the initiation and arrest values are equal propagation takes place by a continuous mode. It was found that although the absolute values of K_{Ic} or G_{Ic} were different for different curing agents when tested under similar conditions, in all cases the difference between the initiation and arrest values decreased as the cross-head speed was increased.[26,32]

As well as the cross-head speed controlling crack stability it has also been found recently that the temperature of testing has an important effect.[26,32,33] Figure 5 shows the variation of K_{Ici} and K_{Ica} with cross-

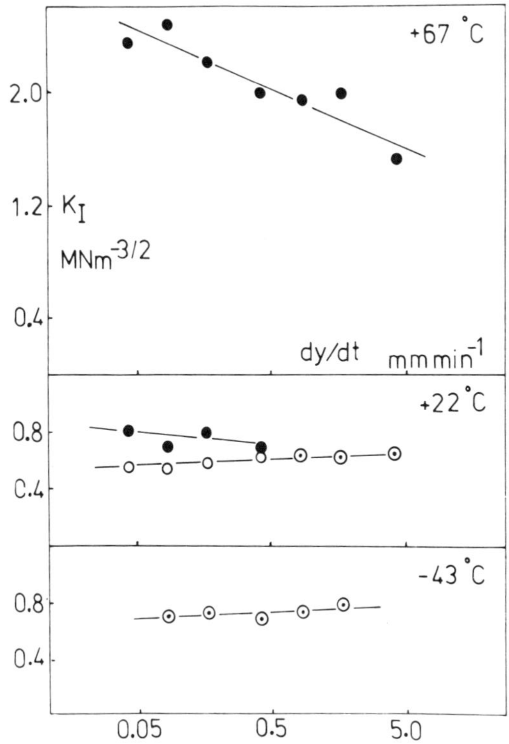

FIG. 5. Variation of K_{Ic} with cross-head speed (dy/dt) for a DGEBA epoxy resin cured with 9·8 phr TETA and tested at different temperatures. (After reference 34.) ● – K_{Ici}, ○ – K_{Ica}, ⊙ – K_{Ic} continuous propagation.

head speed for a DGEBA resin cured with 9·8 phr TETA and tested at different temperatures.[33] At $-43\,°C$ propagation is continuous (K_{Ici} is equal to K_{Ica}) and K_{Ic} increases slightly with increasing cross-head speed. At room temperature (22 °C) propagation is by a stick-slip mode at low cross-head speeds and is continuous at high cross-head speeds. At 67 °C propagation is entirely by a stick-slip mode and the jumps are so large that arrest does not occur within the length of the specimen. Figure 6

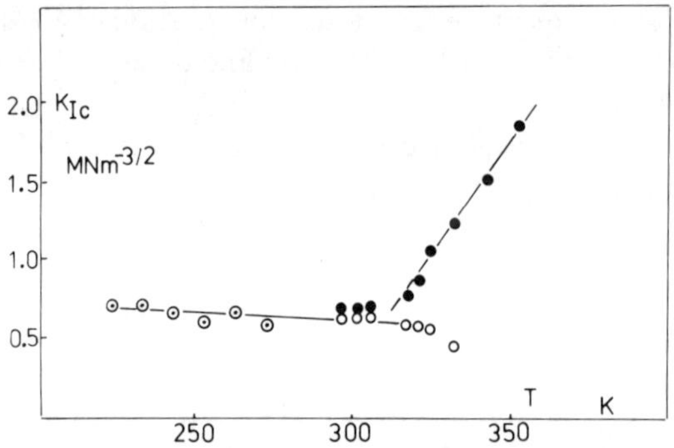

FIG. 6. Variation of K_{Ic} with temperature for a DGEBA epoxy resin cured with 9·8 phr TETA and tested at the same cross-head speed (0·5 mm min^{-1}). Symbols have the same meaning as in Fig. 5. $K = 273 + \theta°C$. (After reference 32.)

shows the variation of K_{Ici} and K_{Ica} with temperature over a large range of temperature, at a fixed cross-head speed.[32] Propagation is entirely stable below 10 °C but jumping occurs at higher temperatures, until very high values of K_{Ici} are obtained as the glass-transition temperature (T_g) of the material ($\approx 90\,°C$) is neared. It appears that jumping may be promoted by the activation of visco-elastic loss processes as T_g is approached.

Another factor which can cause jumping to take place is the presence of water in the vicinity of the crack tip. Investigations by the author and co-workers[27,32] have shown that in epoxy resins cured with both anhydride and amine hardeners, water can promote stick-slip propagation under testing conditions which would cause continuous propagation in laboratory air. Figure 7 shows the variation of K_{Ici} and K_{Ica} with cross-head speed for a DGEBA resin (Epikote 828) cured with 9·8 phr TETA and tested in both air and water. It can be seen that there is a transition between stable and unstable propagation in air which is not present when the test is done in water where propagation is entirely unstable. On the

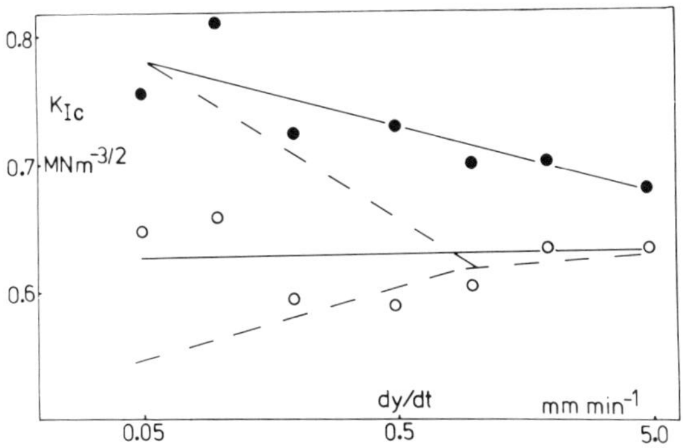

FIG. 7. Variation of K_{Ic} with cross-head speed for a DGEBA epoxy resin tested at room temperature in distilled water. The symbols have the same meaning as in Fig. 5. The dashed lines refer to propagation in laboratory air. (After reference 32.)

other hand it has been found that under vacuum crack propagation can still take place by the stick-slip mechanism.[32]

It is also known that the amount and type of curing agent and the state of cure can control the stability of propagation in epoxy resins. Yamini and Young[34] have recently looked in detail at the effect of changing these variables upon crack stability in the DGEBA/TETA system. Initially they took a constant curing temperature (100 °C) and time (3 h) and showed that at a constant cross-head speed the difference between K_{Ici} and K_{Ica} increased as the amount of hardener was increased. Also the tendency for jumping in resins with a given amount of curing agent decreased as the cross-head speed was increased. This behaviour is illustrated in Fig. 8. Yamini and Young also varied the curing temperature and time for a fixed amount of hardener. It was found that unstable propagation was promoted by using higher curing temperatures and longer curing periods. They concluded that stick-slip behaviour at ambient temperatures was more likely in the more highly cured resins.

Another important factor that has recently been shown to affect the stability of crack propagation in brittle polymers that are prone to crack jumping is specimen geometry. Mai and Atkins[35,36] have pointed out that stability depends upon two factors. One varies with the type of test-piece and is called the geometric stability factor. The other, which depends upon the material, is $dG/d\dot{a}$, which is the rate of change of the fracture toughness with crack velocity and will be discussed in detail in the next section. Mai and Atkins have considered the effect of specimen

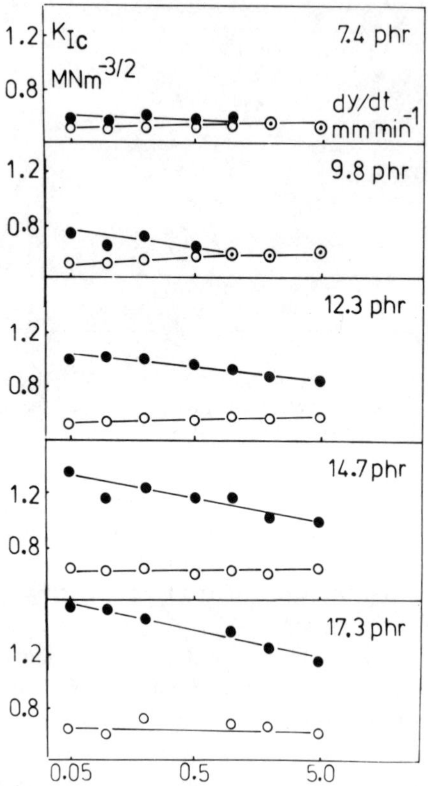

FIG. 8. Variation of K_{Ic} with cross-head speed for a DGEBA epoxy resin cured with different phr of TETA hardener and tested at room temperature in laboratory air. The symbols have the same meaning as in Fig. 5. (After reference 34.)

geometry in detail[35,36] and have mentioned some interesting observations. For instance, it can be shown that with the TDCB specimen stable crack propagation can be promoted by reversing the direction of crack growth.[36]

3.4 Crack Velocity

It is now well established that the value of the critical stress intensity factor K_{Ic} and the strain energy release rate G_{Ic} for crack propagation in brittle polymers depend strongly upon crack velocity. The polymer which has been studied most in this respect is poly(methylmethacrylate), for which the relation between K_{Ic} and velocity has now been very well characterised.[37–39] It is found that K_{Ic} increases with increasing velocity until about 0·1 m s^{-1} when it starts to fall because of the occurrence of an adiabatic–isothermal transition. Eventually K_{Ic} rises with further increase in velocity until the limiting propagation velocity is reached.

It is sometimes possible to obtain stable continuous propagation in thermosets and this enables the crack velocity to be readily measured. Brown and Ward[31] have shown that for a cross-linked polyurethane polymer there is a steady increase in K_{Ic} with increasing crack velocity between 10^{-8} m s^{-1} and 10^{-4} m s^{-1}, which is very similar to the behaviour of poly(methylmethacrylate).[37–39] Gledhill *et al.*[33] have shown for amine-cured epoxy resins that when stable propagation takes place there is also a slight increase in K_{Ic} with increasing crack velocity.

In thermosetting polymers propagation is often unstable with crack growth taking place by a stick-slip mode. In this case the crack velocity is by no means constant and is in fact very difficult to evaluate. The situation in a linear compliance-variation specimen can be shown by considering Fig. 3(b). As the specimen is loaded the crack is stationary until P_i is reached; here it starts to move and accelerates to an undefined speed. Eventually the crack decelerates until it becomes stationary again at P_a. The crack velocity, which varies during the jump, can only be evaluated with the use of sophisticated measuring equipment. Andrews and Stevenson[40] have recently used high-speed photography to measure the velocities of unstable cracks in a DGEBA epoxy resin (Epikote 828) cured with Epikure 114. The test-piece used involved the rapid pressurisation of an internal crack and was designed to assure plane strain conditions. They found that the fracture energy, $2\mathscr{T}$, which is equivalent to G_{Ic} for this material, increased with increasing crack velocity up to a value of 400 J m^{-2} at 37 m s^{-1}. The fracture energy was found to become smaller at higher velocities, a behaviour not unlike the reduction in K_{Ic} at the adiabatic–isothermal transition in poly(methylmethacrylate).[62]

Mai and Atkins[35] have suggested some ingenious methods of promoting stable crack growth in materials such as epoxy resins. They showed that stable cracking could be achieved by putting adhesive tapes across the crack paths in beam-like specimens, the tapes acting as effective elastic springs. Their results are also shown in Fig. 9 along with the data of Andrews and Stevenson.[40] Although the two sets of data are for different systems and so not strictly comparable, they do show the rather complex relationships between G_{Ic} and crack velocity that can occur in epoxy resins. Mai and Atkins[35] have pointed out that continuous propagation will take place when $dG_I/d\dot{a}$, where $\dot{a}$ is the crack velocity, is

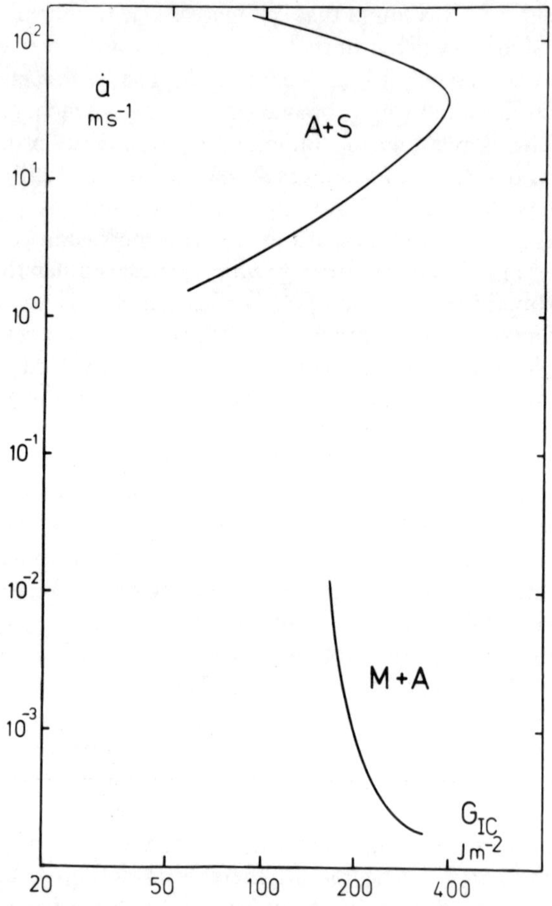

FIG. 9. A log–log plot of crack velocity $\dot{a}$ against G_{Ic} for different DGEBA epoxy resins. A + S—data of Andrews and Stevenson.[40] M + A—data of Mai and Atkins.[35]

positive. When it is negative propagation will become unstable even in specimens which should promote stable crack growth. Figure 9 shows that the value and sign of $dG_I/d\dot{a}$ can vary significantly with crack velocity and indicates how the stability of propagation may vary for these materials.

Another type of specimen to determine the relationship between G_{Ic} and $\dot{a}$ when $dG_I/d\dot{a}$ is negative has been proposed by Mai and Atkins.[35] Under such circumstances an epoxy resin would exhibit stick-slip propagation, but when it is incorporated in a sandwich with poly(methylmethacrylate), which exhibits stable propagation, cracks will grow in a stable manner through the two layers. G_I for the sandwich is given by

$$G_I = \frac{G_{Ip} t_p + G_{Ie} t_e}{t_p + t_e} \tag{8}$$

where G_{Ip} and G_{Ie} are the values of G_I for the poly(methylmethacrylate) and epoxy resin respectively and t_p and t_e are the thickness of the respective layers of material.

A very interesting method of determining the relationship between K_{Ic} and $\dot{a}$ during stick-slip propagation in linear compliance-variation specimens (such as DT or TDCB) has been suggested by Phillips and Scott.[42] It involves measuring the slope of the load–time curve, shown in Fig. 3(b), at various times during the crack jumps. The basic and probably most questionable assumption of the analysis is that by increasing the chart speed of a testing machine such as an Instron it is possible to get an accurate idea of the variation of load with time for the sample. The best response time of conventional chart recorders is of the order of 0·5 s and it is doubtful whether accurate pen movements can be followed with ordinary equipment. However, assuming that this experimental problem can be overcome, the method provides a potentially very useful way of measuring the relationship between K_{Ic} and $\dot{a}$ during unstable propagation. Figure 10 shows a magnified part of the load–time curve in Fig. 3(b) which can be obtained during stick–slip propagation in a linear compliance-variation specimen. For this type of specimen the deflection, y, at the loading point is given by

$$y = P(A_1 a + A_2) \tag{9}$$

where A_1 and A_2 are constants and may be determined from the slope and intercept, respectively, of the compliance calibration curve.[42] Differentiating eqn. (9) with respect to time, t, and rearranging gives

$$\dot{a} = \frac{\mathrm{d}a}{\mathrm{d}t} = \frac{1}{A_1 P} \cdot \left(\frac{\mathrm{d}y}{\mathrm{d}t} - A_1 a \frac{\mathrm{d}P}{\mathrm{d}t} - A_2 \frac{\mathrm{d}P}{\mathrm{d}t} \right) \tag{10}$$

For a hard screw-driven testing machine $\mathrm{d}y/\mathrm{d}t$ is equal to the cross-head speed and is constant. The quantity $\mathrm{d}P/\mathrm{d}t$ can be obtained from the slope of the load-time curve (Fig. 10). When the specimen is first loaded the crack remains stationary until the peak load is reached. At this point it starts to propagate. From Fig. 10 it can be seen that $\mathrm{d}P/\mathrm{d}t$ is negative at point A and becomes increasingly less negative with time until point B is

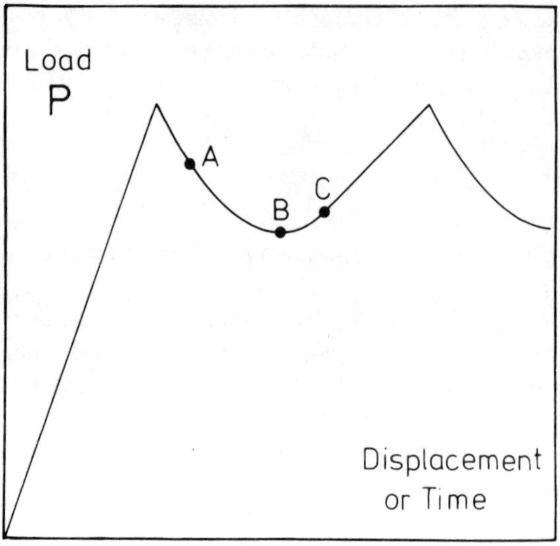

FIG. 10. Schematic load–time (or load–displacement) curves for a specimen undergoing stick–slip propagation in a linear compliance-variation specimen. The curve is similar to Fig. 3(b) but is magnified by increasing the chart speed.

reached, where dP/dt is zero. From eqn. (10) it can be seen that when dP/dt is large and negative the crack velocity is high. However the reduction in the size of dP/dt from A to B shows that the crack is slowing down. At B, where dP/dt is equal to zero, it follows from eqn. (10) that

$$\dot{a} = \frac{1}{A_1 P} \cdot \frac{dy}{dt} \tag{11}$$

This means that at the minimum load, P_a (Fig. 3(b)), the crack has not arrested and is in fact moving at a velocity given by eqn. (11). At first sight this would seem to be a paradox, as P_a is conventionally thought to be the crack arrest load. Hakeem and Phillips[43] have recently solved this problem by looking at the condition in eqn. (10) for $\dot{a}$ to be equal to zero. When $\dot{a}$ is zero then from eqn. (10) it can be seen that

$$\frac{dy}{dt} = \frac{dP}{dt} \cdot (A_1 a + A_2) \tag{12}$$

Using eqn. (9) it follows that when the crack has arrested

$$\frac{dy}{dt} = \frac{dP}{dt} \frac{y}{P} \tag{13}$$

During loading $\mathrm{d}y/\mathrm{d}t$ is always constant (constant cross-head speed) and when the crack has arrested y/P is constant, from eqn. (9). So the condition for the crack to arrest is

$$\frac{\mathrm{d}P}{\mathrm{d}t} = \text{constant} \tag{14}$$

which means that the slope of the loading curve must be constant, a condition which is obeyed at point C in Fig. 10. The implication is that arrest does not occur at the minimum load but at some higher load when the load-time curve becomes a straight line again.

There are also consequences of this observation concerned with the relationship between K_{Ic} and crack velocity $\dot{a}$. In any linear compliance variation specimen (such as DT) K_{I} is proportional to the applied load P. Since the load at crack arrest is higher than the minimum it implies that if K_{Ic} is plotted against $\dot{a}$, the slope of the curve must at some point become negative. Figure 11 shows a schematic log–log plot showing this situation for $\dot{a}$ plotted against K_{Ic} (*cf.* Fig. 9). In this case $\mathrm{d}K_{\mathrm{I}}/\mathrm{d}\dot{a}$ (or more accurately $\mathrm{d}\dot{a}/\mathrm{d}K_{\mathrm{I}}$) is positive above some transition velocity, $\dot{a}_{\mathrm{T}}$, whereas below this velocity it is negative. A similar situation is found in Fig. 9 for

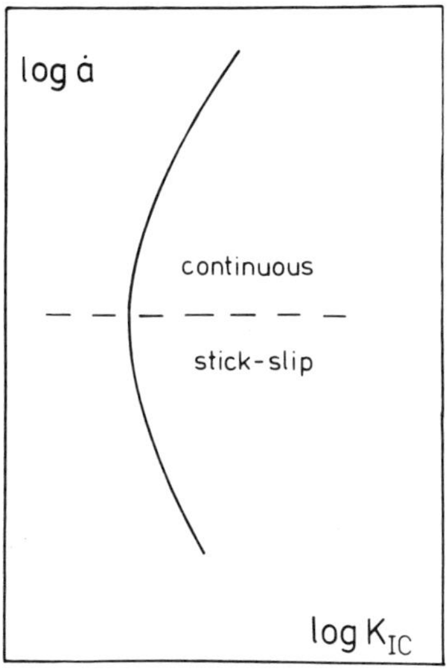

FIG. 11. Schematic log–log plot of crack velocity $\dot{a}$ against K_{Ic} for a material which undergoes a transition between stick–slip and continuous propagation at a critical velocity ($\dot{a}_{\mathrm{T}}$) indicated by the dashed line.

the relationship between G_{Ic} and $\dot{a}$. However, the negative slope in Fig. 11 comes directly from the saw-tooth load–displacement (or load–time) curve, which is caused by the crack jumping behaviour. Earlier investigators have suggested that a negative value of $dK_I/d\dot{a}$ or $dG_I/d\dot{a}$ will cause stick-slip behaviour[35] and it can be seen that this negative value is also a consequence of that behaviour. However although the interpretation of the stick-slip propagation described above explains the phenomenon well it gives no idea of the mechanisms which cause it to occur. These are discussed later (Section 4.2).

A completely accurate idea of the relationship between K_{Ic} and $\dot{a}$ during unstable propagation can only be obtained by direct measurement of K_{Ic} and $\dot{a}$ during each jump. Using sophisticated measuring techniques Kalthoff and co-workers[41] have determined K_{Ic} and $\dot{a}$ directly for unstable propagation in wedge-loaded DCB specimens of epoxy resins. They used notches with different root radii to achieve different initial crack velocities and were able to measure a dynamic stress intensity factor K_{Ic}^{dyn} during propagation. They found that the velocity of the propagating crack remained constant for a considerable period but decreased to smaller values before arrest. They also showed that at the beginning of crack propagation K_{Ic}^{dyn} was smaller than the corresponding static value but at the end of propagation it became larger. It was only after the crack had arrested that K_{Ic}^{dyn} approached the static value and even then it tended to oscillate around the static value because of specimen vibrations.[41] In general, dynamic effects were found to be more pronounced with increasing maximum crack velocity.

3.5 Fatigue

Fatigue failure can take two different forms. The first is failure during a long period of constant loading and is known as *static fatigue*; the second is failure after a period of variable loading and is known as *dynamic fatigue*. In both cases the failure loads or stresses are considerably lower than those registered during normal tensile deformation. Both types of fatigue can be aggravated by environmental attack. There has only been a small amount of work done with this type of failure in thermosetting polymers but it is known that they can suffer to a certain extent from both static and dynamic fatigue failure.

Static fatigue has been studied mostly with respect to the durability of adhesive joints bonded with thermosetting polymers such as epoxy resins.[44,45] Attempts to obtain long-term static fatigue in bulk epoxy resins have had rather mixed success. Brittle polymers such as poly-

(methylmethacrylate) which normally fail by continuous propagation do suffer from static fatigue.[46] However, Phillips *et al.*[26] found that pre-cracked epoxy resin DT samples which were loaded statically to values of K_{Ic} close to those which caused cracks to propagate during dynamic loading did not fail even after loading for periods in excess of 4 months. The explanation of this behaviour probably lies in the tendency of epoxy resins and thermosetting polymers in general to undergo unstable, stick-slip propagation. In Fig. 8, for example, it can be seen that K_{Ic} increases as the cross-head speed is reduced. Now, the static loading tests can be thought of as being equivalent to dynamic tests at very low cross-head speeds and so the value of K_{Ic} for failure by static fatigue is in fact *higher* than that for dynamic crack propagation. This means that epoxy resins have the remarkable ability to become tougher with time during static loading. This point was emphasised by Phillips *et al.*,[26] who performed normal constant cross-head speed tests on specimens which had not failed after several months static loading and found that the value of K_{Ic} for propagation was approximately 20% higher than for similar samples which had not been statically loaded. However, they showed also that the material had only been affected in the vicinity of the original crack as further propagation along the specimen took place at values of K_{Ic} similar to those for normal propagation.

The time-to-failure during static loading, t_f, is in general controlled by two factors through the equation[46]

$$t_f = t_i + t_p \tag{15}$$

where t_i is the time to initiate cracks and t_p is the time to propagate them. For poly(methylmethacrylate) t_f is controlled by t_p[46] whereas it has been shown above that in pre-cracked epoxy resins t_i tends to infinity. There is some indication from work on adhesive joints where failure takes place through the centre of an epoxy resin adhesive layer that it is the time to initiate cracks, t_i, which controls failure in this case.[45] Several years ago Kwei[47] showed that epoxy resin samples which had not been pre-cracked could show static fatigue at loads below the short-term fracture load. He also showed that the failure times were greatly reduced by performing the tests at 100% relative humidity and at elevated temperatures. It appears, comparing the work of Kwei with more recent research, that he was dealing with static fatigue which was controlled by the time to initiate cracks, t_i.

In general it is found that thermosetting polymers are rather resistant to dynamic fatigue failure during cyclically varying loading. For example, although Hertzberg, Nordberg and Manson[48] were able to measure the fatigue behaviour of many different thermoplastics they had difficulty in obtaining fatigue crack propagation in epoxy resins. More recently Sutton[49] has shown that it is possible to measure fatigue crack propagation in an epoxy resin and obtained a relation between the fatigue crack growth rate, $\mathrm{d}a/\mathrm{d}N$, and the range of the stress intensity factor, ΔK_I, during each cycle of the form

$$\frac{\mathrm{d}a}{\mathrm{d}N} = A(\Delta K_\mathrm{I})^m \tag{16}$$

where N is the number of cycles and A and m are experimentally determined constants. This type of relation has been obtained for many metals and polymers and m is typically between 2 and 7 for such materials. Sutton[49] showed that for his epoxy resin m was of the order of 10 which is extraordinarily high and probably explains why Hertzberg *et al.*[48] had problems obtaining fatigue crack growth in their resins. It appears that high values of m are generally found when fatigue crack propagation is obtained in thermosetting resins. Harris and Ward[50] obtained values of m between 4 and 10 for cross-linked vinyl–urethane polymers and Owen and Rose[28] have measured m values of the order of 5 for polyester resins. High values of m are desirable as it means that the fatigue crack growth rate will drop rapidly as ΔK_I is reduced. The high values of m for thermosetting resins shows that they are in general relatively resistant to failure during dynamic fatigue loading.

4 FAILURE MECHANISMS

4.1 Fractography

The fracture surfaces of thermosetting polymers are relatively featureless when compared with the surfaces of other brittle polymers such as polystyrene.[51] The detailed nature of the fracture surface depends upon specimen geometry. In unstable test-pieces, such as SEN or three-point bend types, the fracture surface can show a variety of features because the crack velocity changes greatly across the surface. Using stable test-pieces, such as DT or TDCB types, if the material itself exhibits stable propagation then the appearance of the fracture surface will depend upon the particular crack velocity obtained and should be uniform in appearance. However, as shown in Section 3, even in stable test-pieces propagation in thermosets often takes place by a stick-slip mechanism and this can lead to a variety of different features on the fracture surface.

There has been a considerable amount of work investigating the appearance of fracture surfaces of thermosetting resins using unstable test-pieces and many interesting features have been observed.[10,12,30,52] They have been reviewed by Pritchard and Rhoades[12] and only recent developments will be considered in detail here. The main features that are observed in these specimens in the optical microscope are an initiation region (especially when the specimens are not pre-cracked) and a region of slow growth followed by an area of rapid growth over the remainder of the surface. The direction of crack growth can usually be determined from river marking in the slow growth area, radiating from the initiation region. In most thermosets the region of rapid growth is relatively smooth and featureless.[12] Recently Morgan and O'Neal[53–55] and Morgan, *et al.*[56] have looked at the effect of different structural variables upon the fracture topography of epoxy resins using scanning electron microscopy as their principal investigative technique. They have shown that several different features can be observed and have suggested that some are due to crazing.[53] They have also shown that irregularities in the fracture surfaces can be caused by the original presence of DGEBA epoxy monomer crystals.[55] Higher-magnification micrographs can be obtained by taking replicas of fracture surfaces and examining them in the transmission electron microscope. Fracture surface replicas of both brittle thermoplastics and thermosets show a granular appearance at the 100-Å level.[53,57] This type of surface is shown in Fig. 12 and has been interpreted as being due to a modular structure in epoxy resins, but no consistent relation has been found between composition and cure and the module size or density.[57] The granular appearance can be enhanced by etching the fracture surfaces,[53] as shown in Fig. 12. Morgan and O'Neal have found recently that a domain structure on a similar level (50–100 Å) can be seen in thin films viewed in the transmission electron microscope.[53]

With the use of stable test-pieces it is possible to relate features on fracture surfaces to particular types of crack propagation behaviour (Section 3.3). When failure takes place by continuous propagation the fracture surfaces are relatively smooth, whereas when propagation takes place by a stick-slip mode then crack arrest lines are seen on the fracture surface, each one corresponding to a jump-arrest event. This behaviour was seen by Young and Beaumont[27] in an anhydride-cured DGEBA epoxy resin where DT specimens were tested at different cross-head

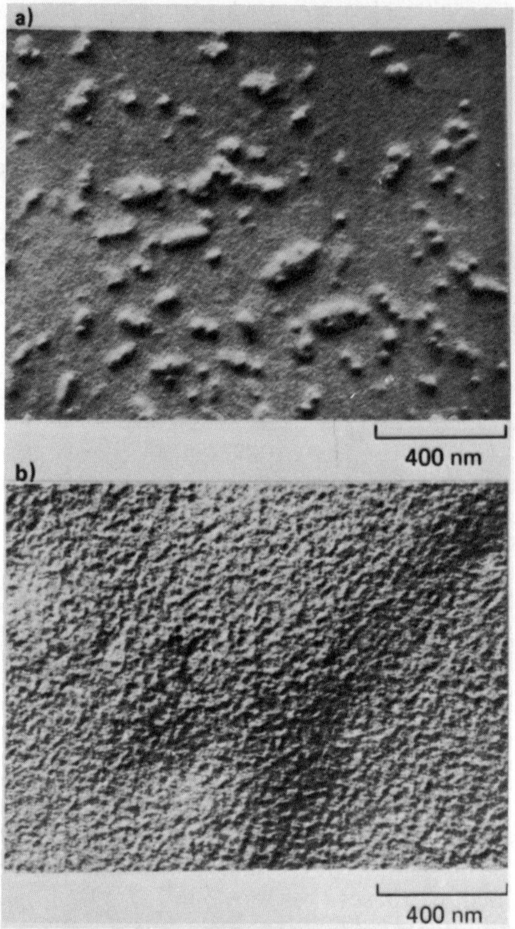

FIG. 12. Transmission electron micrographs of replicas of fracture surfaces of an epoxy resin showing the granular appearance of the surfaces. (After reference 53.) (a) DGEBA/11 phr DETA epoxy surface; (b) DGEBA/11 phr DETA epoxy surface after leaching with acetone for 3 h. (Micrographs courtesy of Dr R. J. Morgan.)

speeds (Fig. 13). Selby and Miller[23] and Phillips *et al.*[26] have also examined the fracture surfaces of a DGEBA epoxy resin (Epikote 828) cured with several different amine hardeners. When propagation takes place by stick-slip propagation they also see crack arrest lines on the fracture surface. In general these lines are not as fine as those seen in

(a)

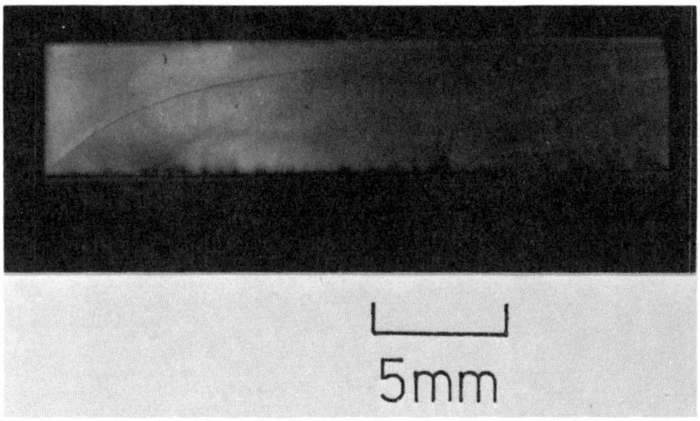

(b)

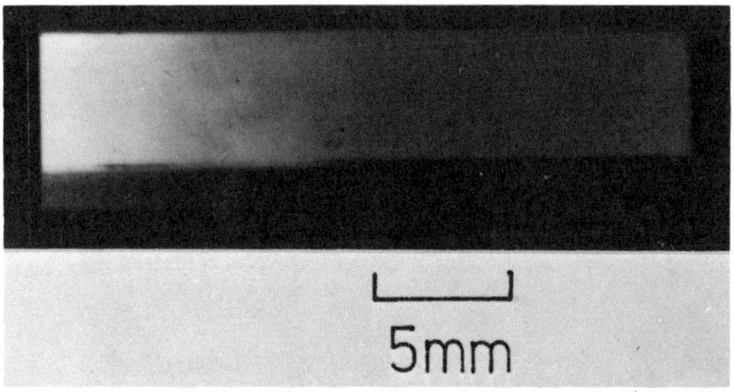

FIG. 13. Optical micrographs of fracture surfaces of a DGEBA epoxy resin (CT 200) cured with phthalic anhydride and tested at room temperature at different cross-head speeds. (After reference 27.) (a) 0·02 mm min^{-1} showing crack arrest lines; (b) 2 mm min^{-1} after continuous propagation.

Figure 13(a), but can be quite broad (up to 1 mm) depending upon the curing and testing conditions. Recently Yamini and Young[34] have looked in detail at the effect of both curing and testing variables upon the fractography of a DGEBA epoxy resin (Epikote 828) which had been cured by the amine TETA. They found that there was a general correlation between the width of crack arrest lines and the size of the jumps as given by the difference between K_{Ici} and K_{Ica}. Factors which lead to large jumps, such as high testing temperatures and a large amount of curing agent or high curing temperatures, tended to lead to broad arrest lines. Figure 14 shows a set of fracture surfaces for the same resin tested at three different

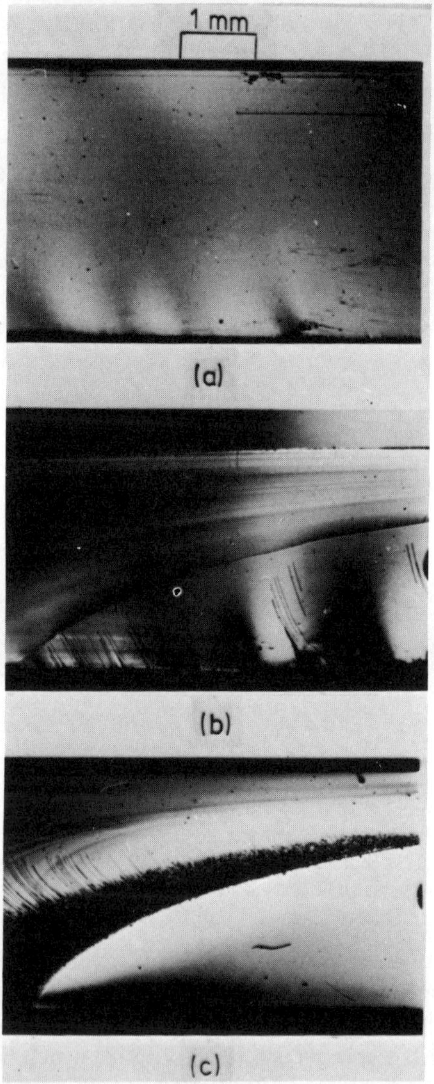

FIG. 14. Optical micrographs of fracture surfaces of a DGEBA epoxy resin (Epikote 828) cured with 9·8 phr TETA and tested at the same cross-head speed (0·5 mm min^{-1}) at different temperatures. (After references 34.) (a) -40 °C, (b) $+22$ °C, (c) $+60$ °C.

temperatures. At around −40 °C propagation is by a continuous stable mode and the surface is smooth. At room temperature fine arrest lines are seen on the surface and at 60 °C the crack arrest line is very broad. Values of K_{Ici} and K_{Ica} for the three samples can be obtained from Fig. 5 and the difference between the two values increases greatly as the temperature is increased (Fig. 6).

The exact nature of the crack arrest lines is of considerable interest. Figure 15 shows a higher magnification view of the lines on each half of a broken DT specimen. Both lines appear to be ridges sticking out of the surface. It is very difficult to tell from visual observation the exact nature of the crack arrest lines and Yamini and Young[34] have recently attempted to determine their form by taking Talysurf traces from the fracture surfaces of both halves of a specimen. They were able to show that the surfaces tended to interlock on a level of above 10 μm leading to the

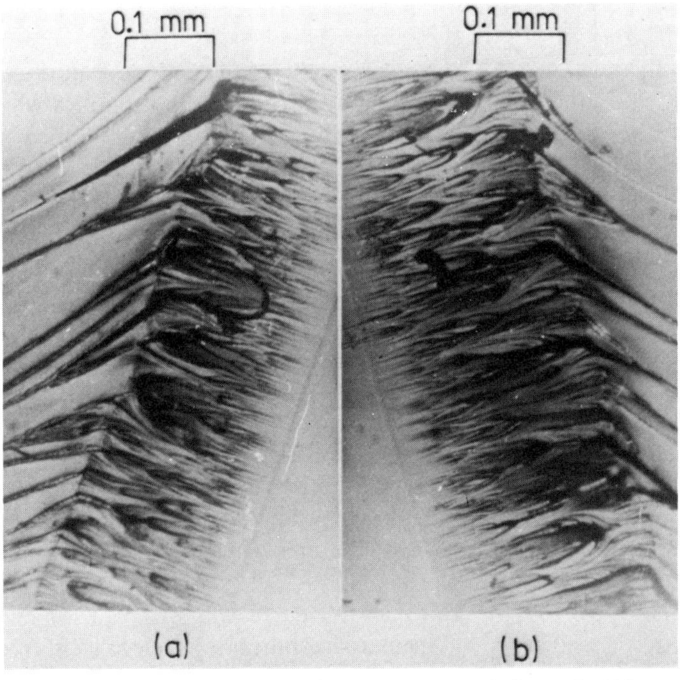

FIG. 15. Optical micrographs of similar areas on two halves of a DT specimen (Epikote 828 cured with 14.7 phr TETA) tested at room temperature. The features correspond to crack arrest lines viewed at a higher magnification than in the previous two figures. (After reference 34.)

shape of the crack arrest lines illustrated schematically in Fig. 16. A similar form was suggested earlier by Phillips *et al.*[26] but they only showed one surface. It appears therefore that during and after the arrest event in epoxy resins there is considerable deviation of the crack from the main fracture plane. Any plastic deformation takes place on a finer scale (below 10 μm).

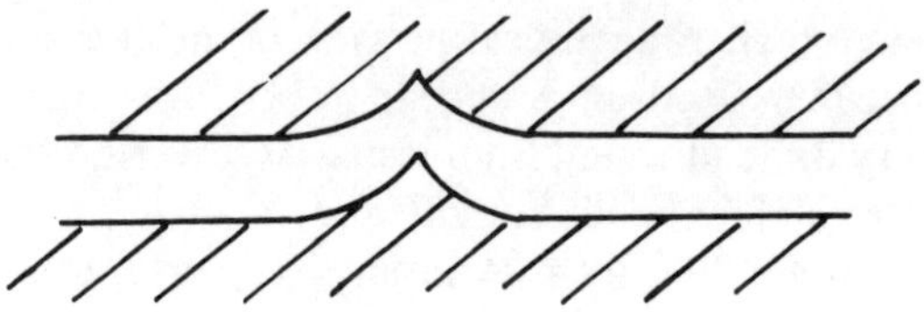

FIG. 16. Schematic section perpendicular to the fracture plane of the two halves of the fractured specimen in Fig. 15 showing the tendency for the features on the fracture surface to interlock. (After reference 34.)

4.2 Crazing

It is well known that glassy polymers will readily undergo deformation by crazing.[58] This is a process whereby small crack-like entities which are normal to the maximum principal tensile stress are formed during deformation.[59] However, crazes differ from cracks in that they are internally bridged by fibrils. Although crazing is found widely in thermoplastics there is little evidence for it in most thermosetting resins. In a vinyl–urethane polymer cross-linked by different amounts of styrene Brown and Ward[31] showed that craze zones were almost certainly formed. They were able to measure a craze stress and showed that it decreased monotonically with increasing styrene content.

There is little direct evidence for crazing in epoxy resins. Lilley and Holloway[60] have reported seeing small crazes in the vicinity of crack tips in some epoxy resin systems. Van den Boogaart[61] has observed crazes at the tips of moving cracks in under-cured epoxy resins and reported that these crazes were identical to those in poly(methylmethacrylate), although no micrographs were shown of the epoxy resin crazes. In fully cured resins the crazes were not present.[61]

Recent work by Morgan and O'Neal,[53] who strained thin films of epoxy resins and examined them in the transmission electron microscope, has shown that craze-like entities similar to those shown in Fig. 17 can be formed. However, they are relatively short ($\sim$ 1 μm) and although they appear to be bridged by material they are by no means as well defined as crazes in thermoplastics, which can be many micrometres long.[58]

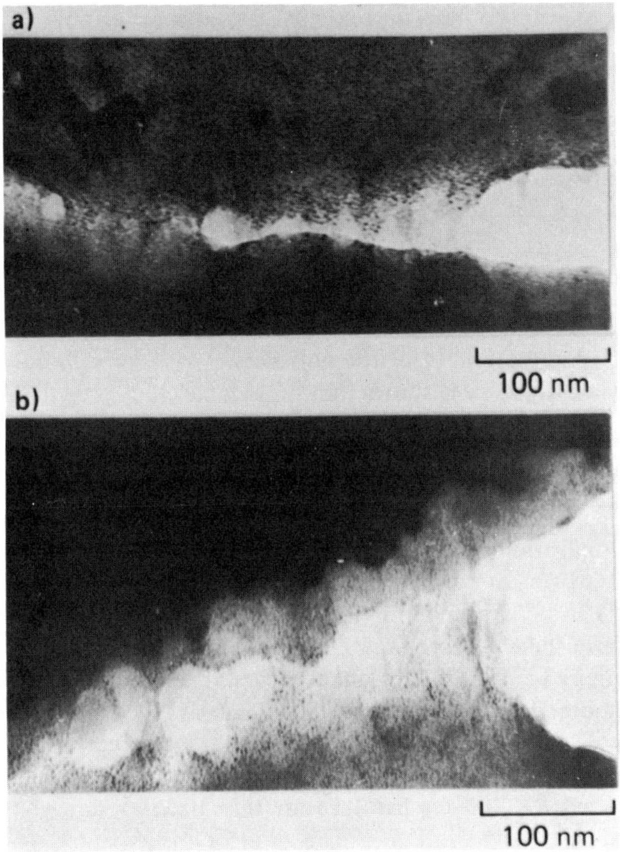

FIG. 17. Transmission electron micrographs of a strained thin film of an epoxy resin showing possible crazes at crack tips. Both are micrographs of a DGEBA epoxy resin cured with 11 phr DETA. (After reference 53; micrographs courtesy of Dr R. J. Morgan.)

One-μm crazes would not be observable at crack tips by optical techniques and this probably explains why they are not normally observed by this method.

The presence of surface colours on the fracture surfaces of brittle polymers has been taken as evidence of plastic deformation taking place during the fracture process.[10–12] In thermoplastics this has been interpreted as giving an indication that crazing has taken place.[58] The coloration is thought to be due to layers of material on the fracture surfaces possessing a different refractive index from the bulk. The surface colours are particularly noticeable on the fracture surfaces of poly-(methylmethacrylate) but the amount of coloration is considerably reduced when the material is cross-linked.[11] Nelson and Turner[10] found surface coloration during their study of the fracture of phenol–formaldehyde resins. However, when the testing temperature was raised the coloration decreased until at 175 °C it became completely absent.

4.3 Failure Criteria

In recent years there have been important developments in the determination of failure criteria in brittle polymers.[31,33,37,46,62] In poly(methylmethacrylate) it has been shown that slow stable crack growth takes place in accordance with a constant crack-opening displacement (δ) criterion.[37,46,62] It has further been shown that in this material propagation takes place though a region of crazed polymer at the tip of the crack which can be accurately modelled by a Dugdale line plastic zone.[37,62] Stable propagation is also found with cross-linked vinyl–urethane polymers and Brown and Ward[31] have shown that crack propagation in these materials can be interpreted in terms of the Dugdale plastic zone model.

In general the crack propagation behaviour of thermosetting resins is rather complex and jumping often occurs. The failure criterion in this case is more difficult to determine. Recently Gledhill *et al.*[33] have determined the crack opening displacement δ for an epoxy resin system (DGEBA/9·8 phr TETA) tested at different temperatures. The relationship between K_{Ic} and temperature for this material at a given rate of testing has been given in Fig. 6. The value of δ is given by[46]

$$\delta = \frac{K_{Ic}^2}{\sigma_y E} \tag{17}$$

where σ_y is the yield stress of the material and E is the Young's modulus, both of which were determined at the appropriate temperatures by a separate experiment. The variation of δ with temperature is given in Fig. 18. Below about 10 °C, δ is approximately constant at about 1 μm. This means that continuous stable propagation appears to takes place at a constant value of δ. Above 10 °C propagation becomes unstable and the jumps become more pronounced as the glass-transition temperature is approached. Because of the rapid rise in K_{Ici} the value of δ increases and it shows that above 10 °C a constant-δ criterion is certainly not applicable. Gledhill *et al.*[33] also showed that δ was constant during continuous

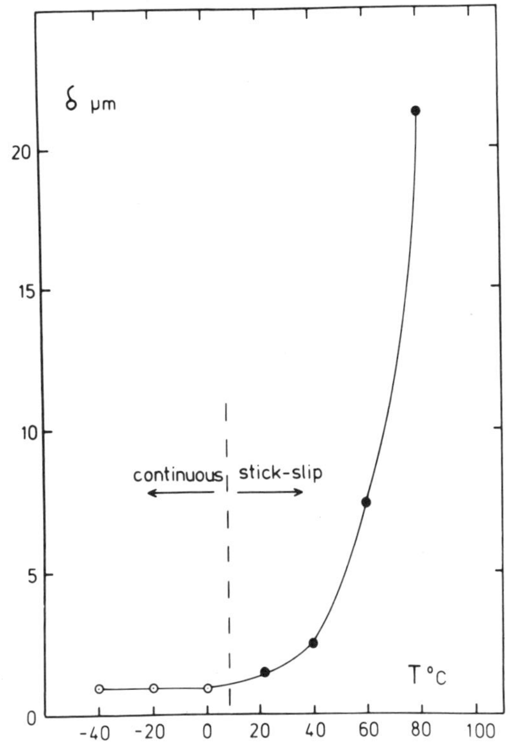

FIG. 18. Variation of crack opening displacement δ with temperature for a DGEBA epoxy resin cured with 9·8 phr TETA. (After reference 33.)

propagation at high rates of testing at room temperature whereas it was not constant during stick-slip propagation at the same temperature.

Although the crack opening displacement approach gives some insight into the phenomenon of crack propagation it has, with some justification, been criticised as being difficult to realise physically. It appears that a constant-δ criterion is appropriate during continuous propagation in thermosets and using the analogy with thermoplastics[46] it would seem likely that this may be consistent with propagation through a single craze at the crack tip. As explained in Section 4.2 Van der Boogaart[61] observed a craze at the tip of a stable crack in an epoxy resin. However, the evidence for crack propagation through crazes in expoxy resins is very tenuous and more work is required before this suggestion can be confirmed.

An important point which still has not been explained is why stick-slip propagation occurs at all and why it is so rate- and temperature-sensitive when other mechanical properties such as modulus and yield stress are not so sensitive to these variables. Consideration of the relation between K_{Ic} or G_{Ic} and crack velocity $\dot{a}$ gives some help in this direction. The onset of unstable behaviour would seem to correspond to a negative value of $dK_{Ic}/d\dot{a}$ or $dG_{Ic}/d\dot{a}$ (Section 3.4). However, although this gives some insight into the phenomenon it does not explain the mechanisms at all. Andrews[63] has noted that both stick-slip and continuous crack propagation can take place in strain-crystallising and filled elastomers at different temperatures and under different rates of testing. He explained this behaviour as being due to a 'stationary stress distribution' at the crack tip caused by the material having high mechanical hysteresis.[64] He also showed that this could lead to considerable deviation of the crack front from the main fracture plane during stick-slip propagation, as is also found in epoxy resins.[34] It appears therefore that the mechanical hysteresis of thermosets may well be worth investigation and could shed some light upon the problems of crack stability.

5 CONCLUSIONS AND FUTURE DEVELOPMENTS

It is clear that over the past decade there have been important developments in the understanding of the fracture of thermosetting resins. The most important feature that all thermosetting polymers have in common is that they are generally brittle. There has been some confusion concerning the main features of fracture because of the wide variety of thermosetting polymers that have been used and, with epoxy resins in particular, the large number of hardeners that have been used. It is only recently that any systematic studies have taken place of the effect of different curing agents and curing conditions upon fracture behaviour.

The most widely studied aspects of fracture have been concerned with crack propagation. In stable test-pieces it is found that cracks generally propagate in an unstable stick-slip manner although under some conditions continuous, stable propagation can be found. Although crack propagation is well-characterised, little is known about the mechanisms which control it and this would appear to be an area where future developments will take place. It is only recently that there have been any attempts to study the local deformation that undoubtedly takes place at the crack tips. As far as is known, crazing appears to be very much more difficult in thermosetting polymers than in thermoplastics and any crazes that may be present are relatively small. There have also been few attempts to relate fracture behaviour to other mechanical properties or loss processes which may affect it. In conclusion, therefore, it appears that there is still much to be done before the fracture of thermsetting resins can be completely understood.

REFERENCES

1. MAY, C. and TANAKA, Y. (eds.). *Epoxy Resins and their Technology*, Marcel Dekker, New York, 1974.
2. HARRIS, B. and PHILLIPS, D. C., in *Polymer Engineering Composites*, ed. M. O. Richardson, Applied Science Publishers, London, 1977.
3. BUCKNALL, C. B. *Toughened Plastics*, Applied Science Publishers, London, 1977.
4. MULVILLE, D. R. and WOLOCK, I. This volume, Chapter 8.
5. KELLY, A. *Strong Solids*, Clarendon Press, Oxford, 1966.
6. GRIFFITH, A. A. *Phil. Trans. Roy. Soc.*, **A221** (1920) p. 163.
7. BERRY, J. P. *J. Polymer Sci.*, **50** (1961) p. 107.
8. BERRY, J. P. *J. Polymer Sci.*, **50** (1961) p. 313.
9. BERRY, J. P., in *Fracture VII*, ed. H. Liebowitz, Academic Press, New York, 1972.
10. NELSON, B. E. and TURNER, D. T. *J. Polymer Sci., Poly. Phys. Ed.*, **11** (1973) p. 1949.
11. BROUTMAN, L. J. and MCGARRY, F. J. *J. Appl. Polymer Sci.*, **9** (1965) p. 609.
12. PRITCHARD, G. and RHOADES, G. V. *Mater. Sci. Engng.*, **26** (1976) p. 1.
13. KNOTT, J. F. *Fundamentals of Fracture Mechanics*, Butterworths, London, 1973.
14. ANDREWS, E. H. *J. Mat. Sci.* **9** (1974) p. 887.
15. ANDREWS, E. H. This volume, Chapter 1.
16. ROOKE, D. P. and CARTWRIGHT, D. J. *Compendium of Stress Intensity Factors* HMSO, London, 1976.
17. BROWN, H. F. and SRAWLEY, J. E. ASTM Spec. Tech. Publ. (1966) p. 410.
18. YOUNG, R. J. and BEAUMONT, P. W. R. *J. Mat. Sci.*, **12** (1977) p. 684.
19. BASCOM, W. D., COTTINGTON, R. L., JONES, R. L. and PEYSER, P. *J. Appl. Polymer Sci.*, **19** (1975) p. 2545.
20. MOSTOVOY, S. and RIPLING, E. J. *J. Appl. Polymer Sci.*, **10** (1966) p. 1351.
21. DIGGWA, A. D. S. *Polymer* **15** (1974) p. 101.
22. MEEKS, A. C. *Polymer* **15** (1974) p. 675.
23. SELBY, K. and MILLER, L. E. *J. Mat. Sci.*, **10** (1975) p. 12.
24. GRIFFITHS, R. and HOLLOWAY, D. G. *J. Mat. Sci.*, **5** (1970) p. 302.
25. EVANS, W. T. and BARR, B. I. G. *J. Strain Anal.*, **9** (1974) p. 166.
26. PHILLIPS, D. C., SCOTT, J. M. and JONES. M. *J. Mat. Sci.*, **13** (1978) p. 311.
27. YOUNG, R. J. and BEAUMONT, P. W. R. *J. Mat. Sci.*, **11** (1976) p. 766.
28. OWEN, M. J. and ROSE, R. G. *J. Phys. D. Appl. Phys.* **6** (1973) p. 42.
29. CHRISTIANSEN, A. and SHORTALL, J. B. *J. Mat. Sci.*, **11** (1976) p. 1113.

30. PRITCHARD, G., ROSE, R. G. and TANEJA, N. *J. Mat. Sci.*, **11** (1976) p. 718.
31. BROWN, H. R. and WARD, I. M. *J. Mat. Sci.* **8** (1973) p. 1365.
32. YAMINI, S. and YOUNG R. J. *Polymer* **18** (1977) p. 1075.
33. GLEDHILL, R. A., KINLOCH, A. J., YAMINI, S. and YOUNG, R. J. *Polymer*, **19** (1978) p. 574.
34. YAMINI, S. and YOUNG, R. J. *J. Mat. Sci.*, to be published.
35. MAI, Y. W. and ATKINS, A. G. *J. Mat. Sci.*, **10** (1975) p. 2003.
36. MAI, Y. W., ATKINS, A. G. and CADDELL, R. M. *Int. J. Fract.*, **11** (1975) p. 939.
37. MARSHALL, G. P., COUTTS, L. H. and WILLIAMS, J. G. *J. Mat. Sci.*, **9** (1974) p. 1409.
38. BEAUMONT, P. W. R. and YOUNG, R. J. *J. Mat. Sci.*, **10** (1975) p. 1334.
39. CHAN, C. S. and YOUNG, R. J. *Mater. Sci. Eng.*, **30** (1977) p. 181.
40. ANDREWS, E. H. and STEVENSON, A. *J. Mat. Sci.*, to be published.
41. KALTOFF, J. F., WINKLER, S. and BEINERT, J. *Int. J. Fract.* **12** (1976) p. 317.
42. PHILLIPS, D. C. and SCOTT, J. M. *J. Mat. Sci.*, **9** (1974) p. 1202.
43. HAKEEM, M. and PHILLIPS, M. G. Private communication.
44. MOSTOVOY, S. and RIPLING, E. J. *J. Appl. Polymer Sci.*, **13** (1969) p. 1083.
45. GLEDHILL, R. A. and KINLOCH, A. J. *Polymer* **17** (1976) p. 727.
46. YOUNG, R. J. and BEAUMONT, P. W. R. *Polymer* **17** (1976) p. 717.
47. KWEI, T. K. *J. Appl. Polymer Sci.*, **10** (1966) p. 1647.
48. HERTZBERG, R. W., NORDBERG, H. and MANSON, J. A. *J. Mat. Sci.*, **5** (1970) p. 521.
49. SUTTON, S. A. *Eng. Fract. Mech.*, **6** (1974) p. 587.
50, HARRIS, J. S. and WARD, I. M. *J. Mat. Sci.*, **8** (1973) p. 1655.
51. HULL, D. *J. Mat. Sci.*, **5** (1970) p. 357.
52. OWEN, M. J. and ROSE, R. G. *J. Mat. Sci.*, **10** (1975) p. 1711.
53. MORGAN, R. J. and O'NEAL, J. E. *J. Mat. Sci.*, **12** (1977) p. 1966.
54. MORGAN, R. J. and O'NEAL, J. E., in *Chemistry and Properties of Cross-linked Polymers*, Academic Press, New York, 1977.
55. MORGAN, R. J. and O'NEAL, J. E. *J. Macromol. Sci.—Phys.*, **B15** (1978) p. 139.
56. MORGAN, R. J., O'NEAL, J. E. and MILLER, D. E. *J. Mat. Sci.*, to be published.
57. RACICH, J. L. and KONTSKY, J. A. *J. Appl. Polymer Sci.*, **20** (1976) p. 2111.
58. KAMBOUR, R. P. *J. Poly. Sci., Macromol. Rev.*, **7** (1973) p. 1.
59. OXBOROUGH, R. J. and BOWDEN, P. B. *Phil. Mag.*, **28** (1973) p. 547.
60. LILLEY, J. and HOLLOWAY, D. G., *Phil. Mag.*, **28** (1973) p. 215.
61. VAN DEN BOOGAART, A., in *Physical Basis of Glassy Polymers*, ed. R. N. Haward, Inst. of Phys., London, 1966.
62. WILLIAMS, J. G. and MARSHALL, G. P. *Polymer*, **15** (1974) p. 251.
63. ANDREWS, E. H. *Fracture in Polymers*, Oliver and Boyd, London, 1968.
64. ANDREWS, E. H. *J. Mech. Phys. Sol.*, **11** (1963) p. 231.

Chapter 7

DEFORMATION AND FRACTURE OF POLYMER CRYSTALS

ROBERT J. YOUNG
Queen Mary College, University of London, UK

SUMMARY

Recent developments in the deformation and fracture of polymer crystals are reviewed. First of all, the structure and properties of conventional semicrystalline polymers are discussed. The mechanical properties of oriented semicrystalline polymers are then described briefly and the recent development of highly oriented ultra-high modulus polymers is covered in depth. Also the recent discoveries in the synthesis of macroscopic polymer crystals are reviewed. These crystals can be prepared both in the form of large lozenges and fibres with dimensions of the order of centimetres, and their mechanical properties have been described. The macroscopic crystals readily deform by chain axis rotation twinning. They also undergo intermolecular cleavage, forming cracks parallel to the chain direction but are very strong and stiff when deformed parallel to the chain axis. The potential that these materials and polymer crystals in general have as light strong materials is discussed in detail.

1 INTRODUCTION

Throughout history man has continually searched for materials possessing high stiffness and strength. In recent years much effort has been applied to obtaining molecular alignment and extension in crystalline polymers in order to exploit the high strength and stiffness of the covalently bonded polymer backbone. In a Royal Society of London discussion meeting eight years ago[1] Frank presented an excellent and stimulating review of the potential that polymer crystals have to offer as materials with high strength and stiffness. He pointed out that a polymer sample with all its molecules fully aligned in one direction could easily have the stiffness of steel in the direction of alignment. He also explained why in 1970, because of the problems associated with producing crystalline polymers having suitable morphologies, the potential of these crystals had not yet been realised. In the last decade there have been important developments in the preparation of crystalline polymers with specific morphologies which have enabled us to start to attain the ultimate strength and stiffness of these materials. The great attraction of crystalline polymers as high-strength materials is that because they generally have low densities their specific strength (ultimate tensile stress divided by density or specific gravity) is better than all other materials. Frank[1] pointed out that it was generally their low stiffness due to morphological defects which excluded them from replacing metals for high-strength applications.

In this article it is the intention to concentrate upon recent developments in the preparation of crystalline polymers with specific morphologies which give rise to specimens possessing high strengths and stiffnesses. The variety of morphologies that can be obtained in crystalline polymers are first of all reviewed with an emphasis upon the recent preparation of highly oriented conventional polymers and the development of new polymers in the form of macroscopic single crystals with dimensions of the order of centimetres. The elastic deformation of polymer crystals is then discussed from both theoretical and experimental viewpoints since it has long been realised that high strength can only be obtained in a material if it already possesses a high elastic stiffness.[1] The occurrence of plastic deformation in crystalline polymers at stresses considerably lower than the theoretical strengths of the materials often means that high values of ultimate tensile strength are not obtained. The understanding of the mechanisms of plastic deformation is therefore essential for the efficient utilisation of crystalline polymers and the ability to exploit their ultimate properties. Following the discussion on plastic deformation, recent developments in the fracture of crystalline polymers are reviewed in detail. In particular the properties of highly oriented crystalline polymers and the fracture of polymer single crystals are covered in depth.

2 MORPHOLOGY OF CRYSTALLINE POLYMERS

2.1 Isotropic Semicrystalline Polymers

The normal morphology of semicrystalline polymers cooled from the molten state is spherulitic, as shown in Fig. 1. This has been recognised

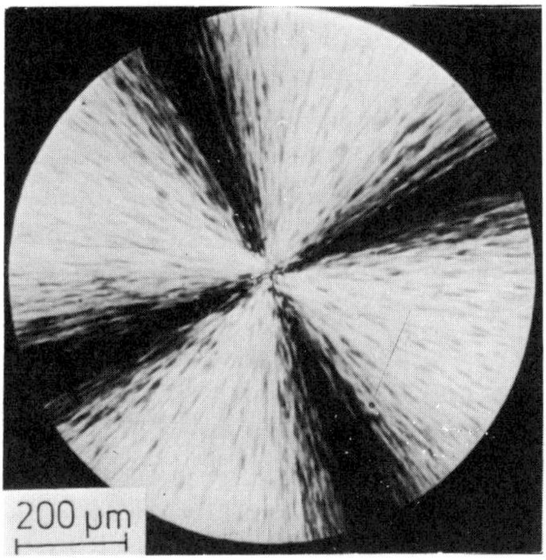

FIG. 1. Optical micrograph of a single spherulite in isotactic polystyrene (courtesy of Dr. D. Tod).

for many years with much of the pioneering work in this area being done by Keller.[2–4] A spherulite is thought to be a composite made up of crystalline lamellar ribbons, in which at least some of the molecules are folded, radiating from a central nucleus. In polyethylene the *b*-axis of the crystals has been shown to be radial.[4] The lamellae have been shown by Keith *et al.*[5] to be joined together by tie molecules which give the structure its internal cohesion. The interlamellar and interspherulitic material is thought to be either low-molecular-weight or non-crystallisable polymeric material which has been rejected by the growing crystals. Spherulites are seen most readily in polarised light in the optical microscope (Fig. 1). In some polymers with certain crystallisation conditions the spherulites can take on a banded appearance which has been attributed to a regular twisting of the lamellar ribbons.[2] However, recent investigations using a scanning transmission electron microscope and microelectron-diffraction techniques have put this model into some doubt.[6] The non-crystalline phase of the composite is generally soft and this, coupled with the three-dimensional orientation of the crystal, means that spherulitic polymers are not particularly strong or stiff.

2.2 Oriented Semicrystalline Polymers

Ever since synthetic high polymers were first discovered it has been realised that good uni-directional strength can be obtained by using them in the oriented state. The universal use of man-made fibres is an excellent example of this. Although polymer fibres are widely used the structure of these materials is still not fully understood.[7] It is known from density determinations and X-ray diffraction measurements that fibres formed by conventional techniques such as solvent spinning or hot drawing are still semicrystalline.[7] The molecular orientation in these materials leads to high tensile strengths but the presence of the amorphous phase leads to a drastic reduction in stiffness compared with that which could be obtained for completely crystalline material.

Using conventional semicrystalline polymers such as polyethylene and polypropylene Ward and co-workers[8–11] have developed techniques to draw the material to high draw ratios (up to 36 times extension). Similar material has been produced by Arridge *et al.*[12] This leads to samples which have a significantly higher modulus than conventionally drawn polymers and the increase in stiffness has been attributed to a higher degree of chain extension.[8–12] Highly aligned crystalline polymer samples have also been produced by melt extrusion. With the combination of high pressure and extensional flow found in a capillary rheometer it is possible to make polyethylene filaments exhibiting high moduli.[14] The various methods of producing highly oriented polymers have been reviewed by Bigg.[13] The presence of extensional flow can lead to fibrous crystallisation from solution. Pennings and Kiel[15] have shown that by stirring supercooled solutions of polyethylene, fibrous crystals with the well-known 'shish-kebab' morphology can be produced. This consists of an extended-chain backbone with an overgrowth of chain-folded lamellar crystals. A similar morphology has been obtained by crystallising a stretched *cis*-polyisoprene melt as reported by Andrews.[16] The crystallisation of polyethylene solutions in the extensional-flows field produced by impinging jets has been shown to again produce fibrous crystallisation[17] not unlike the 'shish-kebab' structures produced by Pennings.

The crystallisation or annealing of polyethylene at pressures in excess of 4 kbar* leads to crystals with thicknesses considerably larger than those

* 1 kbar ≡ 100 MPa

obtained at ambient pressures.[18–21] In this so-called chain-extended polyethylene (CEPE), crystals may be of the order of several micrometers thick but the material is normally very brittle. Bassett and Carder[22] have recently produced highly crystalline oriented CEPE by annealing drawn polymer under pressure but this material does not appear to have properties that are significantly better than the drawn unannealed material.

2.3 Polymer Single Crystals

It is now over 20 years since microscopic single crystals of polyethylene were first isolated by growth from dilute solution.[23–25] Single crystals have since been obtained for most semicrystalline polymers, usually by precipitation from dilute solution.[26] An electron micrograph of a typical polyethylene single crystal is given in Fig. 2. Microscopic polymer single crystals are generally lamellae several micrometres across and a few hundred Ångstroms thick, with the molecules folding back along their length in the short dimension of the crystal.[2,26] The thinness of the crystals in the chain direction means that the strength of the covalent bonding in the chain direction cannot be utilised and the small size of the crystals means that they can only be properly studied in the electron microscope.

Attempts to obtain single crystals of larger dimensions, in which use could possibly be made of the uni-directional strength that the crystals undoubtedly possess, have been made for many years. The bi-axial orientation of polyethylene by rolling[27] or other plane-strain deformation[28]

FIG. 2. Electron micrograph of a solution-grown single crystal of polyethylene (courtesy of Dr. P. Allan).

and subsequent annealing can produce material with a good single-crystal texture.[29] However, these materials are still semicrystalline and the individual crystals are relatively small (~ 200 Å). In terms of realising the ultimate properties of the polymer chains this material is no better than conventionally drawn polymer.

An important advance in producing polymer crystals in the form of large single crystals has been through the solid-state polymerisation of certain monomer single crystals. The first attempts in this area were with the solid-state polymerisation of trioxane to polyoxymethylene by irradiation and subsequent post-polymerisation heat treatment.[30–33] The conversion to polymer in this system depends upon the polymerisation conditions,[33] with the highest degrees of conversion reported being about 90%.[30,33] As well as containing monomer the polyoxymethylene coexists in two crystal modifications[31,32] and has internal boundaries. However, despite these shortcomings the solid-state polymerised polyoxymethylene represented a large improvement in terms of molecular alignment over conventionally oriented polymer.

A great step forward in the preparation of macroscopic polymer single crystals was made by Wegner and co-workers[34–37] in the early 1970s. This work in Germany has been subsequently followed at Allied Chemicals in the USA[38–40] and at Queen Mary College in the UK.[41–44] Using disubstituted diacetylene monomers crystallised from dilute solution in the form of single crystals it is possible to produce large, highly perfect crystalline conjugated polymers by solid-state polymerisation, which can be initiated by either heat, light or high-energy radiation, depending upon the reactivity of the monomer. The diacetylene monomers which give rise to the best polymer crystals have crystal structures in which the monomer molecules are aligned in a similar way to the segments of the chain in the polymer crystals. Polymerisation takes place, probably by means of a free-radical mechanism,[43] with a minimum of movement of the atoms and giving rise to highly perfect crystals which can be remarkably defect-free.[45] The general scheme for the polymerisation of disubstituted diacetylenes is given in Fig. 3 and the different types of R groups that have been used are listed in Table 1 along with abbreviations commonly used to describe the polymers.

The polymer crystals replicate the morphology of the monomer crystals. Toluene sulphonate diacetylene polymer crystals can be obtained in the form of lozenges with dimensions of the order of centimetres, as shown in Fig. 4. On the other hand crystals can also be obtained in the shape of fibres, such as the ethyl urethane diacetylene polymer shown in Fig. 5.

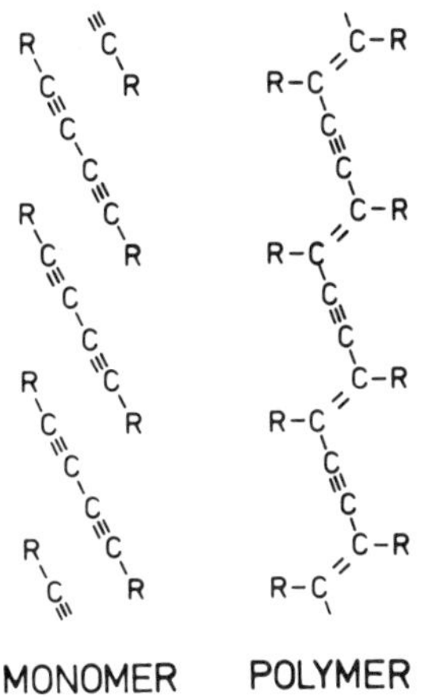

FIG. 3. General reaction scheme for the solid-state polymerisation of disubstituted diacetylenes. Typical R groups are listed in Table 1.

Such fibres have lengths of several centimetres and are of the order of 10–100 micrometers across. Until recently most of the research interest in these materials had been concerned with their electronic and optical properties. It will be shown in the subsequent sections that the mechanical properties of these large crystals are proving to be very promising indeed.

TABLE 1

SOME OF THE DIFFERENT SIDE-GROUPS R THAT CAN BE USED TO OBTAIN POLYDIACETYLENE SINGLE CRYSTALS AS SHOWN IN FIG. 3

Name of polymer	*Side-group R*	*Abbreviations*
Toluene sulphonate diacetylene polymer	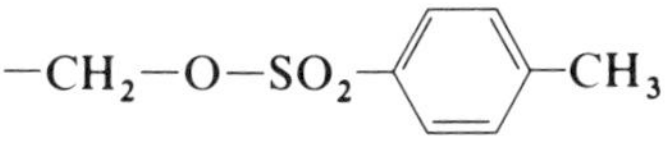	poly(TSHD)
Phenyl urethane diacetylene polymer		poly(HDU) poly(PUHD)
Ethyl urethane diacetylene polymer	$-CH_2-O-CONH-C_2H_5$	poly(EUHD)

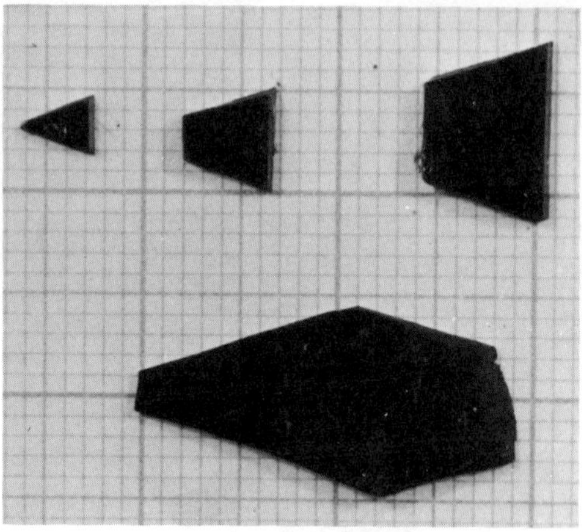

FIG. 4. Macroscopic single crystals of poly(TSHD). The background is graph paper with mm markings. (Crystals supplied by Dr. D. Bloor.)

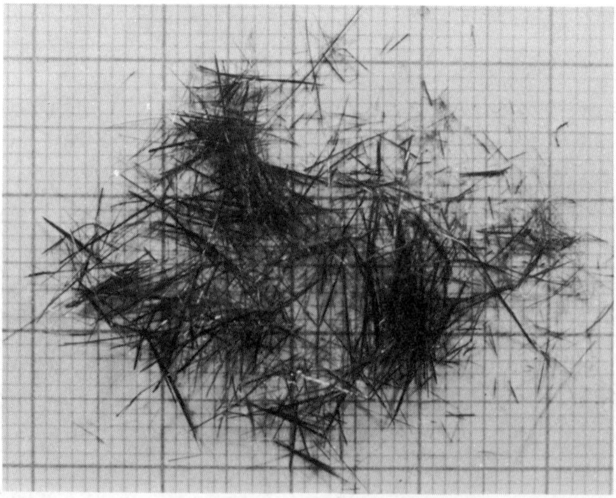

FIG. 5. A bundle of single-crystal fibres of poly(EUHD). The background is again mm graph paper. (Crystals supplied by Dr. D. Bloor.)

Another type of polymer that can be made by solid-state polymerisation is polymeric sulphur nitride $(SN)_x$, which can be made by the polymerisation of vapour-phase-grown crystals of S_2N_2.[46] The crystals are not as perfect as the diacetylenes and tend to be rather fibrous. They can come in the form of lozenges several millimetres across or fibres up to 1 centimetre long. The polymer is a metallic conductor at ambient temperatures and becomes a superconductor at 0·3 K.[47]

3 ELASTIC DEFORMATION

It has long been recognised that there is a broad relationship between stiffness and strength in materials.[48] The strongest materials tend to be those which possess the highest elastic moduli. Vincent[49] pointed out that we cannot expect to reach the theoretical strength of a polymer before we reach the theoretical modulus.

The simplest measure of stiffness is Young's modulus, E, which is just the stress divided by the strain when the deformation is elastic. The modulus of a crystal depends upon the type and direction of the stress to which it is subjected.[50] This leads to the crystal having several elastic constants, the exact number depending upon the crystal symmetry. Polymer crystals are extremely anisotropic because of the covalent bonding in the chain direction. This means that the tensile modulus in the chain direction is generally at least an order of magnitude higher than the other moduli. It is this high value of the modulus that leads to the high unidirectional strength that polymer crystals possess and so has tended to be the one which most workers have calculated theoretically and measured experimentally.

3.1 Theoretical Calculation of Crystal Moduli

Holliday and White[51] have reviewed the attempts at calculating the moduli of a whole range of polymer crystals and the values obtained for several common polymers have been summarised in Table 2. The earliest calculations were of the valence force field type, using the force constants for bond opening and bond stretching obtained from spectroscopic measurements.[52] An improvement upon this method was made by the introduction of a Urey–Bradley force field for the molecule.[51] Shimanouchi *et al.*[53] used this type of force field and took into account interatomic repulsions and bond-twisting as well as stretching and bending modes to calculate crystal moduli. The method used data obtained from Raman scattering experiments and gave rise to a modulus value for poly-

TABLE 2.

CALCULATED AND MEASURED VALUES OF THE CHAIN-DIRECTION MODULI FOR SEVERAL POLYMER CRYSTALS

Polymer	*Calculated*			*Measured*		
	E ($GN\,m^{-2}$)	*Method*	*Ref.*	E ($GN\,m^{-2}$)	*Method*	*Ref.*
Polyethylene	182	Valence force field	[52]	240	X-ray	[56]
	340	Urey–Bradley force field	[53]	358	Raman	[61]
	256	Urey–Bradley force field	[54]	329	Neutron $(CD_2)_n$	[51]
Polypropylene	49	Urey–Bradley force field	[51]	42	X-ray	[51]
Polyoxymethylene	150	Urey–Bradley force field	[51]	54	X-ray	[51]
Polytetrafluoroethylene	160	Urey–Bradley force field	[51]	156	X-ray	[51]
				222	Neutron	[62]
Poly(HDU)	49	Valence force field	[55]	45	Mechanical	[66]

ethylene of the order of 340 GPa, which is in excess of those calculated by the valence force field model (Table 2) and also greater than moduli measured by Raman and neutron scattering.[51] More sophisticated calculations have been carried out by Odajima and Maeda[54] on polyethylene taking into account stretching and bending of C—H bonds as well as deformation of the C—C backbone. They calculated the nine crystal moduli for the orthorhombic polyethylene crystal and obtained a value of 256 GPa for the modulus in the chain direction, which is in good agreement with that measured by X-rays but paradoxically less than that measured by Raman or neutron scattering.[51]

The general picture that emerges from these calculations is that the stiffness of a polymer crystal in the chain direction depends principally upon the conformation of the polymer chain and the density of packing of the chains within the crystal lattice. Chains in a planar zig-zag conformation, such as polyethylene, give rise to the highest moduli through C—C bond bending and stretching. It also follows that if a polymer could be synthesised with completely linear chains, and so only able to deform by bond stretching, it should be very stiff indeed since bond bending is a relatively easy mode of deformation. The density of packing of the chains in the polyethylene crystal is very efficient since it has no side-groups. This also helps to give it a high chain-direction modulus. The presence of side-groups on the polymer chain reduces the modulus for two reasons. Firstly, they tend to reduce the density of chain packing; and secondly, they can often force the molecule into a helix. The helical chain will readily deform by bond bending, and this leads to the moduli of polyoxymethylene and polypropylene, for example, being considerably lower than that for polyethylene (Table 2).

Batchelder[55] has recently calculated the elastic moduli in the chain direction for several different polydiacetylene crystals which can be made in the form of macroscopic single crystals (Section 2.3), using force constants derived from Raman scattering measurements (Table 2). The polydiacetylene chain appears to be stiffer than the single C—C zig-zag chain because it has double and triple bonds and fewer zig-zags (Fig. 3) but the crystal moduli are lower than those of polyethylene because the large side-groups (Table 1) on the polydiacetylene molecules reduce the density of chain packing.

3.2 Experimental Determination of Crystal Moduli

The principal method used to determine the elastic moduli of the crystal phase in semicrystalline polymers is that of X-ray diffraction of deformed samples, developed by Sakurada and co-workers.[56,57] The method depends upon measuring the shift of diffraction peaks in oriented polymer samples when they are stressed in the X-ray beam. Typical values of the crystal moduli in the chain direction for several polymers are given in Table 2. The technique has the advantage of observing changes only in the crystalline regions since the amorphous areas do not give rise to discrete diffraction peaks. However, problems can arise in determining the stress on the crystalline elements from the overall stress applied to the sample. With conventionally oriented polymers it is generally safe to assume that the stress on the crystals is the same as the overall stress on the sample, since the structure usually consists of stacks of alternating layers of crystalline and amorphous material.[58] However, it has recently been shown[59] that when the crystal modulus of ultra-drawn polyethylene is measured at room temperature by X-ray diffraction a low value ($\sim$ 140 GPa) is, rather surprisingly, obtained even though the specimen modulus may itself be relatively high ($\sim$ 68 GPa). If the specimens are tested at low temperatures the measured modulus of the crystalline regions is close to the theoretical value ($\sim$ 240 GPa).[60] It is thought that this is due to the amorphous phase in the highly oriented polymer being in parallel with the crystals rather than in series as in conventionally oriented polymer.[60]

Since the development of lasers many of the problems which were encountered with the Raman scattering of polymer specimens have now been overcome[51] and the full potential of this technique for studying inelastic scattering in polymers is now being exploited.[55] For example, Shauffele and Shimanouchi[61] have been able to determine the crystal modulus of polyethylene in the chain direction by the extrapolation of frequency and wavelength data for the longitudinal modes of long-chain hydrocarbon crystals.

Inelastic neutron scattering is potentially an even more powerful technique than Raman scattering. With hydrogen-containing polymers there is the problem of the high incoherent cross-section of hydrogen,[51] and the greatest success has been with deuteropolyethylene and polytetrafluoroethylene with measured *c*-axis moduli of 329 GPa[51] and 222 GPa[62] respectively. However, these are both larger than X-ray measured values (Table 2).

The most straightforward way to determine the chain-direction modulus of a polymer crystal is to deform a crystal in a direction parallel to the molecules and measure the strain as a function of applied stress. However, until recently this had been impossible to do because of morphological problems (Section 2). Solution-grown lamellar crystals are only of the order of a few hundred Ångstroms thick in the chain direction and cannot be tested in this way. Drawn polymer normally has good crystal orientation, but chain folding and the presence of amorphous material mean that the modulus of the material in the chain direction is usually less than one-tenth of the crystal modulus. Ultra-drawn polymers are thought to have higher degrees of chain extension and can possess sample moduli within 50% of the theoretical crystal moduli.[63]

An important breakthrough in the direct mechanical method of measuring polymer crystal moduli has been made by the preparation of macroscopic single crystals by the solid-state polymerisation of monomer crystals, as described earlier in Section 2.3. Andrews and Martin[64] have recently measured the modulus in the chain direction of partially polymerised single-crystal textured fibres of polyoxymethylene, the structure of which were described in an earlier publication.[33] The fibres were only partly polymerised and, as might be expected, the modulus depends upon the degree of conversion, as shown in Fig. 6. The highest modulus obtained was 38 GPa at 90% conversion, which is close to the X-ray modulus of polyoxymethylene crystals, 54 GPa (Table 2). Attempts have been made to measure the modulus of polymeric sulphur nitride $(SN)_x$ single-crystal fibres but problems have arisen from stress relaxation due to the microfibrillar nature of the material.[65] Baughman *et al.*[66] have measured

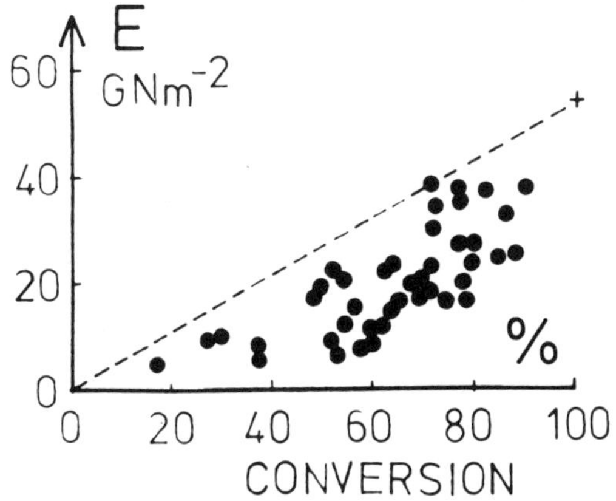

FIG. 6. Young's modulus, *E*, as a function of percentage conversion to polymer for partly polymerised fibres of polyoxymethylene. (After reference 64, with permission.)

the modulus of fibres of solid-state polymerised bisphenylurethane of 2,4-hexadyne-1,6-diol (Table 1). These fibres, which they called poly(HDU) had a high degree of perfection although they did contain 11% by weight of dioxane in solid solution. They measured a static tensile modulus of 42 ± 4 GPa, which compared well with a dynamic modulus of 47 ± 3 GPa that they measured by means of a Vibron dynamic visco-elastometer.[66] The large side-groups in the poly(HDU) crystals lead to a large interchain separation, and mean that each chain supports a cross-sectional area of 97·4 $Å^2$.[66] In contrast each chain in polyethylene supports only 18·4 $Å^2$ and the modulus per chain in the poly(HDU) is then about the same as that for polyethylene. This indicates that in the polydiacetylene crystals the ultimate modulus of the crystal structure is being measured and shows that if the packing of the chains can be improved a new generation of high-stiffness materials could be developed. Even so the stiffness of the existing poly(HDU) crystals compares well with that of other high-modulus materials (Table 3). Any further improvements should produce materials which are very stiff indeed.

3.3 The Stiffness of Semicrystalline Polymers

The presence of the amorphous phase in semicrystalline polymers causes a dramatic reduction in the stiffness of the polymer. In unoriented samples the random orientation of the crystals means that the low transverse and shear moduli of the polymer crystals[54] tend to control the specimen modulus. The random crystal orientation can be represented by Reuss or Voigt averages which may be used to predict the specimen modulus.[67] The two-phase composite nature of semicrystalline polymers was recognised by Halpin and Kardos[68] who showed how it was possible to calculate the stiffness of these materials using theories developed first of all for fibre-reinforced composites. This idea has been extended by Andrews[69] who showed that this approach could be used to predict the mechanical behaviour of semicrystalline polymers with spherulitic, row-nucleated and fibrillar morphologies. There are numerous other examples of composite models that have been used to predict the stiffness of semicrystalline polymers,[70–72] most of which give good quantitative agreement with experimental values.

The high modulus that is possessed by polymer crystals in the chain direction is more closely realised in the new generation of highly oriented ultra-high-modulus polymers.[10] Although they are still semicrystalline there is often a good degree of chain extension, and moduli of up to about 50% of the theoretical value are attainable. The best material in this

TABLE 3

ABSOLUTE AND SPECIFIC VALUES OF STIFFNESS AND STRENGTH FOR VARIOUS HIGH-STRENGTH POLYMERIC AND NON-POLYMERIC MATERIALS

Materials	*Specific gravity*	*Tensile modulus* ($GN\ m^{-2}$)	*Specific modulus* ($GN\ m^{-2}$)	*Tensile strength* ($GN\ m^{-2}$)	*Specific strength* ($GN\ m^{-2}$)	*Elongation to break* (%)	*Ref.*
Highly drawn polyethylene	0·966	68	71	> 0.35	>0·36	> 3	[11]
High-modulus extruded polyethylene	0·97	67	69	0·48	0·49	3·2	[14]
Poly(HDU) diacetylene crystals	1·37	45	33	0·4–1·7	0·29–1·24	< 5	[66]
Glass fibres	2·5	69–138	28–55	0·39–0·69	0·15–0·28	≈ 0.5	[14]
'Kevlar' 49 fibres	1·45	132	91	–	–	–	[11]
Carbon steel	7·9	210	27	0·45–0·62	0·06–0·08	25–30	[14]
Carbon fibre	2·0	200–420	100–210	2–3	1·0–1·5	≈ 1	[14]

respect is polyoxymethylene[13] but the highest absolute values are obtained in ultra-high-modulus polyethylenes.[8–14] Values of moduli for these polymers and other high-modulus materials are given in Table 3. For many purposes it is the specific stiffness (modulus divided by density or specific gravity) that is important, especially when light high-stiffness materials are required. It can be seen from Table 3 that since polymers generally have low specific gravities, their specific stiffnesses are even more favourable than those for non-polymeric materials.

4 PLASTIC DEFORMATION

The ability of semicrystalline polymers to undergo plastic deformation leads to the toughness for which they are renowned. In these materials the amorphous phase, in which most semicrystalline polymers are rubbery at ambient temperatures, allows the crystals to change shape without interacting with each other and causing cleavage since the crystals do not possess the five independent slip systems necessary for general deformation.[50] In polymer crystals, as in other crystalline solids, plastic deformation is found to take place in three general ways. These are slip, twinning and martensitic transformations. Bowden and Young[73] have reviewed the deformation mechanisms in crystalline polymers and only the more important features and recent developments will be discussed here. In polymer crystals which are deformed by a tensile stress parallel to the chain direction, which is the direction in which they are strongest, plastic deformation cannot take place. This is because the only deformation mechanisms which could be activated in this orientation would break molecules. However, in practice polymer crystals are invariably subjected to other types of stress, such as shear or compressive, in all directions which lead to plastic deformation taking place without the necessity of breaking molecules.

4.1 Slip

The ability of polymer crystals to undergo deformation by slip has long been established.[73,74] Since the slip processes do not break molecules the only slip planes that are found are those which contain the chain direction.[73] It is conventional to index the chain direction in polymer crystals as *c*, so this means that slip planes are limited to the $\{hk0\}$ type. As with other crystals, the slip planes are usually close-packed, but slip does not generally take place if there is hydrogen bonding between the planes, as in nylon crystals.[73] Slip has been found to take place either parallel or perpendicular to the chain direction, which leads to the two types of slip in polymer crystals: (*hk*0) [001] chain-direction slip and (*hk*0) [$k\bar{h}0$] transverse slip.[73]

Chain-direction slip is relatively easy to identify during the deformation of a semicrystalline polymer as it manifests itself as a rotation of the chain direction towards the axis of extension.[73,74] It is thought to be an important mechanism in the drawing of semicrystalline polymers such as polyethylene or polypropylene.[75] During the early stages of deformation the original spherulitic structure becomes elongated through chain-direction slip and other mechanisms.[76] Eventually the original lamellar and spherulitic structures break up and a new structure is formed with the molecules approximately parallel to the extension direction.[75,76] The lamellar crystals reform and the long period of the new structure in the extension direction is independent of the initial long period and depends only upon the temperature of drawing.[75] This feature is found with both tensile drawing[77] and plane-strain compressive deformation.[78] In polymers containing thick crystals (> 1000 Å) such as chain-extended polyethylene (CEPE) or polytetrafluoroethylene (PTFE) the occurrence of chain-direction slip can be shown by taking replicas of fracture surfaces before and after deformation. In these materials the chain direction is defined by the direction of striations on the crystal surfaces. Chain-direction slip, which is shown in Fig. 7 for PTFE, occurs at temperatures above −40 °C in PTFE[79,80] and at 80 °C in CEPE.[81] In the undeformed material the molecules are normal to the crystal surfaces but after deformation by chain-direction slip the angle becomes less than 90°. Strong evidence has also been found for chain-direction slip taking place during the deformation of oriented crystalline polymers by shear stresses parallel to the molecular direction.[73,82,83] There is also evidence for chain-direction slip taking place in lamellar single crystals of polyethylene when they are deformed on an extensible substrate.[73,84] The molecules which are initially approximately normal to the crystal surface rotate towards the direction of tensile stress.[84] However, when the molecules are accurately perpendicular to the surface of the crystals the resolved shear stress in the chain direction on the {*hk*0} slip planes is zero and so the crystals either cleave[2] or deform by other mechanisms.[85] Chain-direction slip can also be induced by deforming polyethylene single crystals on copper single-crystal substrates.[86]

Transverse slip has been widely reported during the deformation of spherulitic semicrystalline polymers[74,76] and lamellar single crystals.[86,87]

(a)

(b)

FIG. 7. Transmission electron micrographs of replicas of fracture surfaces of polytetrafluoroethylene. The striations define the chain direction. (a) Undeformed, showing crystals 1 μm thick; (b) compressed in a vertical direction at 21 °C, showing the results of [001] chain-direction slip in the crystals.

It is not as easy to identify as chain-direction slip and usually appears as a rotation of the other crystal axes about the chain direction, *c*.[73] In polyethylene single crystals the slip plane for this deformation mechanism appears to be restricted to the {110} fold planes.[87]

Slip in metals is known to take place by the motion of dislocations but it has not yet been shown convincingly that this is the case for polymer crystals. The existence of dislocations was deduced from the discontinuity of Moiré fringes from overlapping polyethylene crystals viewed in the electron microscope.[88] However, the first direct observation of dislocations was claimed by Petermann and Gleiter.[89] They observed contrast effects in polyethylene single crystals which they interpreted as being due to screw dislocations with Burgers vectors parallel to the chain direction [001]. Such dislocations could be responsible for chain-direction slip and in a later publication they showed that there was a greater density of dislocations in the vicinity of slip steps.[90] The possibility of screw dislocations with [001] Burgers vectors giving rise to chain-direction slip in polyethylene was first suggested by Peterson.[91] The idea has been more recently extended by Young[92] and a screw dislocation in a lamellar crystal is illustrated in Fig. 8. The theories developed by Peterson and Young assume that polymer crystals have their theoretical shear strength[48] at absolute zero. Dislocations are then thermally activated at ambient temperatures under the action of an applied shear stress.[92] The implications of the theory are that chain-direction slip should become easier as the temperature is increased and the crystal thickness reduced. Young[92] has shown that the critical resolved shear stress for chain-direction slip in oriented polyethylene is reduced as the temperature of testing is increased. Also, CEPE is in general brittle at room temperature but

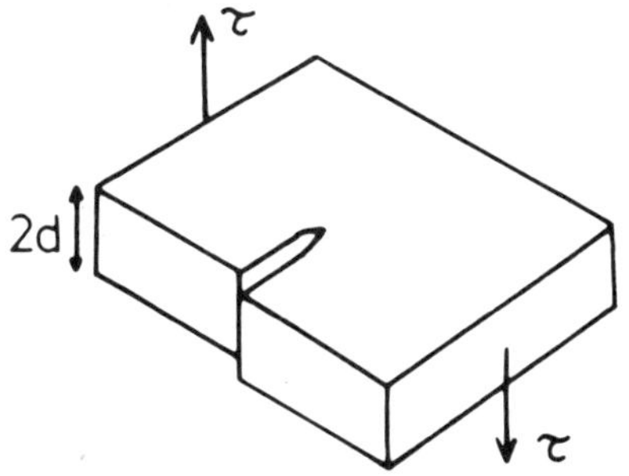

FIG. 8. Schematic representation of a screw dislocation with an [001] Burgers vector in a lamellar polyethylene single crystal of thickness 2*d*. (After reference 92.)

will undergo chain-direction slip at 80 °C[70] and so the theoretical predictions appear to be borne out in practice.

4.2 Twinning

It has been known for many years that twinning is an important deformation mechanism in polymer crystals.[73,74] However, very recently there have been important developments in this area with the recognition by Bevis[93] that there are two general types of twins in polymer crystals. It had been thought previously that the only type of twinning that could occur in polymer crystals was chain-axis invariant twinning (CIT). This is where the twins are formed by means of a shear perpendicular to the chain axis, which remains undistorted.[94] Bevis[93] has shown that in polyethylene the shear strains involved in twinning the crystal structure by shears at angles of less than 90° to the chain direction were in some cases smaller than those needed to form CITs. The new type of twinning causes the molecules to bend across the twinning plane and so has been called[93] chain-axis rotation twinning (CRT).

There is much accumulated evidence supporting the occurrence of CITs in polymer crystals. They were first suggested by Frank *et al.*[74] to explain the deformation of polyethylene and {110} and {310} were thought to be the most likely twinning planes. {110} twinning has been widely reported in deformed oriented polymer[73,95] and in lamellar single crystals[85,87,96,97] following deformation. It is shown schematically in Fig. 9 for polyethylene. {310} twinning is more difficult to find but it has been reported in highly deformed bulk material;[98] there is no strong evidence for it taking place in deformed single crystals.[87,96] {310} twinning is also illustrated in Fig. 9. When polyethylene single crystals are cross-linked by irradiation, {110} twinning is found to be the only deformation mechanism that will occur.[97]

Until very recently it was thought that only CITs could be found in polymer crystals and any twinning modes which distorted the molecules, such as CRTs, would not occur.[73,94] Pietralla[99] invoked CRTs to explain the drawing behaviour of semicrystalline polymers such as polyethylene and suggested {111} as a possible twinning plane. This was later extended by Bevis[93] who considered the idea in more detail and calculated theoretically the shears involved with all low-shear CITs and CRTs in polyethylene. He showed that CRTs could also occur with relatively small shears.[93] The first direct evidence for the occurrence of CRTs in polymer crystals has been obtained by Young *et al.*[100] in deformed macroscopic single crystals of a toluene sulphonate diacetylene polymer

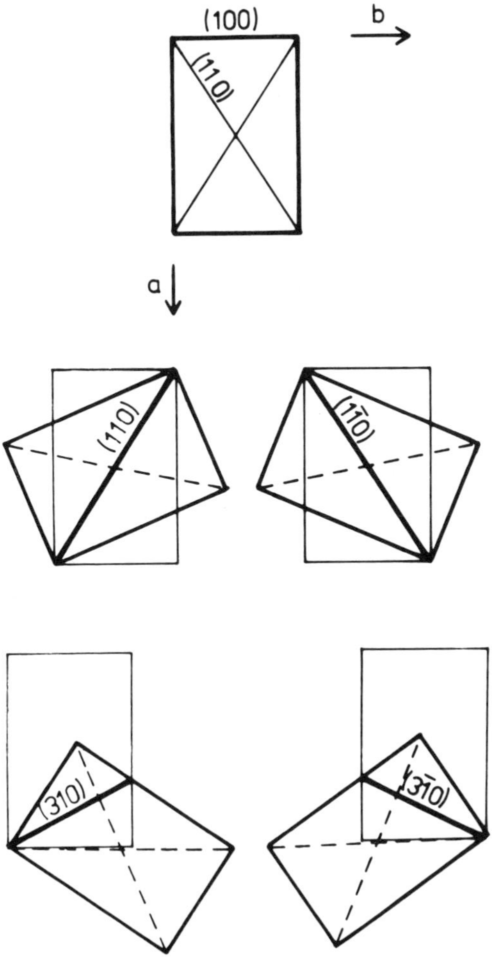

FIG. 9. Chain-axis invariant twinning in the polyethylene unit cell viewed parallel to *c*. {110} and {310} twinning are illustrated. (After reference 74.)

(Table 1). A scanning electron micrograph of the (012) CRT in this material and a schematic diagram of the twin on the molecular level are given in Fig. 10. The micrograph shows clearly the rotation of the molecules across the (012) twin plane. This particular twin is relatively thick (~ 100 μm). Thinner twins of this orientation of twin have also been found. Young *et al.*[100] have shown that CRTs are very common deformation mechanisms in polydiacetylenes and that other orientations of CRTs

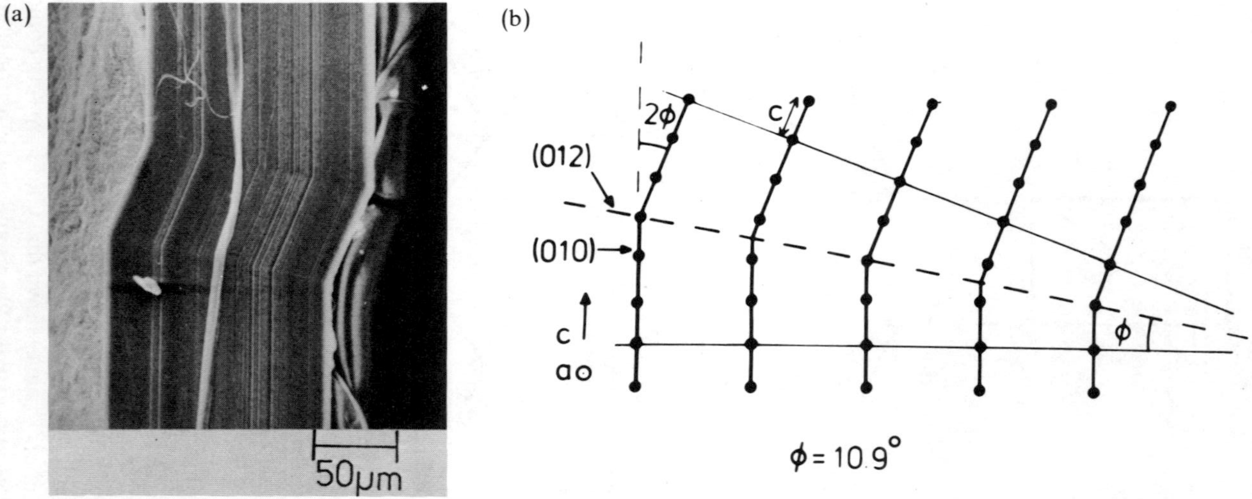

FIG. 10. Chain-axis rotation twinning in poly(TSHD). (a) Scanning electron micrograph of a single crystal containing an (012) twin. Viewing direction parallel to *a* and perpendicular to *c*. Chain direction is defined by striations; (b) schematic diagram of the above twin on a molecular level. (After reference 100.)

can also occur. Some twins are seen to vary in thickness and even to taper to a point when they terminate within the crystal. This behaviour is typical of twins in single crystals of other non-polymeric materials.[50]

Both Pietralla[99] and Young *et al.*[100] pointed out that the kinks often seen within chain-extended polyethylene (CEPE) lamellae may well be CRTs. This suggestion has been supported by Bassett.[101] A good example of possible CRTs in a CEPE lamella is given in Fig. 11.

FIG. 11. Transmission electron micrograph of a replica of a fracture surface of chain-extended polyethylene after deformation. The single lamellar crystal appears to contain chain-axis rotation twins. (Courtesy of Dr. D. C. Bassett, unpublished work.)

However, because of the uncertainty about the orientation of the crystal axes in such a micrograph it is impossible to identify the modes of the CRTs. It is highly likely that when polyethylene crystals are compressed parallel to the chain direction CRTs will be activated and indeed Bassett has evidence to support this.[101] It is also possible, since the result of completely twinning a lamella is indistinguishable from shearing the crystal by means of [001] chain-direction slip, that on many occasions when [001] slip (Section 4.1) has been suggested to have taken place CRTs may have occurred instead. It appears, therefore, that chain-axis rotation twins may be extremely important deformation mechanisms that have been hitherto virtually discounted in polymer crystals.

4.3 Martensitic Transformations

The ability of crystals to undergo stress-induced martensitic transformations has been long recognised. These are changes in crystal structure which are usually induced by the action of shear stresses.[50] In polymer crystals the only transformations known at the moment involve shear stresses perpendicular to the chain direction. When spherulitic polyethylene is deformed extra reflections appear in the X-ray[74] or electron diffraction[96] patterns indicating the presence of a new phase. The orientation relationship between the new phase and the parent phase can only accurately be determined in single crystals[87,96] or in oriented bulk material with a well-defined texture.[95] Bevis and Crellin [94] predicted theoretically the orientation relationships between the normal orthorhombic and stress-induced monoclinic forms in polyethylene. Subsequent work by Bevis and co-workers[87,96] has shown that the observed modes in polyethylene single crystals are close to the predicted ones, with the modes involving the lowest shears being preferred. The modes in single-crystal textured oriented polyethylene are found to be different from those in single crystals, but even so only the modes involving low shears are found.[73,95]

There has been little evidence for well-defined martensitic transformations taking place in other crystalline polymers. However, recently it has been shown that polymeric sulphur nitride $(SN)_x$ (Section 2.3) will undergo a stress-induced martensitic transformation from its normal monoclinic cell to an orthorhombic one.[102] This polymer comes in the form of macroscopic single crystals which allow an unambiguous identification of the transformation mode by X-ray diffraction. The transformation is rather similar to that in polyethylene but takes place in the reverse direction (monoclinic to orthorhombic). Young and Baughman[103] have recently calculated the orientation relationships between the two phases and found good agreement between the measured and predicted modes. They also suggested that the observed mode may involve twinning as well as a martensitic transformation.

5 FRACTURE OF CRYSTALLINE POLYMERS

The fracture of crystalline polymers can be approached from one of two points of view. One is to consider the fracture of semicrystalline polymers and the other is to look at the fracture of individual polymer single crystals. Many of the 'unbreakable' plastic artefacts are made from semicrystalline polymers. In these materials the non-crystalline phase is usually above its glass transition temperature and is therefore rubbery. This means that the polymers are usually extremely tough and normally undergo high degrees of drawing before fracture eventually intervenes. In these cases we are really concerned with the eventual fracture of an oriented polymer with a structure not unlike that of polymer fibres. With the development of microscopic and macroscopic polymer single crystals it has been possible to examine crystalline polymer fracture from an entirely different viewpoint as a fundamental property of the crystals. This aspect will be considered finally.

5.1 Isotropic Semicrystalline Polymers

This chapter is concerned mainly with the deformation and fracture of polymer crystals but it is necessary to consider the mechanical behaviour of semicrystalline polymers to obtain a complete idea of the importance of the properties of the crystals. Semicrystalline polymers are generally very tough materials and often undergo fracture only after large amounts of drawing. This is because in most commercial semicrystalline polymers the amorphous phase is in the rubbery state. This allows plastic deformation of the crystals to take place without cracking and allows high degrees of extension to be achieved.

In some cases semicrystalline polymers can exhibit brittle behaviour and the fracture can be followed using fracture mechanics.[112] Tod[113] has recently used a linear-elastic fracture mechanics approach to examine the simple fracture behaviour of isotactic polystyrene as a function of the degree of crystallinity. Figure 12 shows a plot of the stress intensity factor K_{Ic}[112] as a function of crystallinity for isotactic polystyrene. At low degrees of crystallinity (up to 2%) the value of K_{Ic} and hence the toughness increase slightly but at higher degrees of crystallinity the toughness falls below that of the amorphous material. In isotactic polystyrene the amorphous phase is glassy and brittle; such materials are not normally of any commercial importance because of their inherent weakness and only atactic polystyrene, which cannot crystallise, is normally used.

The more conventional semicrystalline polymers can sometimes exhibit brittle behaviour. For example nylon 6 is sometimes prone to brittle fracture whereas plasticisation by water significantly increases its fracture toughness.[114,115] Many semicrystalline polymers are prone to failure by means of environmental stress cracking. Polyethylene is particularly sensitive in this respect and a great deal of work has been done examining the resistance of this material to environmental stress cracking.[116,117] The resistance appears to depend upon many factors including thermal

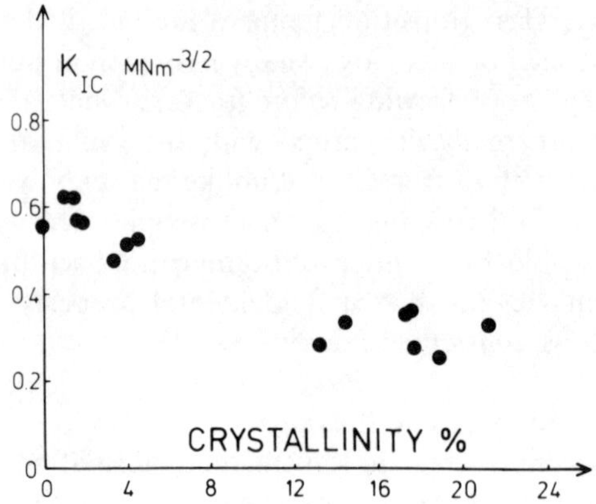

FIG. 12. A plot of critical stress intensity factor K_{Ic} as a function of degree of crystallinity (per cent) for isotactic polystyrene. (After reference 112, with permission.)

history, average molecular weight, molecular weight distribution, chain branching and degree of crystallinity.[116] Marshall *et al.*[117] used a linear-elastic fracture mechanics approach to study brittle crack propagation in low-density polyethylene immersed in different alcohols. They found that there was a steady increase in stress-intensity factor K_{Ic} with crack velocity although the exact value of K_{Ic} for a particular velocity depended upon the grade of polymer and the type of alcohol. A linear-elastic fracture mechanics approach[112] has also been used to study the dynamic fatigue behaviour of crystalline polymers such as nylon and polyethylene by Hertzberg *et al.*[118] For these materials they found a unique relationship between the fatigue crack propagation rate and the stress intensity factor range ΔK_I prevailing at the tip of the crack.

Great care must be exercised when applying linear-elastic fracture mechanics to semicrystalline polymers. Their load–displacement curve is rarely linear-elastic and they are more amenable to investigations using approaches developed for more ductile materials, such as the theory of Rivlin and Thomas,[119] which was first applied to rubbers, or the generalised theory of fracture of Andrews.[120] The tearing of thin sheets of polyethylene[121] and polypropylene[122] has been recently investigated using the approach pioneered for rubbers by Rivlin and Thomas. In the polyolefin sheets the tearing energy is found to depend upon both the testing temperature and any orientation present in the sample. Non-classical fracture mechanics has been used to study the propagation of fatigue cracks in polyethylene by Andrews and Walker.[123] In this case a unique relationship was found between the fracture mechanics parameter, $\mathscr{T}$[112,120] and crack propagation rate, not unlike the relation found by Hertzberg *et al.*[118] between ΔK_{I} and the crack propagation rate for a similar material. However, in the work of Andrews there is no assumption of linear-elastic deformation, which would appear to be more satisfactory for a visco-elastic solid such as polyethylene. Teh *et al.*[124] have recently extended this work and have devised a method of performing creep-compensated fatigue testing on polyethylene under reverse-loading conditions.

When semicrystalline polymers are deformed and undergo plastic deformation rather than brittle fracture they extend parallel to the tensile axis to give a drawn structure with the molecules aligned parallel to the draw direction.[75–78] This is thought to be due to a combination of inter- and intralamellar slip and twinning,[73] as described in Section 4. The details of this deformation depend also upon the morphology of the polymer. For instance, in polypropylene cracks may be seen within the spherulites after deformation.[125] However, it appears that the tendency for polypropylene spherulites to crack depends upon the temperature at which they are grown. Vadimsky *et al.*[126] studied the deformation of spherulitic polyethylene films by electron microscopy and found that the intercrystalline links between the chain-folded lamellar crystals, both within the same spherulite and across interspherulitic boundaries, behaved as strong inextensible elements and played an important part in the deformation of the structure. The links were seen to bridge cracks both within spherulites and between them and so transfer the stress between various elements in the structure.

5.2 Oriented Semicrystalline Polymers

At sufficiently high strains a drawn polymer must break, although it may be at very high extensions, perhaps in excess of 25 times. This is, of course, exactly the same as the fracture of a polymer fibre.[127–129] It has been shown using electron spin resonance (ESR) spectroscopy that when highly oriented polymers or polymer fibres are deformed to failure free radicals are formed, indicating that molecular fracture occurs. A detailed account of ESR and molecular fracture is given in Chapter 2 by Andrews and Reed.

Chain-extended polyethylene (CEPE), which has been described earlier (Section 2.2), is generally brittle, failing on $\{hk0\}$ planes. The fracture of oriented CEPE has been studied recently by Bassett and Carder.[70] They examined the behaviour of drawn polyethylene after annealing at 5·3 Kbar* between 220 and 260 °C. Material annealed below 240 °C with a crystal thickness of less than 1000 Å yielded and underwent cold drawing. However, the CEPE annealed above 240 °C was found to be brittle, with failure taking place across {001} planes because of the segregation of low-melting polymer between the chain-extended lamellae. Ductility was restored by removal of the low-melting material. However, even with the best samples both the stiffness and strength of this material are relatively low for oriented polymers, probably because of the presence of the interlamellar amorphous material and because of some loss of chain orientation during high-pressure annealing.[70,81]

There has been much interest recently in the preparation of ultra-high-modulus semicrystalline polymers[13] (Section 2.2). However, as well as possessing moduli that are relatively high for oriented semicrystalline polymers (Table 3) they also have high ultimate tensile strengths. Capaccio and Ward[11] have reported an ultimate tensile strength in excess of 350 MPa and an elongation-to-break of more than 3% for their ultra-high-modulus polyethylene. Similarly Capiati and Porter[14] find that their ultra-drawn high-density polyethylene filaments prepared by solid-state deformation in a capillary rheometer had strengths of up to 480 MPa with elongations-to-break of 3·2%. In addition they noted that the fractured filaments had a characteristic 'brushlike' appearance and they concluded that rupture took place by slippage between bundles of microfibres rather than by rupture of the polymer chains. The ultimate tensile strengths of these ultra-high-modulus polyethylenes and their specific strengths are given in Table 3. The same parameters for other high-strength materials are also given in the table. It can be seen that the polyethylene samples compare well with the other materials in terms of absolute strength but that their specific strengths are also extremely good and point to the potential that these oriented polymers undoubtedly possess as light materials with high one-dimensional strengths.

5.3 Single Crystals

With the development of microscopic single crystals over the last 20 years and the discovery more recently of macroscopic single crystals, many advances have been made in the study of the fracture of polymer crystals. In simple terms, polymer crystals can fracture in one of two basic ways. If they are stressed in directions not parallel to the chain direction they will

* 1 Kbar ≡ 100 MPa

cleave with cracks running parallel to the molecules, breaking secondary bonds between the molecules. On the other hand if they are stressed parallel to the chain direction, molecular fracture will occur with cracks running across the molecules and breaking the primary covalent bonds. This is, of course, the direction in which the crystals should exhibit their highest strengths.

Single crystals of polymers obtained from dilute solution are invariably in the form of lamellae with the molecules folded so that the chain direction is perpendicular to the lamella surface (Section 2.3). These crystals are microscopic and too small to be gripped in the conventional way so that they are usually deformed only after being deposited upon an extensible substrate and then only in directions perpendicular to the chain axis. When they are deformed in this way the crystals are seen to undergo slip, twinning and martensitic transformations (Section 4). However, they are also found to show cleavage cracks, especially at high strains; the cleavage cracks are on planes containing the chain axis because of the molecular orientation within the lamellae. Figure 13 shows a deformed single crystal of polyethylene containing cracks. The detailed nature of the cleavage cracks in single crystals has been used to advantage to show the fold planes, and to give strong support to the theory of adjacent

FIG. 13. Transmission electron micrograph of a single crystal of polyethylene after tensile deformation in the direction of the arrow. Note the cracking and compare with Fig. 2. (Courtesy of Dr. P. Allan.)

re-entry during folding, in single crystals of polyethylene[2,104] and polypropylene.[105] Cracks between fold planes are clean and do not have fibres pulled across them whereas the sides of cracks between fold planes are connected by fibres. Geil[104] showed that polyethylene crystals grown from certain solvents and having a truncated diamond shape with {010} growth faces were very brittle and tended to crack readily. He also observed[106] that crystals of nylon 6 and polyoxymethylene invariably cracked perpendicular to the straining direction. The tendency for cracking in polyethylene crystals is also increased by deforming them without a substrate. Gleiter and Petermann[107] deformed polyethylene single crystals which had been deposited so that they bridged cracks in a carbon film and found that the crystals cracked at relatively low strains with fibres pulled across the cracks. They concluded that the affine deformation which is normally found when crystals are deformed on extensible substrates occurs because the crystals are forced to deform uniformly when the substrate deforms uniformly. In a later publication[108] they described a detailed examination of the structure of the fibres drawn across the cracks in crystals deformed at different temperatures. The formation of the fibres appears to take place by a two-step process. The first step is the breaking off of blocks of folded chains from the crystals giving a 'string-of-pearls' structure. If the temperature is high enough there appears to be a thermally activated rearrangement of the molecules in the fibres resulting in the formation of a 'bamboo' structure. The irradiation of polyethylene single crystals has a profound effect upon their deformation.[97,109] At low doses fibres are still pulled across cracks but above 35–40 Mrad fibre formation is inhibited[97] and material is drawn across the cracks in the form of a continuous thin (15 Å) film.[109]

Although the fracture of lamellar single crystals is interesting, and informative about the structure of these materials, the potential strength of polymer crystals is realised only when they are stressed parallel to the chain direction. This can only be readily achieved using crystals in a fibrous form with the molecular chains lying along the length of the fibres. However, until about 10 years ago crystals with this particular morphology were not known to exist. An important breakthrough was made with the discovery that it is possible to polymerise certain monomers in the solid state directly to polymer (Section 2.3). Andrews and Martin[33] have shown that trioxane single crystals can be polymerised to polyoxymethylene with a conversion of up to about 90%. These crystals can be produced in the form of needles with moduli approaching the theoretical value (Fig. 6). Andrews and Martin[64] also measured the fracture stresses of these needles and Fig. 14 shows a plot of fracture stress as a function of modulus for the partially polymerised crystals. The best line through the data gives a slope of 0·06 and the upper limit has a slope of 0·1. This indicates that the average strain at rupture is about 6% but that some crystals can fracture at a strain of 10%. The clear implication of this observation is that the needles are exhibiting their theoretical strengths.[1,48,64] It must also be noted that the needles are not true single crystals but are thought to be a composite aggregate of fibrils embedded in a monomer matrix.[64]

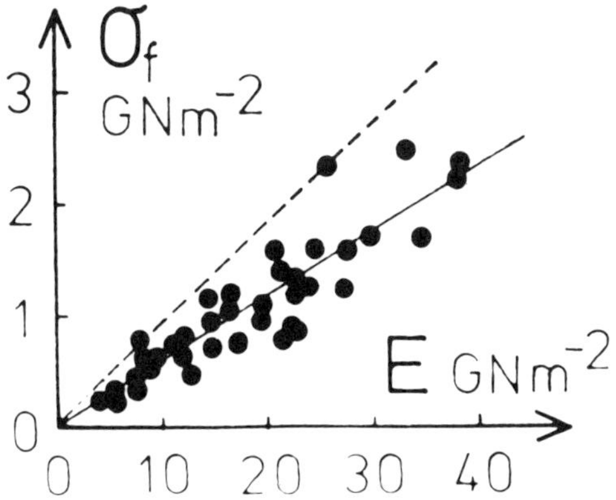

FIG. 14. Fracture stress, σ_f, as a function of modulus, E, for partly polymerised, single-crystal textured polyoxymethylene fibres. (After reference 64, with permission.)

True macroscopic single crystals can be obtained in the form of fibres by polymerising certain diacetylene monomers (Section 2.3). Baughman *et al.*[66] have recently reported some exciting measurements of the fracture behaviour of single crystals of phenyl urethane diacetylene polymer, poly(HDU). These fibrous crystals appear to have their theoretical modulus (Section 3.2) and it has been shown that they also appear to have strengths which are close to their theoretical limit. The poly(HDU) crystals contain a small amount of dioxane solvent in solid solution and have diameters of the order of 10–100 μm and lengths of up to 50 mm. When tested in uni-axial tension they fracture by molecular cleavage at strains of up to 5%. The ultimate tensile strength of such crystals was found to increase as the crystal diameter, d, was reduced, as shown in Fig. 15. The scatter of the data was found to be rather large but the data clearly show

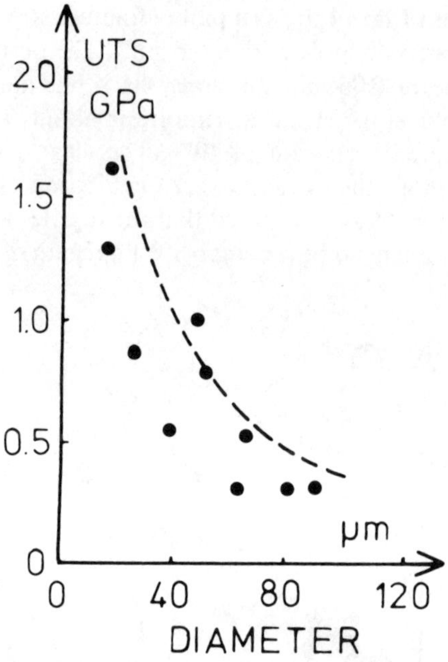

FIG. 15. Ultimate tensile strength as a function of crystal diameter for poly(HDU) single-crystal fibres. (After reference 66.)

the characteristic $1/d$ dependence that has been found for other high-strength fibres[48] due to the presence of surface defects. Galiotis and Young[110] have found a similar relationship for fibrous crystals of an ethyl urethane diacetylene polymer (Table 1). When the diacetylene crystals fail they tend to fracture by splintering parallel to the molecules and by molecular cleavage as shown for the ethyl urethane diacetylene crystals in Fig. 16. The maximum fracture stress obtained for polydiacetylene crystals so far is about 1·7 GPa.[66] This compares with an ultimate tensile strength of about 3 GPa for carbon fibres[111] and betters that of other polymeric materials (Table 3). However, as explained by Frank,[1] high-strength materials must also have a high modulus and the modulus of the poly(HDU) crystals of Baughman *et al.*[66] is relatively low at about 45 GPa (Table 3). If the polydiacetylene crystals can therefore be synthesised with smaller side-groups so that the molecular packing is more efficient, giving a higher modulus, values of ultimate tensile strength well in excess of those of carbon fibres could very well be obtained. Since they

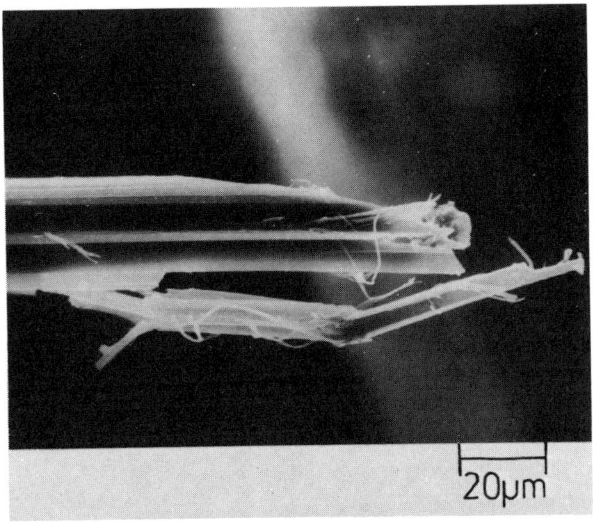

FIG. 16. Scanning electron micrograph of one-half of a fractured poly(EUHD) single-crystal fibre viewed perpendicular to the chain direction.

would probably have a lower density than carbon fibres (Table 3) their specific strengths would be even more favourable than those of carbon fibres.

A common defect in the polydiacetylene crystals which gives rise to reduction in strength is twinning.[66] In Section 4.2 the tendency for these crystals to deform through chain-axis rotation twinning was clearly demonstrated. The twins can give rise to surface steps and hence stress concentrations. They can also interact with each other and lead to cleavage cracks parallel to the chain direction, as shown in Fig. 17(a). This type of cracking is commonly found in other materials where twins interact with each other.[50] Cleavage also readily takes place when the macroscopic polydiacetylene crystals are compressed perpendicular to the chain direction. Figure 17(b) shows cleavage cracks on {*hk*0} planes in a toluene sulphonate diacetylene polymer crystal which had been compressed perpendicular to the (010) face. This type of intramolecular cleavage is analogous to that in the deformed lamellar single crystals (Fig. 13) but in the case of the polydiacetylenes there are no fold-plane restrictions and so few fibres pulled across the cracks.

(a)

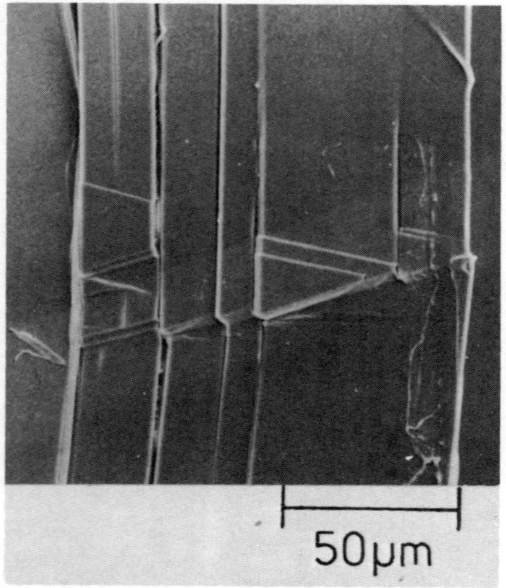

(b)

FIG. 17. Scanning electron micrographs of deformed poly(TSHD) crystals containing cleavage cracks. (a) Cracks parallel to *c* on an (010) face, possibly caused by the interaction of twins; (b) crystal viewed parallel to *c*, containing cracks on {hk0} after compression in the vertical direction.

6 CONCLUSIONS AND FUTURE DEVELOPMENTS

It has been shown that over the last decade there have been important developments in the preparation of crystalline polymers with special morphologies which have strengths and stiffnesses well in excess of those which had been previously obtained from polymeric materials. The two principal methods of preparation have been either by producing ultra-highly oriented conventional polymers or through the synthesis of new polymers in the form of macroscopic single crystals. Both of these methods have produced materials with extremely promising mechanical properties and because polymers normally have low densities the new materials have specific strengths and stiffnesses which are higher than those of most other materials.

Future developments would appear to be in the improvement of both the ultra-highly oriented polymers and the synthesis of better macroscopic crystals. Orientation in conventional polymers can be improved by obtaining higher degrees of chain extension and better internal cohesion. The most promising developments would appear to be in the area of macroscopic single crystals. These materials were developed initially to investigate their optical and electronic properties but they also appear to have mechanical properties which are potentially very interesting as well. The macroscopic polymer crystals which have been prepared so far have a relatively low degree of chain packing when compared with polyethylene, for example. It appears therefore that there is plenty of scope to improve their mechanical strengths and stiffnesses by the synthesis of new crystals with higher degrees of chain packing approaching that of polyethylene.

REFERENCES

1. FRANK, F. C. *Proc. Roy. Soc. (London)*, **A319** (1970) p. 127.
2. KELLER, A. *Rep. Prog. Phys.*, **31** (1968) p. 623.
3. KELLER, A. *J. Polymer Sci.*, **17** (1955) p. 291.
4. KELLER, A. *J. Polymer Sci.*, **17** (1955) p. 351.
5. KEITH, H. D., PADDEN, F. J. and VADIMSKY, R. G. *J. Polymer Sci. A2*, **4** (1966) p. 267.
6. LOW, A., VESELY, D., ALLAN, P. and BEVIS, M. *J. Mat. Sci.*, to be published.
7. HEARLE, J. W. S. and PETERS, R. H. *Fibre Structure*, Butterworths, London, 1968.
8. CAPACCIO, G. and WARD, I. M. *Polymer*, **15** (1974) p. 233.
9. CANSFIELD, D. L. M., CAPACCIO, G. and WARD, I. M. *Polymer Eng. Sci.*, **16** (1976) p. 721.

10. Capaccio, G., Chapman, T. J. and Ward, I. M. *Polymer*, **16** (1975) p. 469.
11. Capaccio, G. and Ward, I. M. *Polymer Eng. Sci.*, **15** (1975) p. 219.
12. Arridge, R. G. C., Barham, P. J. and Keller, A. *J. Polymer Sci., Polymer Phys. Ed.*, **15** (1977) p. 389.
13. Bigg, D. M. *Polymer Eng. Sci.*, **16** (1976) p. 725.
14. Capiati, N. J. and Porter, R. S. *J. Polymer Sci., Polymer Phys. Ed.*, **13** (1975) p. 1177.
15. Pennings, A. J. and Kiel, A. M. *Kolloid-Z*, **205** (1965) p. 160.
16. Andrews, E. H. *Proc. Roy. Soc. (London)*, **A227** (1964) p. 562.
17. Frank, F. C., Keller, A. and Mackley, M. R. *Polymer*, **12** (1971) p. 467.
18. Anderson, F. R. *J. Appl. Phys.*, **35** (1964) p. 64.
19. Geil, P. H., Anderson, F. R., Wunderlich, B. and Arakawa, T. *J. Polymer Sci. A*, **2** (1964) p. 3707.
20. Wunderlich, B. and Melillo, L. *Makromol. Chem.*, **118** (1968) p. 250.
21. Bassett, D. C., Khalifa, B. A. and Turner, B. *Nature Phys. Sci.*, **239** (1972) p. 106.
22. Bassett, D. C. and Carder, D. R. *Phil. Mag.*, **28** (1973) p. 513.
23. Keller, A. *Phil. Mag.*, **2** (1957) p. 1171.
24. Till, P. H. *J. Polymer Sci.*, **24** (1957) p. 301.
25. Fischer, E. W. *Z. Naturforsch.*, **12a** (1957) p. 753.
26. Geil, P. H. *Polymer Single Crystals*, Interscience, New York, 1963.
27. Hay, I. L. and Keller, A. *J. Mat. Sci.*, **2** (1967) p. 538.
28. Young, R. J. and Bowden, P. B. *J. Mat. Sci.*, **8** (1973) p. 1177.
29. Gray, R. W. and Young, R. J. *J. Mat. Sci.*, **9** (1974) p. 521.
30. Hayashi, K., Ochi, H. and Okamura, S. *J. Polymer Sci. A*, **2** (1964) p. 2929.
31. Colson, J. P. and Reneker, D. H. *J. Appl. Phys.*, **41** (1970) p. 4296.
32. Reneker, D. H. and Colson, J. P. *J. Appl. Phys.*, **42** (1971) p. 4606.
33. Andrews, E. H. and Martin, G. E. *J. Mat. Sci.*, **8** (1973) p. 1315.
34. Wegner, G. *Makromol. Chem.*, **145** (1971) p. 85.
35. Wegner, G. *Makromol. Chem.*, **154** (1972) p. 35.
36. Kaiser, J., Wegner, G. and Fischer, E. W. *Israel J. Chem.*, **10** (1972) p. 157.
37. Schermann, W., Wegner, G., Williams, J. O. and Thomas, J. M. *J. Polymer Sci., Polymer Phys. Ed.*, **13** (1975) p. 753.
38. Baughman, R. H. *J. Appl. Phys.*, **43** (1972) p. 4362.
39. Baughman, R. H. and Chance, R. R. *J. Polymer Sci., Polymer Phys. Ed.*, **14** (1976) p. 2037.
40. Baughman, R. H. and Chance, R. R. *J. Appl. Phys.*, **47** (1976) p. 4295.
41. Bloor, D., Koski, L., Stevens, G. C., Preston, F. H. and Ando, D. J. *J. Mat. Sci.*, **10** (1975) p. 1678.
42. Bloor, D., Koski, L. and Stevens, G. C. *J. Mat. Sci.*, **10** (1975) p. 1689.
43. Stevens, G. C. and Bloor, D. *J. Polymer Sci., Polymer Phys. Ed.*, **13** (1975) p. 2411.
44. Batchelder, D. N. and Bloor, D. *Chem. Phys. Lett.*, **38** (1976) p. 37.
45. Schultz, J. M. *J. Mat. Sci.*, **12** (1976) p. 2258.
46. Baughman, R. H., Chance, R. R. and Cohen, M. J. *J. Chem. Phys.*, **64** (1976) p. 1869.
47. Greene, R. L., Street, G. B. and Suter, L. J. *Phys. Rev. Lett.*, **34** (1975) p. 577.

48. KELLY, A. *Strong Solids*, Clarendon Press, Oxford, 1966.
49. VINCENT, P. *Proc. Roy. Soc. (London)*, **A282** (1964) p. 113.
50. KELLY, A. and GROVES, G. W. *Crystallography and Crystal Defects*, Longman, London, 1970.
51. HOLLIDAY, L. and WHITE, J. W. *Pure and Appl. Chem.*, **26** (1971) p. 545.
52. TRELOAR, L. R. G. *Polymer*, **1** (1960) p. 290.
53. SHIMANOUCHI, T., ASAHINA, M. and ENOMOTO, S. *J. Polymer Sci.*, **59** (1962) p. 93.
54. ODAJIMA, A. and MAEDA, T. *J. Polymer Sci. C*, **15** (1966) p. 55.
55. BATCHELDER, D. N., private communication.
56. SAKURADA, I., ITO, T. and NAKAMAE, K. *Makromol. Chem.*, **75** (1964) p. 1.
57. SAKURADA, I., ITO, T. and NAKAMAE, K. *J. Polymer Sci. C*, **15** (1966) p. 75.
58. GRUBB, D. T., DLUGOSZ, J. and KELLER, A. *J. Mat. Sci.*, **10** (1975) p. 1826.
59. BRITTON, R. N., JAKEWAYS, R. and WARD, I. M. *J. Mat. Sci.*, **11** (1976) p. 2057.
60. WARD, I. M., 1977, paper presented at IOP conference *Polymeric Materials at Low Temperatures*, London. (Private communication.)
61. SHAUFFELE, R. F. and SHIMANOUCHI, T. *J. Chem. Phys.*, **47** (1967) p. 3605.
62. LA GARDE, V., PRASK, H. and TREVINO, S. *Disc. Farad. Soc.*, **48** (1969) p. 15.
63. CAPIATI, N. J. and PORTER, R. S. *J. Polymer Sci., Polymer Phys. Ed.*, **13** (1975) p. 1177.
64. ANDREWS, E. H. and MARTIN, G. E. *J. Mat. Sci.*, **9** (1974) p. 1507.
65. BRISCOE, B., private communication.
66. BAUGHMAN, R. H., GLEITER, H. and SENDFELD, N. *J. Polymer Sci., Poly. Phys. Ed.*, **13** (1975) p. 1871.
67. KAMBOUR, R. P. and ROBERTSON, R. E., in *Polymer Science*, ed. A. D. Jenkins, North-Holland, London, 1972.
68. HALPIN, J. C. and KARDOS, J. L. *J. Appl. Phys.*, **43** (1972) p. 2235.
69. ANDREWS, E. H. *Pure and Appl. Chem.*, **39** (1974) p. 179.
70. BASSETT, D. C. and CARDER, D. R. *Phil. Mag.*, **28** (1973) p. 535.
71. BUCKLEY, C. P., GRAY, R. W. and MCCRUM, N. G. *J. Polymer Sci., Polymer Lett. Ed.*, **8** (1970) p. 341.
72. TAKAYANAGI, M., IMADA, K. and KAJIYAMA, T. *J. Polymer Sci. C*, **15** (1966) p. 263.
73. BOWDEN, P. B. and YOUNG, R. J. *J. Mat. Sci.*, **9** (1974) p. 2034.
74. FRANK, F. C., KELLER, A. and O'CONNOR, A. *Phil. Mag.*, **3** (1958) p. 64.
75. PETERLIN, A. *J. Mat. Sci.*, **6** (1971) p. 490.
76. HAY, I. L. and KELLER, A. *Kolloid-Z.*, **204** (1965) p. 43.
77. PETERLIN, A. and CORNELIUSSEN, R. *J. Polymer Sci. A2*, **6** (1968) p. 1273.
78. BESSELL, T. J. and YOUNG, R. J. *J. Polymer Sci., Polymer Lett. Ed.*, **12** (1974) p. 629.
79. SPEERSCHNEIDER, C. J. and LI, C. H. *J. Appl. Phys.*, **34** (1963) p. 3004.
80. YOUNG, R. J. *Polymer*, **16** (1975) p. 450.
81. ATTENBURROW, G. E. and BASSETT, D. C. *J. Mat. Sci.*, **12** (1977) p. 192.
82. YOUNG, R. J., BOWDEN, P. B., RITCHIE, J. M. and RIDER, J. G. *J. Mat Sci.*, **8** (1974) p. 23.
83. SHINOZAKI, D. and GROVES, G. W. *J. Mat. Sci.*, **8** (1973) p. 1012.
84. KIHO, H., PETERLIN, A. and GEIL, P. H. *J. Polymer Sci. B*, **3** (1965) p. 257.

85. Kiho, H., Peterlin, A. and Geil, P. H. *J. Appl. Phys.*, **35** (1964) p. 1599.
86. Gleiter, H. and Argon, A. S. *Phil. Mag.*, **24** (1971) p. 187.
87. Allan, P. and Bevis, M. *Proc. Roy. Soc.* (*London*), **A341** (1974) p. 75.
88. Holland, V. F. *J. Appl. Phys.*, **35** (1964) pp. 1351, 3235.
89. Petermann, J. and Gleiter, H. *Phil. Mag.*, **25** (1972) p. 813.
90. Petermann, J. and Gleiter, H. *J. Mat. Sci.*, **8** (1973) p. 673.
91. Peterson, J. M. *J. Appl. Phys.*, **37** (1966) p. 4047.
92. Young, R. J. *Phil. Mag.*, **30** (1974) p. 85.
93. Bevis, M. *Colloid and Polymer Sci.*, to be published.
94. Bevis, M. and Crellin, E. B. *Polymer*, **12** (1971) p. 666.
95. Young, R. J. and Bowden, P. B. *Phil. Mag.*, **29** (1974) p. 1061.
96. Allan, P., Crellin, E. B. and Bevis, M. *Phil. Mag.*, **27** (1973) p. 127.
97. Andrews, E. H. and Voigt-Martin, I. G. *Proc. Roy. Soc.* (*London*), **A327** (1972) p. 251.
98. Preedy, J. E. and Wheeler, E. J. *J. Mat. Sci.*, **12** (1977) p. 810.
99. Pietralla, M. *Colloid and Polymer Sci.*, **254** (1976) p. 249.
100. Young, R. J., Bloor, D., Batchelder, D. N. and Hubble, C. L. *J. Mater. Sci.*, **13** (1978) p. 62.
101. Bassett, D. C., private communication.
102. Baughman, R. H., Apgar, P. A., Chance, R. R., MacDiarmid, A. G. and Garito, A. F. *J. Chem. Phys.*, **66** (1977) p. 401.
103. Young, R. J. and Baughman, R. H. *J. Mat. Sci.*, **13** (1978) p. 55.
104. Geil, P. H. *J. Polymer Sci. A*, **2** (1964) p. 3813.
105. Cerra, P., Morrow, D. R. and Sauer, J. A. *J. Macromol. Sci.—Phys.*, **B3(1)** (1969) p. 33.
106. Geil, P. H. *J. Polymer Sci. A*, **2** (1964) p. 3857.
107. Gleiter, H. and Petermann, J. *J. Polymer Sci., Polymer Lett. Ed.*, **10** (1972) p. 877.
108. Petermann, J. and Gleiter, H. *J. Polymer Sci., Polymer Phys. Ed.*, **10** (1972) p. 2333.
109. Petermann, J. and Gleiter, H. *J. Polymer Sci., Polymer Phys. Ed.*, **11** (1973) p. 359.
110. Galiotis, C. and Young, R. J., to be published.
111. Watt, W. *Proc. Roy. Soc.* (*London*), **A319** (1970) p. 5.
112. Andrews, E. H., this volume, Chapter 1.
113. Tod, D. Ph.D. thesis, Queen Mary College, University of London, 1978.
114. Carbuglio, C., Ajroldi, G., Casiraghi, T. and Vittadini, G. *J. Appl. Polymer Sci.*, **15** (1971) p. 2487.
115. Bessell, T. J., Hull, D. and Shortall, J. B. *J. Mat. Sci.*, **10** (1975) p. 1127.
116. Martin, J. R., Johnson, J. F. and Cooper, A. R. *J. Macromol. Sci.*, **C8(1)** (1972) p. 57.
117. Marshall, G. P., Culver, L. E. and Williams, J. G. *Plastics and Polymers* (1970) p. 95.
118. Hertzberg, R. W., Nordberg, H. and Manson, J. A. *J. Mat. Sci.*, **5** (1970) p. 521.
119. Rivlin, R. A. and Thomas, A. G. *J. Polymer Sci.*, **10** (1953) p. 291.
120. Andrews, E. H. *J. Mat. Sci.*, **9** (1974) p. 887.

121. Anderton, G. E. and Treloar, L. R. G. *J. Mat. Sci.*, **5** (1971) p. 562.
122. Sims, G. L. A. *J. Mat. Sci.*, **10** (1975) p. 647.
123. Andrews, E. H. and Walker, B. J. *Proc. Roy. Soc. (London)*, **A325** (1971) p. 57.
124. Teh, J. W., White, J. R. and Andrews, E. H. *J. Mat. Sci.*, **10** (1975) p. 1626.
125. Way, J. L. and Atkinson, J. R. *J. Mat. Sci.*, **7** (1972) p. 1345.
126. Vadimsky, R. G., Keith, H. D. and Padden, F. J. *J. Polymer Sci. A2*, **7** (1969) p. 1367.
127. Peterlin, A. *J. Polymer Sci. A2*, **7** (1969) p. 1151.
128. Peterlin, A. *J. Macromol. Sci.—Phys.*, **B6(4)** (1972) p. 583.
129. Lloyd, B. A., De Vries, K. L. and Williams, M. L. *J. Polymer Sci. A2*, **10** (1972) p. 1415.

Chapter 8

FAILURE OF POLYMER COMPOSITES

D. R. MULVILLE and I. WOLOCK
U.S. Naval Research Laboratory, Washington DC, USA.

SUMMARY

A large number of studies have been conducted over the past five years on the failure of polymer composites. They have demonstrated that a number of variables affect the failure behaviour of composites, and in addition, the fracture process itself is quite complex. As a result, it is difficult to conduct comprehensive investigations of composite failure encompassing all of the variables and to derive broad generalisations regarding failure behaviour. With the complexities of the failure process, failure criteria developed for composites are also complex, as are the failure modes.

In this chapter fracture mechanics and delamination models are reviewed for describing failure of initially debonded or notched composites. First ply failure and probabilistic models are discussed for unnotched composites, as well as empirical multiaxial failure criteria for assessing the effect of complex loading states.

Failure modes observed experimentally for polymer composites under both static and fatigue loading are described. The residual strength of laminates after fatigue loading and the effects of cycling frequency are discussed.

1 INTRODUCTION

This chapter is a review of work reported over approximately the past five years in certain areas of the broad field of 'Failure of Polymer Composites'. The areas covered are those that are considered important or that bear a close relationship to important topics and in which significant progress has been reported. There are a number of fairly recent publications which provide information on additional aspects of the chapter topic, to which the reader is referred.[1–12]

The fracture behaviour of polymer composites is affected by many variables. These include fibre, matrix, fibre/matrix bond, fibre orientation, stacking sequence, void content, mode and rate of loading, and environment. Because of this very large number of variables, it is difficult to make broad, generalising statements about the fracture behaviour of composites. Most investigators conduct a relatively limited number of tests and study only a few of these variables. The resulting conclusions are often reported without specifying all of the limitations. Their results may conflict with the results of other investigators in which some of the more subtle unreported conditions, such as stacking sequence, are different. Therefore the comparison of data on the failure of composites must be done with care.

In reviewing the fracture behaviour of fibre-reinforced polymers, one must keep in mind that the failure process in composites is far more complex than that of isotropic materials such as metals, ceramics and plastics. Certain differences have been observed in various aspects of the failure process of composites compared to the more conventional and more familiar isotropic materials.

Metals, for example, fail by the initiation and growth of a single crack. However, there are several possible damage modes in composites. These include matrix cracking, fibre-matrix interfacial bond failure, fibre breakage, void growth, and delamination. As a result, it is usually difficult to identify and observe the growth of a single crack that controls strength in a composite. A multi-directional composite under complex loading may exhibit one or all of these damage modes, and it is currently difficult to predict in advance which will dominate and cause failure.

Since fibre reinforced composites are highly anisotropic, complex local stress states will exist in the composite even under simple uniaxial loading. This will influence the direction of failure propagation, which may not be perpendicular to the applied load.

The complex nature of the fracture process usually makes it difficult to develop a clear unambiguous general definition of failure in composites. There may not be a clean physical separation of the specimen into two or more pieces such as is usually observed in metals, ceramics and plastics. Some composites can carry a significant load even after a failure zone has propagated completely across the specimen. The complex failure process also makes the use of a single failure criterion difficult.

There may be some additional factors to be considered when one reviews the fracture behaviour of composites under cyclic loading, rather than static or monotonic loading. In metals, much of the fatigue life is spent before cracks appear. Then a dominant crack propagates fairly rapidly at a rate that can be predicted. In composites, cracks usually appear after relatively few cycles, even at low stress levels. However, the damage is quite complex and grows slowly, and as stated previously, one cannot usually identify a single dominant crack that controls the strength. Thus in composites much of the fatigue life is spent after the appearance of the first crack. In addition, the mode of failure in composites under cyclic loading may be significantly different from that observed for monotonic loading.

2 THEORETICAL FAILURE CRITERIA

2.1 First Ply Failure

Failure initiation in composite laminates can be characterised in some cases by failure within a single ply. Analysis methods used to describe this behaviour are known as first ply failure criteria (FPF), and have been reviewed recently by Tsai and Hahn.[13] Generally, cross-plied glass reinforced laminates subjected to uniaxial tension along one of the fibre directions will exhibit a change in stress-strain response as the failure strain of the 90° plies is approached. Microcracking in the matrix of a 90° ply (FPF), occurs at this point, reducing the stiffness of the composite laminate. While failure of a single ply may occur well below the ultimate failure of the laminate, Tsai and Hahn point out that environmental degradation and reduced fatigue life may eventually result.

Sendeckyj, Richardson and Pappas[14] conducted an investigation of the basic stress-strain response and strength of graphite/epoxy composites under tensile loading. Their theoretically predicted response curves based on non-linear analysis of balanced laminates gave excellent results up to first ply fracture. Lack of agreement beyond first ply failure was related to failed ply unloading mechanisms which redistribute failed ply strains.

Recent developments in application of FPF address the mechanism by which loading in the failed ply is transferred to adjacent plies. Yeow and Brinson[15] describe various stiffness reduction methods used to unload the failed ply and/or uncouple the failed lamina from load-displacement relations of the laminate. Generally these reduced stiffness methods include using a tangent modulus to unload the failed ply and then modifying the laminate stiffness relations, for example, by assigning zero stiffness to appropriate terms in the stress-strain relations. Chou *et al.*[16] have recently proposed reducing the modulus of the 90° plies in a quasi-isotropic glass reinforced composite to one third of its initial value when first ply failure occurs. Their subsequent numerical analysis shows that this model can be used to compute load-strain relations which agree well with experimental results obtained under uniaxial loading. Yeow and Brinson[15] proposed a staggered first ply failure model in which failed plies are assumed to be capable of carrying a fraction of the applied stress and are not totally ineffective. Fracture studies conducted on graphite epoxy laminates, (0°)8s and (0/±30/0°)2s, indicate that their model provides a more conservative strength prediction for uniaxial loading than reduced stiffness methods.

McLaughlin *et al.*[17] have also investigated load transfer between laminae resulting from first ply failure. They point out that an increase in load carrying capability is possible if the failed ply does not induce a large stress concentration or if large scale delamination between laminae does not occur. These investigators used an inter-laminar stress analysis in conjunction with laminated plate theory to analyse first ply failure as well as second ply and subsequent failure mechanisms in graphite/epoxy composites, (0/90°), (±45°) and (0/±45°) laminates. Predictions of second ply failure were reasonably accurate as long as axial tension (fibre breakage) was the failure mode of the second ply. If other failure modes occurred, such as delamination or transverse failure (matrix failure), their layered plate theory could not accurately predict second ply failure. Their results demonstrate the importance of interlaminar stress effects in prediction of ultimate laminate failure.

Francis *et al.*[18] have recently conducted an investigation of FPF in graphite/epoxy laminates under biaxial loading. Tubular specimens, (0/90°) and (±45°), were subjected to axial and torsional stress in addition to internal pressurisation. A FPF envelope was predicted on the basis of a maximum strain criterion, and results of experimental studies were in general agreement with predictions. Ultimate failure of the tubular samples ranged from 15% to 400% of predicted FPF depending on mode of loading; consequently the degree of conservatism in a structural design based on FPF can be quite variable. Tsai and Hahn[13] point out that under biaxial loading ultimate failure stress is not always guaranteed to be higher than the FPF stress. Their data on graphite/epoxy composites indicate that under some conditions FPF may trigger ultimate failure, and that the degree of conservatism predicted can be changed substantially by the failure criteria used to determine FPF.

Reifsnider[19] points out that FPF is rather poorly defined since an entire ply does not fail completely at a specific load level. Instead, a small number of cracks initiate within the ply. His observations indicate that the transverse cracks, which initiate within the ply, develop a characterisitic spacing, and that the crack spacing depends on the laminate stacking sequence and the material properties of the ply. He also notes that when the characteristic spacing is attained further cracking no longer occurs within the ply, although delamination and cracking may spread into adjacent plies. Reifsnider[20] postulates that the characteristic spacing can be predicted by analysis of the stress distribution within the laminate, and his numerical simulation studies show qualitative agreement with experimental failure studies in graphite/epoxy composites.

2.2 Empirical Multiaxial Failure Criteria

In 1972 Sandhu[21] and Sendeckyj[22] presented brief surveys of failure theories with particular emphasis on empirical multiaxial strength criteria for composite materials. Sendeckyj discussed the basis for many of the present failure criteria and stated that Gol'denblat-Kapnov and Tsai-Wu criteria provided the best fit to experimental data. Subsequently, Wu[23] described a method for conducting optimal experimental measurements of a biaxial interaction tensor to improve the accuracy of the Tsai-Wu failure criterion. Experimental results were obtained under biaxial loading using axial force, torsion and internal pressure on composite tubular specimens. Later Rowlands[24] presented a review of various failure criteria including maximum strain, Petit, Waddoups, von Mises, Tsai-Wu, Sandhu, Chamis, Puppo-Evensen and Wu-Scheublein analyses. He compares theoretical predictions with experimental data obtained on biaxial tubular specimens and off-axis tensile coupons. Based on the correlation with experimental data he concludes that none of the present analysis methods represents observed results sufficiently accurately to be employed confidently by itself in design. The two biggest deficiencies are the lack of laminate theory to realistically account for interlaminar and non-linear effects and an insufficient base of reliable experimental results.

In his review of failure criteria, Wu[25] defined a phenomenological failure criterion as a mathematical model of the material transfer function relating external excitation to the material response. Failure was interpreted as a significant change in material response. This approach to failure prediction is not intended for extrapolation of data or for interpretation of failure mechanisms. Its intent is rather to provide aid in experimental design, interpolation and correlation of experimental observations, and to provide allowed quantitative values for strength for rational engineering design, fabrication and materials formulation. Wu has shown that various phenomenological failure criteria for anisotropic materials, such as maximum strain criterion, maximum stress criterion, von Mises-Hill criterion and numerous other quadratic formulations such as Ashkenazi, Gol'denblat-Kapnov, Hoffman, Marin, Norris, Fischer, Franklin, and Chamis are special cases of a general tensor polynomial failure criterion. Experimental results were presented in this study which show the application of the tensor polynomial and other theories to describe failure in graphite composites. Wu and Ruhmann[26] extended the tensor polynomial strength criterion to permit a description of transient failure surfaces, which result from the materials' time-dependent loss of strength. The modified criterion uses time/environment dependent strength tensor components as compared to the static strength components used in the original formulation. Strength vectors, which define the radial distance to the failure surface, can be used with a vectorial shift parameter to describe time-dependent changes in the strength of composites subject to complex stress states. Recently Wu[27] has described the application of a three dimensional tensor polynomial failure criterion to strength characterisation of thick composite plates. The potential areas of application are the effects of free edges, notches and cutouts on the strength of thick composite structures. Wu's three dimensional analysis employs a number of assumptions concerning the symmetry of the tensor polynomial strength coefficients. These assumptions reduce the number of experiments which must be performed; however, even with his simplifications, additional experimental data is required on the transverse shear strength and transverse biaxial strength in the thickness direction.

Hashin[28] has proposed a modification of the tensor polynomial failure criterion by requiring that the failure criterion be piecewise continuous. Consequently, his model can take different forms in each of the loading quadrants, where the form of the failure criterion could conceivably be based on observed failure modes for each loading quadrant. This model is similar in some respects to a higher order polynomial representation of the data, although it is composed of a number of discrete lower order polynomials. In this failure criterion as well as those previously discussed, the question of path independence has not been raised. There is some evidence suggesting that the resultant failure stress levels may depend on the path taken in the loading space. If experimental studies demonstrate that the mode of failure and the magnitude of the loading at failure are dependent on loading history, then it will be necessary to modify these criteria to account for path dependence. Wu and Scheublein [29] have discussed formulation of an anisotropic failure criterion using higher order polynomials. Tennyson *et al.*[30] have recently shown that the cubic form of the tensor polynomial strength criterion provides a more accurate failure prediction of pressurised graphite/epoxy and glass/epoxy tubes, four ply ($\pm\theta°$), than a quadratic formulation. Recently Guess and Gerstle[31] have used filament wound tubular specimens of graphite/epoxy and Kevlar/epoxy composite to study the effects of combined stress states produced by internal pressure and axial loading. They contend that a Norris-Ashkenazi failure criterion provides a better prediction of failure than a maximum stress criterion. Other polynomial formulations were not considered in this study. In another study, Narayanaswami and Adelman[32] have shown that a Hoffmann or Tsai-Wu tensor polynomial formulation can predict failure in a composite under general biaxial load to within 10%.

An extension of the polynomial failure criterion to cyclic fatigue has been reported by Sims and Brogdon.[33] They modified the Tsai-Hill failure criterion for static strength by replacing the static strength parameters in the formulation with fatigue functions. The three necessary fatigue functions for prediction of fatigue strength are determined from the shape of the longitudinal, transverse, and in-plane shear fatigue curves, i.e., the S-N or stress v. number of cycles to failure curves. Sims and Brogdon used S-N data for longitudinal, transverse, and ($\pm 45°$) laminates to compute the in-plane shear S-N curve shape. First ply fatigue failure and corresponding S-N curve shapes for any other laminate were then computed, based on the fatigue functions. Predicted S-N curves obtained using this formulation showed good agreement with fatigue test data obtained on ($\pm 30°$) and ($\pm 60°$) glass reinforced epoxy laminates. They also used a Weibull model, similar to that of Hahn and Kim[34] for ($\pm 45/90°$)s glass reinforced epoxy, to compute probability of survival S-N curves for cross-plied glass/epoxy systems.

In order to facilitate the use of quadratic and tensor polynomial failure criteria, Belytschko *et al.*[35] and Collins and Crane[36] have developed graphical methods for visual presentation. The former graphics package has been developed to illustrate failure surfaces in both stress and strain space. Hahn and Tsai[37] have proposed a graphical representation for failure surfaces based on maximum strain theory, tensor polynomials or other lamina failure criteria. Their method relies on an orthogonal transformation of strain components to construct a failure surface for each lamina of the composite.

2.3 Fracture Mechanics

Applications of linear elastic fracture mechanics to composite materials has met with only limited success. This is in part due to the complex nature of the failure processes in most composite laminates and the assumptions made in formulating fracture mechanics parameters. Kanninen *et al.*[38] presented a critical assessment of the applications of fracture mechanics to fibre-reinforced composites. They divide composite materials fracture research into two broad categories. The first is a continuum mechanics analysis for homogeneous anisotropic linear elastic materials containing a flaw of known geometry, and the second is a micromechanical analysis of the crack tip region in unidirectional or laminated composites.

Numerous experimental studies have attempted to demonstrate the validity of linear elastic fracture mechanics based on a continuum approach. Durchlaub and Sacco[39] studied fracture of unidirectional and cross-plied glass reinforced composites, as did Owen and Bishop,[40] who also formulated an empirical correction factor to obtain constant stress intensity factors independent of the crack length. Slepetz and Carlson[41] investigated the fracture of graphite/epoxy composites and Pines *et al.*[42] studied the effects of fibre/matrix interface combination on fracture toughness. The conclusions of all of these researchers was that classical fracture mechanics is not valid when a zone of damage occurs rather than a planar crack. In general, planar cracks in composites result from loading which initiates crack growth parallel to the fibres in unidirectional composites or causes delamination between plies in composite lamination. Consequently, most successful fracture mechanics applications have addressed these types of problems.

Other investigators who have reported successful application of fracture mechanics in describing transverse fracture of glass reinforced unidirectional laminates and of (0/90°) graphite fibre composites include Barnby and Spencer,[43] and Phillips,[44] respectively; although in some cases the mode of failure was perpendicular to the notch rather than self-similar crack propagation. Barnby and Spencer[43] used an experimental calibration curve to compute a stress intensity factor based on transverse fracture, and subsequently they applied these results to describe mixed mode failure of a glass reinforced composite (Spencer and

Barnby[45]). Slepetz[46] conducted experimental fracture studies on single-edge notched cross-plied S-glass/epoxy and graphite/epoxy laminates. He observed that a significant damage zone developed at the crack tip, which resulted in an increase in compliance and fracture toughness for graphite/epoxy specimens. Critical values of the stress intensity factor reported by Slepetz are based on an experimental compliance calibration, which was correlated with a finite element calibration using a crack length correction factor based on damage zone site. Glass/epoxy composites, on the other hand, did not exhibit crack growth, but instead the damage zone continued to grow with increased load. He concludes that failures observed in cross-plied glass laminates appear to be too complex to be predicted by the usual fracture mechanics approach.

Despite the complex nature of the failure process, many investigators have continued to develop formulations for stress intensity factors for composites assuming self-similar crack growth. Stress intensity factors have been obtained using analytic analyses by Sani and Stern,[47] Sih[48] and others, and using numerical methods by Papaioannou *et al.*,[49] Mandell *et al.*[50] and Atluri *et al.*[51] Other fracture parameters such as the strain energy density theory have been applied to composites by Sih and Chen[52] and Sih *et al.*[53] This theory was used to investigate mixed mode failure in filament reinforced composites with particular emphasis on predicting direction of crack growth based on minimum energy density. As with formulation for the stress intensity factor the material is treated as a continuum. Wu[54] has also predicted crack initiation trajectory by combining his formulation of a tensor polynomial failure criterion with an analysis of the stress distribution in the vicinity of a notch tip. Experimental observations of crack initiation in laminated glass reinforced epoxy composites were in good agreement with theoretical predictions.

Path-independent integral concepts, such as the J integral, have also been used to study the toughness of plane orthotropic materials. Tirosh and Berg[55] described an experimental method to evaluate the J integral via strain gauge measurements along an appropriate path. These results for a unidirectional glass reinforced epoxy composite were then related to an associated stress intensity factor. Tirosh and Berg also discussed the formulation of J integral components for mixed-mode loading. Mixed-mode fracture of graphite composite laminates, $(0/\pm45°)$s and $(0/\pm45/90°)$s, has also been studied by Morris and Hahn[56] using an effective normal crack length to fit experimental data.

A brief summary of some of these fracture mechanics approaches applied to composites has recently been compiled by Smith and

Mullinix[57] as a fracture mechanics design handbook for composite materials. The limitations and restrictions discussed in their work as well as those cited here demonstrate that the application of fracture mechanics methods to composite materials requires considerable caution. In fact, most studies conducted to date have applied fracture mechanics methods to relatively simple laminate configurations under unidirectional loading, which have limited applications to complex composite structures.

Whitney and Nuismer[58,59] have presented two stress criteria for predicting variation in tensile strength for various size discontinuities, such as a hole in a composite plate, without applying principles of linear elastic fracture mechanics. Their point stress criterion predicts failure when the stress over some distance (d_0) away from the discontinuity is greater than the strength of the unnotched material. In their analysis, (d_0) is a characteristic dimension independent of laminate geometry and stress distribution, and is determined empirically from experimental data. Physically this approach analyses the distance over which material must be critically stressed to find sufficient flaws of critical size to initiate failure.

Their second approach, the average stress criterion, states that failure occurs when the average stress over a distance (a_0) exceeds the unnotched laminate tensile strength. The distance (a_0) is also assumed to be independent of laminate geometry and stress distribution and is considered a first approximation to the distance over which failure has taken place. As with (d_0), (a_0) is determined from a best fit of experimental data. Although these criteria were developed without resorting to classical concepts of linear elastic fracture mechanics, a relationship between unnotched laminate tensile strength and mode I fracture toughness was determined.

An approach similar to the average stress criterion was proposed by Waddoups *et al.*[60] based on a fracture mechanics formulation. He analysed the problem of a crack emanating from a circular hole in a composite plate under tensile load by assuming a region of intense energy at the hole boundary. The intense energy region was defined in terms of a characteristic length, a, evaluated by fitting experimental data for graphite epoxy $(0/\pm 45°)2s$ laminates. Waddoups' data was analysed by Cruse,[61] who presented a formulation in which a notch of half length lambda was modelled as an ideal crack of a greater half length lambda*. His results show that values of a* = lambda* – lambda are similar to values for characteristic length obtained by Whitney and Nuismer[58] and Waddoups.[60] According to this model an equivalent notch size can be used to predict fracture initiation from notches in composite laminates under tensile loading. Cruse[61] also postulated a strength criterion for laminates based on lamina failure, by assuming the energy released by co-planar crack growth at fracture was the sum of the energies released by each angle-ply constituent. This formulation is subject to the restriction that crack growth is co-planar, failure is by cleavage mode and that the zone of failure generated by matrix cracking is small. This analysis does not apply for delamination type of failure or for failure by other than opening mode. Konish and Cruse[62] applied a failure model for angle-ply laminates based on lamina strength properties to predict failure in angle-ply graphite/epoxy laminates. They assumed that crack growth occurs when either the ply stress parallel to the fibres or transverse to the fibres reaches its ultimate value at some specified distance ahead of the initial flaw. Consequently, their lamina model yielded two strength predictions, one for fibre-dominated failure and another for matrix-dominated failure. Agreement between experimental results and theoretical predictions are generally good as long as the fracture is fibre-dominated. Konish and Cruse also used Cruse's model described above for general orthotropic laminates to predict failure in angle ply as well as $(0/\pm 45°)$s and $(90/\pm 45°)$s laminates. As previously mentioned, self-similar crack growth is assumed, and inter-ply cracking or independent matrix cracking are not accounted for in the formulation for fracture toughness. Strain energy release rate values obtained on edge-cracked and centre-cracked plates agreed with predicted values provided that the crack length was of sufficient size. Their results indicate that specimen configuration and thickness may have less of an effect on fracture toughness than the size of the crack-tip damage zone, notch-tip width and ply thickness relative to the crack length.

Recently, Tirosh[63] has analysed fracture initiation near an inhomogeneity, i.e., cut-out or inclusion, in a laminated plate. He assumes that a Griffith crack of length l interacts with the elevated stress field generated by the embedded inhomogeneity. Using this analysis along with a uniaxial tensile strength and a characteristic flaw dimension, the critical tensile or compressive load can be predicted for a given inhomogeneity. These results can be viewed as using an equivalent crack size to predict the stress at fracture for a circular hole subjected to tension or compression.

Of associated interest is the work reported by Bhatia and Verette,[64] Waddoups[65] and Sendeckyj[66] on developing concepts for crack arrest in composites. Bhatia and Verette conducted an experimental study of crack arrest in panels of graphite/epoxy with buffer strips of the same material. They found that fracture can be controlled under both static and fatigue loading; the crack arrest capability however, depending on the initial flaw size. They also reported that linear elastic fracture mechanics can be used to predict initial fracture strength. In the latter study, Sendeckyj uses parameters proposed by Waddoups *et al.*[60] to account for the effect of crack size. Three crack-arrestment criteria are presented for different locations of crack-arrestment buffer strips relative to position of the initial crack. The comparison between predicted and observed loads are in good agreement for boron/epoxy laminates with buffer strips of the same material, but not as good for graphite/epoxy laminates. This lack of agreement is attributed to extensive crack branching and delamination which occurred in the graphite/epoxy buffer strips.

A number of empirical laws have been proposed to describe crack growth rate in fatigue as a function of the stress intensity factor. Many authors have applied models of this type to composites, for example, Kunz and Beaumont,[67] Mandell and Meier,[68] Underwood and Kendall,[69] and others. In general, they have found that relations of this type apply as long as the dimensions of the damage zone are small. An additional point is that the restrictions which pertain to the application of fracture mechanics models apply to these interpretations of fatigue crack growth as well. Mandell and Meier in their investigation combined a crack growth rate formulation with a Miner's cumulative damage relationship to predict crack growth in successive ligaments at the crack tip in cross-plied glass/epoxy laminates. They concede that a Miner's equation does not agree with experimental results on cumulative fatigue of composites. Broutman and Sahu[70] have developed a modified Miner's law analysis to predict fatigue behaviour of fibre reinforced composites. They propose a residual strength model which assumes a linear strength reduction in the composite due to cyclic loading. Residual strength of the composite is then determined at a given stress level as a function of the number of cycles or fraction of life spent. They also assume that the slope of the strength reduction curve at a given stress level is not affected by pre-cycling at a different stress level. Predicted residual strengths for tensile loading show good agreement with limited experimental results obtained with cross-plied glass reinforced laminates.

Micromechanical analysis of crack tip regions in fibrous composites has been studied by Piggott[71] and Sanchez-Moya and Pipkin.[72] In these analyses, a regular array of fibres is assumed, and the fracture energy of the composite is determined as an initial crack propagates obliquely or transversely through the composite. Models of a composite system have been fabricated and tested to simulate crack growth in a fibrous composite as, for example, by Conrad.[73] These studies and those of McCall and

Morley[74] on crack growth in hybrid composites provide some insight into toughening mechanisms in composite materials.

An extension beyond the micromechanics analysis of crack tip regions in composites was presented by Zweben[75] in the analysis of notched unidirectional composites. He adapted an engineering approach considering average stress states near notch tips to predict fracture. Analytical methods such as the analysis by Wang *et al.*[76] of a through crack in a ($\pm 45°$) laminate have been used to compare computed stress distributions with those predicted by classical fracture mechanics. They have found in their analysis a deviation from elastic fracture mechanics prediction. In all of the studies involving micromechanics analysis near the crack tip region, a problem arises in relating these results to macromechanics analysis. Kanninen *et al.*[77] have proposed a hybrid model which includes a local heterogeneous region near a notch tip embedded in a linear elastic anisotropic continuum. The two regions are assumed to be uncoupled by setting the displacements at the boundary of the local heterogeneous region to be the same as if the entire cracked body were in an elastic continuum. Nuismer[78] used a continuum model of microdamage growth to develop ply constitutive relations for both damaged and undamaged states of a lamina. Ply constitutive relations were then incorporated in a non-linear incremental finite element model to predict damage growth and ultimate failure in a laminated plate containing various size holes. In his formulation, Nuismer used the concept of internal state variables to represent the current damage state of the material. He assumed two different types of damage occurred: one normal to the applied load and the other parallel to the loading axis. Changes in these damage states give rise to changes in the internal state variables, which are expressed as coefficients in the ply constitutive relations. Using this analysis the spread of both fibre and matrix failure in a laminated plate containing a hole was predicted. Despite the assumptions concerning the type of damage, interaction between damage types, and neglect of three dimensional effects, the comparison between observed and predicted ultimate failure strength of laminates has been good. Further studies of this type which accurately model the complex failure modes in composites may permit the successful application of fracture mechanics to general laminate failure analysis.

Another approach has been taken recently by Beaubien *et al.*[79,80] which differs in several respects from classical fracture mechanics and empirical failure criteria. These theories are based on strain at a point, or, in the case of fracture mechanics, in an infinitesimal neighbourhood of a crack tip. The theory advanced by Beaubien *et al.* requires that the strain field in the entire region in question be used as the failure parameter. This approach is based on the observation that the failure process in a composite is far more complex than the concept of a crack or a crack front would provide. Hence it is necessary to focus on a zone of failure and the strain field about this zone as opposed to focusing on a crack tip or the strain field about that tip. Based on this concept a failure prediction technique for composite structures has been demonstrated using correspondence of the strain field in a failure zone as a failure criterion. First a finite element analysis was used to compute the notch tip zone strain field for failure initiation in laboratory coupons. Then the strain field in the notch tip region of a composite structure was approximated by the strain fields produced in the laboratory coupons, and the corresponding point of failure initiation was predicted. This technique differs from other analyses in that it is not based on an analytic failure theory, but rather on experimental data obtained over a broad range of loadings. Smith[81] in his assessment of the limitations of fracture mechanics applied to composites, points out the complexities associated with modelling multi-layered composites. Smith also discusses in detail a number of micromechanical and macromechanical theories of composite failure and the nature of the approximations used in each.

2.4 Delamination Failure

Interlaminar stresses generated in composite laminates under unidirectional loading have been related to lamina stacking sequence by Pipes and Pagano.[82] They demonstrated that the free-edge effect on the interlaminar stresses could be a significant factor in determining the strength of laminated composites. Recently Wang and Crossman[83] have addressed the same problem for five different graphite/epoxy composites using an improved numerical analysis procedure. They determined that (+45/−45/0/90°)s laminates develop considerable tensile stress normal to the thickness, whereas (90/0/−45/+45°)s laminates develop compressive stresses. Subsequent experimental studies demonstrated that the former laminates were approximately 20% weaker than the latter, suggesting that the tensile edge effects reduced the strength of the laminate. These authors also noted that the edge effects on the in-plane stresses may be as physically significant as the interlaminar stresses, and that the 'singular' behaviour of both in-plane and interlaminar stresses was highly localised. Hsu and Herakovich[84] investigated edge effects in (±45°)s angle-ply graphite/epoxy composites, and found that normal stresses remained finite under tensile loading; however interlaminar shear stresses exhibited a 'singular' behaviour. Hsu and Herakovich[85] also studied cross-ply symmetric laminates subjected to uniaxial extension and determined that the normal stress at the free edge in (0/90°)s graphite/epoxy composites is tensile, whereas in (90/0°)s composites it is compressive. A point worth noting is that while these and other studies of free edge stress distributions under uniaxial tension have demonstrated the variation due to stacking sequence and offered some correlation with tensile strength, there has not been a corresponding study of the effect of more complex loading states. Consequently the conclusions reported for uniaxial loading tests may not necessarily be applicable to the complex loading states normally encountered in composite structures.

Stresses developed along the free edge in composite laminates together with applied mechanical and thermal loading can cause delamination to initiate from an interlaminar debond. Analysis of delamination originating from a surface crack has been conducted by Wang and Mandell.[86] This problem for interlaminar defects has also been investigated by Kulkarni *et al.*[87] using an analytical model based on energy balance to predict crack growth. Estimates of fracture energy were obtained for graphite/epoxy $(0^4/\pm 45°)4s$ laminated cantilever beam specimens containing an interlaminar defect. A strain energy release rate formulation for stable crack growth in free edge delamination was formulated by Rybicki *et al.*[88] They used a finite element analysis to model crack initiation and stable crack growth. Their observations of delamination in boron/epoxy laminates $(\pm 30/90°)$s were that fracture energy was nearly constant during propagation of delamination.

Delamination mode of failure in composites is particularly important under impact loading. Moon *et al.*,[89] Chamis,[90] and Mallick and Broutman[91] have shown that significant normal and shear stresses develop during impact, which can initiate delamination. The latter authors have also shown that impact properties in glass composites are significantly affected by fibre orientation. Off-axis composites fail by brittle interfibre cleavage with little delamination, whereas unidirectional and cross-plied composites fail by interfacial fibre splitting and delamination. Of particular interest is the residual static and fatigue strength of a composite laminate following impact which initiates an interlaminar flaw.

Wang[92] has analysed delamination failure initiating from a surface notch in $(\pm 45/90/0°)$s graphite/epoxy composites. His results are based on a hybrid-stress finite-element model, which includes a crack tip element with singular terms. Fracture mechanics parameters, $K_{\rm I}$, $K_{\rm II}$ and $G_{\rm I}$,

G_{II} are obtained for given crack geometry and laminate properties. (K_I, K_{II} are stress intensity factors in modes I and II respectively and G_I, G_{II} are the corresponding energy release rates.) Stacking sequence, for three different laminate configurations, was demonstrated to have little effect on the computed values for K_I and K_{II}. However, the effect due to the notch depth was found to be significant.

2.5 Probabilistic Failure Models

Various probability models have been proposed to describe failure of unidirectional composite materials. Phoenix[93] has reviewed a number of these theories ranging from single-fibre tensile strength models to fibre bundle strength analysis, and finally failure models for unidirectional composites. These latter models are generally based on 'weakest link' failure theories applied to monotonically loaded composites. When a fibre breaks at its weakest point, load is transferred along the fibre and into the neighbouring fibres. The degree of sophistication of the probability models depends on whether fibres are considered to fail independently of each other, a non-interacting model, or whether failure of a single fibre can be related to subsequent failure of neighbouring fibres, an interacting model. The probability models for unidirectional composites discussed by Phoenix involve partitioning the composite in different ways. The dispersed fracture model divides the composite into a series of layers each of which consists of a bundle of fibre elements loaded in parallel. Failure of the composite occurs when any one bundle fails, based on a 'weakest link' failure hypothesis. Hitchon *et al.*[94] use a non-interacting probability model to describe tensile failure of unidirectional carbon fibre reinforced plastics and found that this simple hypothesis was not a good description of material behaviour. They suggest that interacting models may be more representative.

The Zweben-Rosen cumulative weakening model partitions the composite into fibre elements, and determines the probability that an arbitrary fibre element will fail. This failure is termed a zero-order break. If interaction between zero-order breaks is neglected, the probability of failure in adjacent fibre elements, first-order breaks, can be determined. Second-order breaks are determined in the same manner, and final failure is based on a selected number of failures occurring at a zero-order failure site. A number of limitations of these methods are discussed by Phoenix,[93] and an alternative analysis is proposed with somewhat less restrictive assumptions.

A cumulative weakening type of model has also been formulated by

McMahon.[95] In this model stress enhancement due to a broken fibre is postulated as a mechanism leading to failure of neighbouring fibres. A failure criterion is then based on a 50% probability of an additional fibre failure in a specified neighbourhood of approximately 100 fibres. Sun and Yamada[96] have combined a statistical formulation of strength parameters based on Weibull distributions with three failure criteria, maximum stress, maximum strain and maximum work. Their results on single-ply glass reinforced epoxy composite indicate that the Weibull distribution depicts the data quite well and that a probabilistic maximum work failure criterion adequately predicts the strength distribution for off-axis loaded unidirectional composites.

Recently, Harlow and Phoenix[97] have investigated chain-of-bundles probability models for the strength of fibrous composite materials. They raise several key questions and conjectures regarding tensile failure and then compare two assumptions concerning distribution of loading after fibre failure. The first assumption is the equal load sharing rule in which all non-failed fibre bundles share the load equally, and the second is the local sharing rules in which fibres adjacent to broken fibres carry a significantly larger portion of the load than remote fibres. Harlow and Phoenix point out that although both of these models can be formulated and analysed in theory, computational difficulties are significant.

Awerbuch and Hahn[98] used a two parameter Weibull distribution to determine the probability of survival for unidirectional graphite/epoxy composites under static loading. They applied a similar model to describe the residual strength and fatigue life distribution for the same material. In the studies and in the investigations by Hahn and Kim,[34] a two parameter Weibull model was used to analyse the scatter in fatigue life data. Based on these analyses, fatigue life curves can be determined corresponding to any desired degree of reliability. Hahn and Kim[99] also investigated the application of a Weibull probability model to describe proof testing in composite materials. Using this analysis they were able to compute a minimum expected lifetime guaranteed by proof testing for unidirectional glass reinforced epoxy. Awerbuch and Hahn[98] attempted to relate the fatigue life and residual strength to the static strength through proof testing. Based on their analysis they conclude that a minimum life can be guaranteed through proof testing with approximately 90% probability for unidirectional graphite/epoxy, if there is no visible damage after the proof test. Yang[100] has also used a two parameter Weibull distribution to represent the static tensile strength distribution of notched graphite/ epoxy laminates. He modelled residual strength and fatigue life with a three parameter Weibull distribution and his studies on tension-compression and tension-tension fatigue show reasonably good correlation between experimental data and the predicted residual strength and fatigue life distributions.

Halpin *et al.*[101] developed a probability model for residual strength, where the shape of the static and fatigue failure distributions are materials' variables independent of structural geometry and scale. In addition, his results imply that for a constant mode of failure, the fatigue life shape parameter is independent of history of loading. This analysis provides an analytical method for prediction of wear-out in structures subjected to random or constant amplitude loading, and has been applied with some success to glass/epoxy composites by Halpin[102] and to boron/epoxy composites by Waddoups.[65]

3 EXPERIMENTAL FAILURE MODES

In reviewing the experimental studies of the fracture behaviour of polymer composites, one should keep in mind the comments made in the introduction, of the large number of variables that affect the behaviour. In comparing the results of different investigations, these variables, such as the fibre, matrix, fibre orientation, type of loading used, and whether notched or unnotched specimens were used, all should be kept in mind and compared.

One of the major limitations to the usefulness of work reported on failure modes in polymer composites is that almost without exception, the studies are conducted under conditions of uniaxial loading. However, most applications of composites will be in structures in which the material is subjected to complex loading. The question then is how much of the results already reported and still being obtained under uniaxial loading can be utilised in predicting the behaviour or failure modes under complex loading. With the complexity of the failure process in composites, the answer may be, 'very little'. Unless there are grounds to believe the opposite, it would appear that studies of failure modes of polymer composites under complex loading are very much in order.

The term 'notch sensitivity' is used frequently in this section. If a material is notch sensitive, then it will fail in the presence of a notch when the stress on the net cross-section is less than the ultimate strength of the material. There are several precautions one should be aware of in reviewing the results of notch sensitivity studies. Most of the work reported on notch sensitivity of composites is performed under uniaxial tensile loading only. Yet the results are reported as 'notch sensitivity' without the caveat of tensile loading only. The same conclusion may not apply for shear loading or compressive loading or combinations of these loadings. In addition, the results of notch sensitivity tests may differ, depending on whether the tests are conducted under load control or displacement control, and will probably also be influenced by loading rate.

Unless otherwise stated, for a $(0, \pm\theta°)$ laminate, the load is applied parallel to the 0° fibres, whereas for a $(\pm\theta°)$ laminate, θ is also the angle between the reinforcement and the applied load axis.

For the sake of conciseness and generalisation in this section, the presentation of a composite configuration as, for example, (0, ±45, 90°) refers to any or all stacking sequences with that general orientation, unless indicated otherwise.

3.1 Static Failure Modes

3.1.1 Glass Reinforced Composites

One broad aspect of the failure of composites deals with the behaviour of angle-ply laminates without notches. Rotem and Hashin[103] investigated the failure of balanced angle-ply glass/epoxy laminates under uniaxial tension using unnotched specimens. The study revealed three distinct failure modes: for reinforcement angles smaller than (±45°); for the (±45°) angle-ply; and for angles larger than (±45°).

The onset of failure observed for a (±30°) laminate was caused by high interlaminar shear stresses at the edges of the specimen. Delamination between the layers propagates slowly with increasing stress level, which is in qualitative agreement with results of an analysis of the interlaminar shear stress distribution showing rapid decrease of shear stress toward the specimen centre.[82] As the stress level increases, the failure condition is reached in a location nearer to the centre. The interlaminar delamination does not affect the apparent load and elongation as long as the laminae themselves do not fail, i.e., no cracks appear through the lamina. When through-the-thickness lamina cracking starts, the specimen fails instantaneously. Similar failure mechanisms were observed for the (±35°) and (±40°) laminates.

By changing the lamination angle by a few degrees, from 40° to 45°, the behaviour changes from a brittle type (close to linear stress-strain curve up to fracture) to a ductile type (stress-strain curve has a plateau of elongation under constant load). The onset of failure for this laminate is not due to interlaminar shear but to lamina cracking, starting just before the plateau. The specimen does not fail but continues to sustain the load, unlike the other angle-ply laminates. Cracking occurs progressively in various domains, criss-crossing the specimen. The number of cracks increases with continued elongation of the specimen. After a certain elongation, interlaminar yielding starts, causing bundles of fibres to move and change orientation, and thus increases the applied load necessary to continue the elongation.

When the lamination angle is changed again by just another few degrees, from ($\pm 45°$) to ($\pm 50°$), the failure mode again changes radically. The ($\pm 50°$), ($\pm 55°$) and ($\pm 60°$) laminates show the same type of failure mode, but different from the two previously described. In this case, initial failure occurs by a lamina crack in the fibre direction, followed by progressive parallel cracks at close distance until ultimate failure of the laminate. The failure mode is of local nature, the damaged region being limited to a strip consisting of parallel cracks. The width of the crack region diminishes with increasing lamination angle. These laminates also exhibit brittle behaviour.

To summarise, for laminates with angles greater than ($\pm 45°$), catastrophic failure is observed when the first layer fails and the laminate cannot support additional load. For lamination angles less than ($\pm 45°$), but greater than ($\pm 30°$), the failure mode is one dominated by interlaminar shear behaviour. At a lamination angle of ($\pm 45°$), lamina cracking parallel to the fibres occurs. However, delamination does not initially occur to any great degree, thus permitting transfer of lamina stress to fibres in adjacent laminae.

Mandell *et al.*[104] investigated the size and character of the damage zone at the tip of sharp notches in fibre reinforced plastic laminates under various loading conditions. The damage zone is a region of stable crack growth at the tip of the main crack or precut notch, consisting of subcritical cracks extending parallel to the fibres of each ply (subcracks) and in some cases delaminations between plies. The size of the damage zone is characterised by the length of the subcrack in each ply (there is usually only one subcrack per ply) and the extent of any delamination zone between plies. The damage zone is thought to be related to the crack propagation resistance of composites, with larger damage zones leading to high values of fracture toughness. It has been suggested that the damage zone size may play an analogous role in the fracture of composites to that played by the plastic zone in a high strength metal.

When specimens of graphite/epoxy cross-ply laminates containing a sharp precut notch were subjected to increasing tensile loads, the following sequence of events was observed. At very low loads, the specimen deforms elastically with no damage. At a threshold value of load, subcracks form parallel to the fibres of each ply at the notch tip. As the load increases above the threshold value, the subcracks extend, and some region of delamination may form between plies. At some critical maximum load the notch extends as a main crack, and the specimen is separated into two pieces.

The subcracks propagate parallel to the fibres of the ply in a planar fashion, and penetrate only a single ply thickness. The subcrack is a simple separation between the fibres of the ply, at some points splitting off fragments of the anisotropic fibres.

In ($\pm\theta°$) cross-ply laminates, where $\theta < 45°$, only a single subcrack is typically observed in each ply if the precut notch is sufficiently sharp. For $\theta \geq 45°$, the subcracks reach an instability point whereupon they may multiply in number and extend rapidly across the entire width of the specimen prior to catastrophic failure. Laminates with larger damage zones and laminates having certain ply orientations tend to exhibit interply delaminations as well as intraply subcracks.

For ($\pm 15°$), ($\pm 30°$), and (0, 90°) laminates, all of which failed in a notch-sensitive manner, the subcrack length increased with increasing load. Subcrack length also increased with load for the ($\pm 45°$) laminates but these exhibited notch-insensitive behaviour.

A strong dependence of damage zone size and fracture toughness on ply thickness, fibre orientation and fibre properties is demonstrated for the cases studied.

Data obtained appear to support the use of damage zone size to characterise the fracture toughness of composites of similar composition, analogous to the situation with metals.

Im *et al.*[105] studied the growth of surface cracks in glass/epoxy composites under tensile loading. The configurations studied were ($\pm 15°$), ($\pm 45°$) and (0, 90°), using specimens with surface cracks halfway through the thickness and of various lengths. Damage did not occur until the applied stress reached 20–30% of the ultimate strength. First damage was in the form of delamination at the interface between the last cut layer and the first uncut layer, extending eventually only to the precut crack tips. Splits developed at each end of the crack, somewhat parallel to the applied load, depending on the orientation of the longitudinal fibres. The delamination grew within the bounds established by the splits. In most cases, damage which extended under higher loads was dominated by central delamination between the precut plies and the remainder of the specimen. No significant regions of broken fibres were present in four-ply laminates prior to fracture, but significant fibre fracturing occurred in the precut plies of multi-ply laminates having more layers.

The rate of damage growth was sensitive to ply configuration, materials and specimen geometry. In similar tests conducted on graphite/epoxy specimens with a surface notch, the damage was similar to that for glass/epoxy but less massive in the (0, 90°) configuration, and almost identical in the (±15°) configuration.

Specimens with surface cracks and through-cracks were tested for notch sensitivity under tensile loading. For glass/epoxy, the specimens with surface cracks having the configuration of (0, 90°), (±15°) and (±45°) were all found to be notch insensitive. For through-cracks, the (0, 90°) were notch sensitive but the (±45°) were insensitive. For graphite/epoxy specimens with surface cracks, the (0, 90°) were notch sensitive but the (±15°) and (±45°) were notch insensitive. Thus notch sensitivity of a composite under tensile loading is influenced by the type and orientation of the reinforcement and the type of notch.

Underwood and Kendall[69] conducted fracture studies of glass/epoxy composites using single-edge notch tensile specimens and two layups, (0, 90°) and (0, ±60°).

In monotonic loading tests, crack growth in the (0, 90°) material is accompanied by extensive delamination above and below the notch. Damage in the (0, ±60°) material is restricted to a triangular region ahead of the notch with the crack extension tending to be parallel with one or the other 60° fibre directions. The more concentrated damage zone in the (0, ±60°) material is believed to contribute to the lower crack growth resistance in this material.

For both orientations, the static fracture damage zones were relatively concentrated as compared to the long, narrow damage zones observed under fatigue loading.

Chaplin[106] studied compressive failure in unidirectional glass/epoxy composites to investigate the low compressive strength of unidirectional composites. The approach was that low compressive strength can be attributed to sensitivity to defects from which local failure can propagate across the material. It is shown that under an axial compressive load parallel to the fibres, once initiated, a shear mode of failure can propagate across the material. One of the reasons for glass-reinforced plastics not reaching theoretical predictions for compressive strength must be sensitivity to defects. The failure itself is essentially a shear instability similar to that which occurs in the unreinforced resin. Based on observation of failures in unidirectional composites, Chaplin concludes that the term 'microbuckling' often used to describe the compressive failure mode seems inappropriate. In the failure mechanism described, shear deformation is restricted to a band of material inclined to the plane normal to the fibre axes. The relationship between the orientation of the failed band of material and the limiting shear deformation in the band is explained in terms of volumetric strains. Tests are described which demonstrate that in glass/epoxy composite, this type of failure can propagate from a notch, and this notch sensitivity is proposed as an explanation for the apparent inadequacy of the theoretical model. It is found that fibre fracture at the boundaries of the band and interlaminar failure within the band follow as a result of increasing shear deformation in the band. In the failure band, as the shear strains increase, the increase in volume dictated by the geometry will lead to the development of transverse tensile stresses within the band. These, coupled with the shear, bring about the extensive failure of adhesion.

Several investigations were conducted on the formation of transverse cracks in cross-ply laminates. Garrett and Bailey[107] studied the formation of transverse cracks in cross-ply laminates by testing specimens of a (0, 90°) glass/polyester laminate in tension in a direction parallel to one of the directions of reinforcement. Extensive cracking occurred in the transverse ply at strains much lower than the resin failure strain. The cracks formed in a direction parallel to the transverse reinforcement and perpendicular to the applied stress and showed a remarkably even spacing. The spacing decreased with increasing applied stress and increased with increasing transverse ply thickness. There was no evidence of debonding between the plies during cracking and a multiple cracking theory in which the plies remain elastically bonded is presented.

Stevens and Lupton[108] conducted an analysis which indicated that the number of transverse cracks per unit length at a given stress level in a cross-ply laminate may be reduced by: (1) increasing the transverse ply thickness, which is in agreement with the conclusions of Garrett and Bailey;[107] (2) increasing the transverse ply fracture stress; (3) increasing the elastic modulus of the laminate; (4) decreasing the interply shear strength. It was shown that the relation between stress and crack spacing can be explained in terms of shear-stress transfer from the adjacent longitudinal plies.

Garrett and Bailey[109] also investigated the effect of resin properties on the transverse cracking behaviour of a glass/polyester three-ply (0/90/0°)

laminate. The polyester resin was modified by adding flexibiliser to increase the failure strain from 2% to 11%. Generally, transverse cracking initiated at higher strains for the higher strain resins. No cracking was observed in the most flexibilised resin laminate up to failure. The onset of transverse cracking is associated with a 'knee' in the stress-strain curve and a sharp increase in acoustic noise. Laminates made with the lower failure strain resins cracked in a brittle fashion and the cracks propagated rapidly across the entire specimen cross-section. Occasionally longitudinal cracks formed in the outer plies of the laminate made with the most brittle resin. For the more ductile resins, damage was more diffuse, less catastrophic, and the cracks were less distinct and less straight. The results are explained by the Kies strain magnification theory.

Parvizi *et al.*[110] studied further the effect of inner-ply thickness on the apparent failure strain of the inner ply of a cross-ply laminate. Specimens of a (0/90/0°) cross-ply glass/epoxy laminate were tested in tension parallel to the direction of reinforcement in the outer plies. The thickness of the inner ply was varied and cracking constraint was observed at small thicknesses. At large inner-ply thicknesses, the specimens showed uniform transverse cracking, and at very low inner-ply thicknesses, this transverse cracking could be suppressed completely prior to total specimen failure.

Cracking constraint was observed at inner-ply thicknesses less than about 0·4 mm. Complete suppression of inner-ply failure was observed at inner-ply thicknesses of 0·1 mm. From 0·4 to 0·1 mm, small cracks, predominantly edge cracks, were observed which propagate slowly under increasing load, finally triggering total inner-ply failure.

3.1.2 Graphite Reinforced Composites

Mandell *et al.*[111] reported on fracture modes for glass/epoxy and graphite/epoxy composites of different configurations, in most cases using notched specimens. The initial response of a notched graphite/epoxy laminate to a low tensile stress is the formation and extension of a damage zone at the notch tip. This damage zone consists of splits penetrating through the thickness of each ply and propagating parallel to the fibres, with possible delamination between plies. Four types of ply cracking within a damage zone were observed. These were single splits, multiple splits, propagation of a crack with fibre fracture, and formation of a fibre kink band.

The two most commonly observed fracture modes for catastrophic failure are: opening mode type of specimen displacement with fracture of fibres; shearing type of displacement with buckling of some plies and little tensile fracture of fibre.

In the case of opening mode displacement, and depending somewhat on the reinforcement configuration, splits form at the tip of the notch, defining ligaments of fibres, and the crack propagates ligament by ligament, breaking the fibres in each ligament as it progresses. The length of the splits, which define the ligaments and blunt the crack, is sensitive to the ply thickness and ply stacking sequence.

In the shear-buckling mode, the fracture originally propagates from the notch tip as a split and a delamination zone which subsequently buckles. This mode is usually constrained to fibre angles of greater than 45° to the load. The splits initially extend in a stable manner, and then propagate more rapidly across the entire specimen above some critical load. For these fibre orientations, the damage zone spreads across the specimen prior to fracture and the materials become notch sensitive.

McLaughlin *et al.*[17] conducted an analytical investigation into the behaviour of graphite/epoxy composites under tensile loading after the outer lamina has failed. For (0, 90°) and (0, $\pm 45°$) laminates under the stresses considered, debonding of the failed lamina was not found to be an important failure mechanism. The main mode of failure is overstressing of laminae adjacent to an initially failed layer. Because of this, interlaminar shear stresses are an important feature of failure analysis of these laminates. Delamination effects in (0, $\pm 45°$) graphite/epoxy laminates are not predicted to play a significant role in laminate ultimate failure under tensile loading. However, delamination effects are important for ($\pm 45°$) laminates and may also be important for ($\pm \theta°$) composites over a significant range of θ.

Scott and Phillips[112] studied the effect of increased fracture toughness in an epoxy resin achieved by the addition of carboxyl terminated acrylonitrile-butadiene copolymers on the toughness of an adhesive bonded joint and a graphite fibre composite made using these resins. Fracture tests were conducted using double cantilever beam specimens. In the adhesive tests, an aluminium adherend was used. The tests show that by adding 9% copolymer, the fracture energy, G_c, increased by a factor of 20 for the resin, a factor of 10 for the adhesive joint, and a factor of 2 for a unidirectional composite with pure opening mode force applied. It is postulated that the suppression of the rubber toughening mechanism in the composite is associated with the very small resin film thickness in the composite.

Hancox[113] studied the effects of flaws and voids on the shear properties of graphite/epoxy composites, using a pultruded tube, filament-wound tubes and solid rods, all loaded in torsion. Critical lengths were determined for surface flaws for both types of tubes, but the rods were not very sensitive to surface flaws. Voids had a serious degrading effect on shear strength and modulus, reducing properties to 30% of their void free value at 5 volume % voids.

3.1.3 Boron Reinforced Composites

Durchlaub and Freeman[114] conducted studies of the strength and failure modes of notched boron/epoxy composites, using plate specimens having a centre hole and under tensile loading. Static failures for (±45°), (0, ±45°), (0, 90°), and (0, ±45, 90°) laminates all occurred through the notch at 90° to the load axis. All of these laminates exhibited notch sensitivity.

Kulkarni *et al.*[115] reported on static tensile tests of boron/epoxy laminates using specimens with a centre hole. The specimens failed in a transverse crack propagation mode originating at the centre hole, similar to that reported by Durchlaub and Freeman.[114]

3.2 Fatigue Failure Modes

In a summary of a recent symposium on Fatigue of Filamentary Composite Materials, Lauraitis[116] expressed some relevant comments on that subject. She pointed out that although the amount of work done in fatigue of high-modulus fibre-reinforced composites has increased exponentially in the past several years, progress appears slow and uncertain. She also stated that there are apparent inconsistencies in the conclusions drawn by the authors. This comment also applies in general to the published literature on fatigue of composites. Some researchers observed an increase in compliance with fatigue cycling while others have strongly maintained that no such change occurs. Residual strength has been observed to increase, decrease and remain unchanged for both notched and unnotched laminates. As she pointed out, residual strength results generate considerable confusion. Failure modes due to static and fatigue loading were assumed to be the same by some authors, while others claim that they are different. Only limited summarisation of the results was possible, because of the inconsistencies in the conclusions drawn by the authors.

The apparent contradictions may be caused by different materials, fibre orientations, stacking sequences, specimen size or geometry, test conditions, or other such variables. Often inadequate details are given to permit an exact comparison of different investigations.

These comments serve well as an introduction to a section reviewing recent experimental studies on failure modes in polymer composites under fatigue loading. Summarisation is still quite difficult, not only because of inconsistencies and lack of adequate details of test conditions, but also because of the problem of lack of standardisation in reporting descriptions of morphological behaviour. Finally, in some cases, alarmingly few specimens are tested and examined, making the reliability of the results reported somewhat doubtful.

3.2.1 Glass Reinforced Composites

Dharan[117] conducted fatigue studies of unnotched unidirectional glass/epoxy composites. In a monotonic tension test, he found that first detectable damage started at a stress level of 172 MN m^{-2} where fibre breaks initiated. At a stress of 242 MN m^{-2} matrix microcracks appeared, mainly at the surface and perpendicular to the applied stress. The number of cracks increased as the stress was increased. He stated that these cracks could be crazes.

The response to tensile fatigue loading was divided into three ranges. The first range was at stress levels high enough to produce failure within 100 cycles. In this region, damage was produced on the first cycle, significantly weakening the material and often producing large cracks and a high density of fibre failures. Local fibre breaks coalesced and propagated to connect to other such regions until the specimen failed catastrophically in a manner similar to a static failure. There was small dependence of fatigue life on the number of cycles in this region; fibre mean strength and strength distribution appeared to affect the behaviour profoundly. In the second region, the stresses were high enough to produce failure in 10^2 to 10^6 cycles. The number of cycles to failure increased linearly with decreasing stress on a log-log scale. Matrix microcracks still formed at these stress levels and fibre breaks were apparent, but not on the scale of the first region. The surface microcracks propagated perpendicular to the tensile axis, breaking fibres in the path of propagation until they were large enough to satisfy the criterion for shearing type propagation parallel to the fibres. Delamination began, leading to loss of stiffness and failure. In the third region, the maximum stress was below that required for microcrack initiation under static loading. Unlike the behaviour exhibited by the composites in the first two regions, where defects were formed in the first cycle and subsequently propagated, most of the cycles in this region were used in crack nucleation. After 10^6 cycles, the only obvious damage was crack nucleation; none of the specimens failed.

Kim and Ebert[118] also conducted fatigue studies of unnotched unidirectional glass-reinforced composities under tension-tension axial loading, but they used a vinyl ester matrix. The events in axial fatigue failure are described as fibre cracking, matrix crack propagation and delamination. They stated that the onset of fatigue damage occurs by fibre surface flaw propagation during the first loading cycle. Continued loading results in the flaws propagating and coalescing to produce fibre surface cracking which leads to interfacial failures and matrix cracking. The specimen finally fails completely by delamination parallel to the loading direction. They found that the surface condition of the composite specimen greatly affects the fatigue performance.

Tanimoto and Amijima[119] studied fatigue damage in composites made of a satin weave glass cloth with a polyester resin using unnotched reduced section specimens. The fatigue loading was tensile at a frequency of 30 Hz. They found that cracks form after a very few cycles even at a stress level of 20 to 30% of ultimate. The cracks formed first in the resin phase near fibres perpendicular to the load direction where the stress concentration is high. Cracks in this region may relieve the stress concentration without reduction in strength of the composite. This appears to be the main reason that initial cracks not only fail to decrease the static strength but sometimes increase it initially. These cracks propagate up to the longitudinal fibres and cracks then start to propagate parallel to the load along the longitudinal fibres and the residual strength starts to decrease rapidly. After about one-half of the fatigue life, the crack density remains almost constant in both directions, and the residual strength does not change greatly. The fatigue damage can be classified in three stages: (1) the formation of cracks in one or two cycles with no reduction in static strength; (2) for up to half of the specimen life, cracks propagate and the residual strength decreases continuously; and (3) the rate of propagation of cracks and reduction in static strength decreases until the fatigue life is reached and the specimen fails suddenly.

Hahn and Kim,[34] in tests conducted on a (0, $\pm$45, 90°) configuration of glass/epoxy composite, found that in both static and fatigue tests under tensile loading, cracks appeared first in the 90° plies, which have the lowest strain-to-failure. Cracks appeared in the 45° plies next, and these cracks eventually coalesce into a partial delamination. The ultimate fracture strain of the laminate is about equal to the longitudinal failure strain of the 0° plies.

Smith and Pascoe[120] studied the role of shear deformation and progressive damage in the monotonic and cyclic loading of a woven roving glass-fibre reinforced polyester composite. In-plane shear fatigue tests, using cruciform specimens, and uniaxial fatigue tests were conducted at 0°, 22·5° and 45° to the weave and show that three mechanisms of damage can occur: (1) rectilinear cracking aligned with the fibres; (2) delamination; (3) shear degradation of the matrix and the fibre/matrix interface. The appearance of one or all three mechanisms depends on the fibre angle and the components of normal stress and in-plane shear stress acting on the fibre plane. In-plane shear tests show that the extent of shear degradation is not simply determined by the shear stress magnitude but the fibre angle plays a principal role.

Mandell and Meier[68] described the mode of fatigue crack growth under tensile loading in an E-glass/epoxy laminate with a (0, 90°) configuration and a cycling rate of 5 Hz, using double-edge notched specimens. Crack growth appears to occur in a stepwise fashion. A longitudinal split is formed at the end of the notch on the first loading cycle. The split grows during many cycles of loading and a ligament of longitudinal 0° material is fatigued by the local stress field. The ligament fails and the crack extends by one ligament width. A split in the 0° fibres forms again at the tip of the crack and the process repeats. The 90° plies crack between fibres as the crack extends and are not believed to contribute substantially to the fracture resistance except through the effect of their constraint on the deformation of the 0° plies. Local delamination between plies at the crack tip is also observed. The effect of this on the crack growth rate is not clear.

Mandell *et al.*[111] pointed out further that for the (0, 90°) laminates, the split length, which represents a blunting effect of the notches, increased with stress level and number of cycles under fatigue loading. For (0, ±45°) laminates, the mode of crack propagation under fatigue loading was identical to that for the monotonic test, i.e., the shear-buckle mode, which originally propagates as a split and delamination zone and then subsequently buckles, as previously described for static fracture.

Underwood and Kendall[69] reported on fracture studies of glass/epoxy composites using single-edge notched tensile specimens. Two layups were used, (0, 90°) and (0, ±60°). However, the specimens were cut from the sheets in two ways, so that the load direction was parallel to and perpendicular to the 0° fibres in the outer layer. With regard to the load direction, then, the orientations were (0, 90°), (0, ±60°), (90, 0°), and (90, ±30°). Fatigue tests were conducted at a frequency of 30 Hz and a stress ratio $R = 0{\cdot}1$.

The fatigue damage zones in both the (0, 90°) and the (0, ±60°) materials consisted of relatively constant shaped, semi-elliptical regions of delamination or failure of the epoxy bond between layers. Away from the notch, only the outer two layers and occasional inner layers delaminated, whereas near the notch most of the layer bonds failed and some fracture within the layers occurred. The damage zones were similar to the successive, longitudinal splitting type of fracture observed previously by Mandell and co-workers under monotonic loading in a (0, 90°) glass/epoxy laminate and in another study under fatigue loading in a (0, 90°) glass/epoxy composite. However, apparently due to the difference in layer orientation, the current tests more often displayed a single longitudinal split and associated delamination zone. The fatigue damage zones were long and narrow compared to the more concentrated damage zones developed under monotonic loading.

Measurements indicated that most of the damage zone growth occurred during the early cycles and the rate decreased markedly with number of cycles, and growth tended toward arrest but never ceased entirely.

The (90, 0°) and (90, ±30°) materials with 90° outer plies exhibited lower fatigue life than the first two materials tested. The fatigue damage zone in the (90, 0°) material was considerably more concentrated than those observed in the (0, 90°) material. The monotonic fracture of the same specimen after fatigue loading occurred with almost no damage zone formation. The damage zones in the (90, ±30°) material were also quite concentrated and essentially the same for fatigue and monotonic loading. These concentrated damage zones in the 90° outer layer materials, as opposed to the long semi-elliptical zones observed in the 0° outer layer materials which blunt and retard the crack, are a controlling factor in their poor fatigue life.

3.2.2 Graphite Reinforced Composites

Ramani and Williams[121] reported on failure modes under tensile fatigue loading of graphite/epoxy composites of various configurations. For (0, ±45°) specimens with a centre hole, typical of the (0, $\pm\theta$°) configuration, at 10^3 cycles, a zone of damage occurs around the hole, consisting of parallel longitudinal splits at the edges of the hole, and some fine localised failure at 45°. The splits propagate longitudinally and the failure at 45°, originating at the hole boundary, intensifies. The diffuse zone of damage around the hole tends to spread predominantly in the direction parallel to the loading. At 10^5 cycles, fine delamination appears between the 0°, +45° and −45° plies. Fatigue damage at 10^6 cycles is still confined to the region between the splits and includes patches of delamination. Transverse spread of damage is slight, if any. The delamination cracks suggest interlaminar weakening which will result in a reduction in transverse, shear and compressive properties. Even at stress levels of 70% of the unnotched strength, which is much greater than the static net section strength for the specimens used with a centre hole, intense fatigue damage in the (0, $\pm 45°$) specimens is contained in the longitudinal band bounded by the splits originating at the hole edges. The results suggest the relative insensitivity of these laminates to tension-tension fatigue in the notched condition.

With double-edge notch specimens of (0, $\pm 45°$) laminate, damage was in the form of axial splitting at the notch roots, and fatigue damage is contained between the specimen edges and the notch root splits. In effect, the net section remains virtually undamaged.

The failure of notched (0, $\pm \theta°$) laminates in tensile fatigue is governed by the strong anisotropy introduced by the presence of 50% 0° plies. The high ratio of longitudinal to transverse strength of the 0° lamina encourages splitting at notch roots in 0° plies. It is significant that fatigue damage is well contained between the parallel splits at the hole edges; therefore longitudinal propagation of damage, parallel to the loading direction, constitutes an important mode of failure in situations where stresses act predominantly in the fibre direction. The small but finite amount of transverse cracking in the $\pm \theta°$ directions observed around stress concentrators in the notched fatigue tests could be an important factor at long lives on the order of 10^9 cycles. Moreover, the occurrence of fine-edge delamination cracks in this layup is likely to affect transverse, shear and compressive properties. Also, the authors state that the fact that edge effects occur under cyclic tensile loading in fibre-dominated laminates (containing 50% or more 0° plies) not containing 90° plies demonstrates that analytical predictions of edge delamination criteria in static loading do not directly apply to cyclic loading.

For the (0, ± 45, 90°) laminates with a centre hole, a zone of damage develops around the hole as early as 10^3 cycles, again mainly in the form of longitudinal splits. Some fibre-matrix separation along the 45° directions occurs within the region bounded by the splits, and several 90° ply failures also occur; these extend across the width of the specimen. These 90° failures are first observed in the lamina located in the mid-thickness of the specimen. At 10^4 cycles, the notch root splits are more pronounced and the extent of failure along 90° directions is noticeably increased. In the central region of the specimen, numerous closely spaced 90° failure lines can be observed, and edge delamination between plies close to the specimen surface is apparent. The damage around the hole has spread longitudinally to a considerable extent at 10^5 cycles. Failure along 45° directions intensifies, extending outward into the specimen net section from the hole boundary. A further increase in density of failure in 90° plies is evident and zones of delamination around the hole can be perceived. Beyond 10^5 cycles, the edge delamination effect becomes dominant and is the principal cause of eventual fatigue failure of the laminate. The propagation of edge delamination from the specimen edges into the interior coincides with the sharp decrease in residual tensile strength. Severe delamination between plies and kinks in delamination cracks are apparent, both normal and oblique, which appear to switch from between one set of plies to another set of plies. Eventually, the edge delaminations traverse the entire specimen width to effect failure. The apparent mechanism of failure suggests that the role of a notch in effecting fatigue failure is, at most, only secondary.

Thus the tendency of the (0, ±45, 90°) family of laminates to delaminate at the edges under static and cyclic loading conditions led to failure of the laminates in this study containing 90° plies. In the case of centre notched laminates, damage around the hole spreads longitudinally rather than laterally with cycling, so that the hole does not appear to play a significant role in effecting fatigue failure. This is corroborated by the observations that unnotched specimens when cycled at the same stress level as notched specimens undergo failure through the same edge delamination mechanism.

The fatigue mechanism for unnotched specimens in the (0, $\pm\theta$°) configurations appears to be one of general failure along 0° directions, leading to failure of 0° plies in later stages when load is transferred to them by the failed θ° plies. The dominant fatigue failure mechanism for both notched and unnotched (0, ±45, 90°) laminates is edge delamination propagating across the specimen to effect failure.

Experimental results demonstrate that it is possible to relate notched fatigue behaviour to unnotched fatigue behaviour for laminates under tensile cycling. For the (0, ±45, 90°) laminates, the notched and unnotched fatigue behaviour are virtually the same, since in either case, failure occurs by edge delamination. In the case of (0, $\pm\theta$°) laminates with a centre hole, the notched situation reduces to unnotched fatigue behaviour, since the stress can be increased to the unnotched strength level, with resultant notch insensitivity. However, the longitudinal zone of damage occurring above and below the hole is a potential source of failure.

Chang *et al.*[122] studied damage growth in graphite/epoxy composite specimens with a centre notch under constant amplitude tensile fatigue loading at a frequency of 5 Hz. For ($\pm 45°$) specimens, the failure mode under fatigue loading was similar to that for tensile ramp loading. The predominant damage was matrix cracks in the $\pm 45°$ fibre directions. The matrix cracks started at the ends of the machined slits and continued transversely to the edges of the specimen. Delaminations initiated from the areas of the matrix separation and grew catastrophically as the cracks progressed along the fibre direction. As matrix cracks in one direction occurred, fibres in the perpendicular direction took up the majority of the load until they failed. The fibre fracture lines extended from the slit tips at angles between 45° and 90°.

For (0, $\pm 45°$) specimens, the predominant damage was along the 0° fibre direction, with matrix cracks emanating from the slit ends. There were some matrix cracks in the $\pm 45°$ directions and delaminations started to grow along the 0° and 45° fibre directions as well as along the edges of the slit. Massive delaminations originated from matrix cracks along the 0° direction, and damage was largely confined within the slit area until failure. The specimen failed in the 90° direction by extension of the slit. In contrast to the $\pm 45°$ specimens, there was a difference in the failure modes exhibited by this material under tensile ramp loading and under fatigue loading. Under ramp loading, the predominant damage line was along the $\pm 45°$ fibre direction, whereas under fatigue loading it was along the 0° direction.

For the (0, ± 45, 90°) material, damage growth behaviour was similar to that for the (0, $\pm 45°$) material. The predominant damage line was along the 0° direction, with minor damage lines in the 45° and 90° directions. However, the damage areas extended to the edges of the specimens from the ends of the slit, instead of being confined within the 0° damage lines within the slit area. The 90° damage lines appeared early in the test, but did not appear to contribute significantly to the failure mechanism. Delaminations were mainly between the 0° and 45° plies. The important role of edge delaminations in initiating fatigue failures in this configuration, as described previously by Ramani and Williams,[121] apparently was not observed by these investigators.

Damage growth rate plots indicate that there is a predominant damage length that stabilises at the expense of increased damage growth in other damage lines. The specimen fails when the damages reach a critical stage. Under ramp loading, damage in areas other than the predominant damage grew much faster than under fatigue loading.

Sturgeon[123] conducted fatigue tests on graphite/epoxy composites using (0°), (0, ±45°), and (90, ±45°) orientations. Zero-tension fatigue tests were made at 20 Hz for (0°) and (0, ±45°) laminates and 10 Hz for (90, ±45°) laminates. Monotonically tested specimens with a centre hole failed through the section containing the hole. Damage in the cross-ply specimens initiated in the form of cracks at the hole which appeared early in fatigue life and ran parallel to the outer 45° fibres to the specimen edges where local delamination occurs. Generally final separation was through the section containing the hole. It was also observed that when the 45° cracks reached the specimen edge, initiating local delamination, another series of cracks started through the orthogonal 45° ply propagating across the specimen to the opposite edge. This formed what the author calls a zig-zag ladder. This behaviour has not been reported in other papers.

In the (90, ±45°) specimens, most specimens failed in fatigue through the section containing the hole. Extensive inter-ply delamination occurred before the final break.

The (0, ±45°) laminates exhibited a significant notch sensitivity under monotonic loading, which was reduced by fatigue loading. This reduction in sensitivity was accompanied by cracking of the composite which reduced the effectiveness of the hole as a stress raiser. It was stated that the degree of notch sensitivity would be affected by the level of surface treatment of the graphite fibres.

Porter[124] reported on constant amplitude tensile fatigue tests of graphite/epoxy composites at a frequency of 10 Hz and a stress ratio $R = 0{\cdot}05$. The layups used were (0, ±45, 90°), (0, ±80°), and (0, ±30°), the latter with one 0° ply of S-glass. Specimens were tested with holes and slits, both half penetration and full penetration. The (0, ±45, 90°) showed little visual damage, consisting of a longitudinal split of the outer 0° plies parallel to the loading, and transverse fracture with a small amount of delamination. The (0, ±80°) laminates displayed greater splitting, also penetrating the outer plies, with delamination in the plane of the 80° plies. The (0, ±30°) showed minimal damage, consisting of extensive delamination on the planes of the S-glass and extensive fibre pull-out of S-glass.

Sendeckyj *et al.*[125] studied the failure behaviour of surface-notched (0, ±45°) graphite/epoxy composites under static loading and spectrum fatigue loading. In static tests, failure initiated by matrix cracking and delamination in the vicinity of the notch, accompanied by local buckling of the delaminated surface. Typically, the failure surface consisted of two regions of failed plies, from the central delaminated area to each of the specimen edges, connected by the delamination zone. The primary failure occurs along the 45° directions from the surface notch to each edge of the specimen, while the secondary failure is straight across from edge to edge at the ends of the delaminated region away from the notch.

When load is first applied, the intact fibres in the vicinity of the surface notch are under a complex load distribution arising from (1) the local bending due to the eccentricity of the specimen cross-section and (2) the load concentration due to load transfer through the intact matrix in shear. The matrix not only serves as a binder stabilising the fibres under compressive loading but also provides the mechanism by which loads are transferred from the fibres cut by the surface notch to the adjacent intact ones. The load transfer, occurring by a shear mechanism, results in load concentrations in the fibres adjacent to the notch. The magnitude of the load concentration, which is a function of the number of cut fibres and shear capacity of the resin, dictates the strength reduction due to the surface notch. As the applied load is increased, the matrix above and below the surface notch fails. This results in the observed surface ply buckling pattern and a reduction in the load concentration due to shear load transfer. Fatigue loading causes extensive observable matrix damage in the vicinity of the surface notch. As long as the damage is confined to the matrix and is similar to that observed in static tests, there is no physical reason to expect the residual strength to differ from the static strength.

Awerbuch and Hahn[98] investigated the tensile fatigue behaviour of unidirectional graphite/epoxy composites at a frequency of 30 Hz using unnotched rectangular specimens. At high stress levels, most specimens failed suddenly in a brush-like manner typical of the static failure mode. At lower stress levels, damage initiation was in the form of longitudinal cracking along the edges of the specimen, which indicated the presence of cut fibres. Such damage was also observed in static tests. In some cases, more longitudinal cracks followed, starting from the original cracks, until complete failure occurred. In other cases, the original damage was arrested.

Ryder and Walker[126] studied the fatigue behaviour of graphite/epoxy composite in the (0, ±45, 90°) configuration, using unnotched specimens. Fatigue failure modes appeared to differ from static tension and compression failure modes. In static tension, failures were brittle-like, without any discernible plastic-like flow at the end of the stress-strain curve; specimens exhibited limited delamination. In tension-tension fatigue, specimens usually failed due to extensive delamination which started between two 45° plies at the free edge and progressed inwards. In static compression, failure was sudden and again brittle-like. Specimens tested in tension-compression failed similarly to those tested in tension-tension, but large out-of-plane buckling also occurred under tension-compression.

Nevadunsky *et al.*[127] conducted studies on tubular specimens of (±45°) glass/epoxy and (0/±45°) graphite/glass/epoxy composites loaded in axial fatigue at 30 Hz. In preliminary static tests, acoustic emission was detected for the (0/±45°) specimens at approximately half of the ultimate load at which point a secondary linear modulus was evident. The sound was attributed to 0° fibre breakage and fibre/matrix debonding. By contrast, (±45°) specimens did not emit sounds until 90% of the maximum load was reached.

During the fatigue tests, the specimen temperature increased and the axial modulus decreased correspondingly. This was interpreted as evidence that both are indicating the same phenomenon, i.e., the rate of accumulation of fatigue damage. A change in stiffness due to fatigue damage can be detected very early in the total life to fracture, depending on the magnitude of the load. The change occurs continuously over a large portion of the fatigue life.

However, damage in composites is anisotropic in direction. The torsional modulus did not change to the same extent as the axial modulus. This is related to the specimen fibre orientation and the type of loading. Orientations of (±45°) and (0, ±45°) would tend to sustain matrix damage during axial loading. As a result, the axial modulus will decrease. However, (±45°) fibres will continue to carry the load in torsion even though matrix damage exists. Thus one would expect less change in the torsional modulus. By contrast, 0° specimens would exhibit little change in axial stiffness, but large changes in torsional stiffness.

Phillips and Scott[128] studied the shear fatigue of unidirectional composites of graphite, glass and Kevlar® with epoxy resin by low frequency torsion cycling. During either constant shear stress or shear strain amplitude cycling, there is initially a slow decrease in shear modulus, the strain amplitude increasing and the stress amplitude decreasing approximately linearly with the log of the number of cycles. This behaviour is attributed to fibre/matrix debonding. Eventually macroscopic cracks initiate and propagate along the specimen axis, leading to rapid failure.

The change in compliance prior to cracking decreases to zero at a low shear strain amplitude, implying that debonding does not occur at a low strain. This may not define a fatigue limit; instead it may mean that the dominating fatigue mechanism changes to initiation of matrix cracking. This has not yet been proved. Fatigue in torsion results in a more rapid breakdown in properties than does short beam interlaminar shear fatigue.

Phillips and Scott[129] conducted cyclic torsion experiments on solid rods of graphite/epoxy composites. The salient features of the torsional fatigue of this material are: (1) under constant angular amplitude cycling the torque amplitude decreases linearly with the logarithm of the number of cycles (N), while under constant torque amplitude cycling the angular amplitude increases approximately linearly with log N; (2) the rate of decrease of torque amplitude during constant angular cycling and the rate of increase of angular amplitude during constant torque cycling increases with increasing torque and angular amplitudes; (3) after a given number of cycles, cracks form and propagate roughly radially in the matrix and matrix/fibre interface, resulting in failure.

The events that are proposed to occur in a small volume element of composite during shear stressing are: (1) plastic deformation of the matrix; (2) debonding at the fibre/matrix interface; (3) crack propagation at the fibre/matrix interface and in the matrix.

Sumison and Williams[130] investigated the torsional and flexural fatigue behaviour of (0°) and (±45°) graphite/epoxy composites. Both the unidirectional and cross-ply laminates undergo fatigue damage as evidenced by reduction in torsional stiffness when subjected to cyclic torsional stress. For constant torsional deflection tests, the fatigue damage curves are characterised by: (1) an initial incubation period before damage begins; (2) an accumulation of damage as evidenced by a reduction in stiffness; (3) a decrease in the rate of damage accumulation as evidenced by an approach to a constant stiffness value. Exposure to water during testing decreases the incubation period and increases the rate of damage accumulation; at 74 °C, water also decreases the limiting stiffness. Cross-ply specimens are more prone to fatigue damage than unidirectional.

Reduction in torsional stiffness caused by stress cycling is accompanied by reduction in failure strength in bending. In contrast to torsion fatigue results, flexural fatigue studies showed a greater effect of water on unidirectional specimens than on cross-ply specimens.

Kunz and Beaumont[67] reported on axial compressive fatigue studies of edge-notched bars of (0°) and (0, 90°) graphite/epoxy composites. Axial cracking was found to be the primary fracture mechanism in the unidirectional composites. On the other hand, splitting within and between transverse plies is instrumental in promoting crack extension in cross-ply composites. The cross-ply material has fewer fibres in the loading direction than the unidirectional, resulting in less resistance to crack extension by way of a fibre buckling process. Axial cracking along the fibre/matrix interface took place extensively in both materials but the cross-ply has a lower 0° fibre/matrix interface area and consequently a lower susceptibility to axial cracking. In addition, the alternating orientation of plies provides discontinuities across the specimen and prevents the linking up of microbuckling paths across an entire specimen by means of axial cracks.

Composites with higher interfacial shear strength are less susceptible to axial cracking along the fibre/matrix interface. High interlaminar shear strength minimises splitting between and within the transverse plies in a cross-ply laminate. There is an optimum bond strength, measured by fibre pull-out length, for which axial cracking is minimised and trans-fibrile crack extension prevented.

3.2.3 *Boron Reinforced Composites*

Kulkarni *et al.*[115] conducted analytical and experimental studies of the fatigue behaviour of boron/epoxy laminates in the 0° and (0, ±45°) configuration. There was considerable scatter in the 0° laminate fatigue data. Tensile fatigue tests were conducted on (0, ±45°) specimens with a centre hole with maximum stresses of 80% and 67% of the ultimate static notched strength, R = 0·1, and frequency = 30 Hz. The residual tensile strength increased and the residual tensile modulus decreased slightly as the number of fatigue cycles increased. Fatigue damage initiated with axial cracking in the surface 0° layers emanating from the two sides of the centre hole. This was followed by substantial interlaminar damage propagation, with the 0° surface layers separating from the 45° subsurface layers in the region of the projection of the notch diameter. No through-the-thickness axial cracking was observed.

During fatigue loading of a notched laminate, high axial and shear stresses exist near the notch root. These stresses will cause the material in the vicinity of the notch to be degraded much more rapidly than throughout the rest of the laminate.

Roderick and Whitcomb[131] studied the fatigue behaviour of boron/epoxy specimens with centre holes, with (0, ±45°) and (0, ±45, 90°) configurations, and two stacking sequences for each. The loading was tension-tension (R = 0·05) and compression-compression (R = 20) at a frequency of 10 Hz.

The damage sequence for the (0, ±45°) specimens was: intralaminar cracks developed around the edge of the hole, transthickness (through-the-thickness) and local intralaminar cracks in the ±45° plies developed in narrow bands parallel to the loading axis and tangent to the hole; in these bands, the 45° fibres loaded in maximum tension broke; in the same area, the other 45° fibres loaded in compression broke.

For the (0, ±45, 90°) laminates, intralaminar cracks developed around the edge of the hole; intralaminar cracks in the ±45° plies developed from the edge of the hole along the transverse centreline of the specimen; near the hole, 0° and 45° fibres broke where intralaminar cracks in the 45° plies had occurred; further away from the hole, the 45° fibres broke where intralaminar cracks in the 45° plies had occurred; catastrophic failure followed. There were no transthickness cracks.

Initially, failure damage in both the (0, ±45°) and (0, ±45, 90°) laminates occurred as intralaminar cracks around the edge of the hole. Then, intralaminar cracks in the 45° plies propagated from the edge of the hole. For the (0, ±45°), these cracks occurred simultaneously with transthickness cracks and propagated along lines parallel to the loading axis whereas for (0, ±45, 90°), no transthickness cracks occurred and damage propagated along lines perpendicular to the loading axis. Finally, in both types, fibres broke where intralaminar cracks in the 45° plies had occurred. Damage in the (0, ±45°) laminates was most pronounced under compressive loads and had little effect on either residual strength or stiffness. In contrast, damage in the (0, ±45, 90°) laminates occurred primarily under tensile loads, reduced the residual strength, and frequently resulted in catastrophic failure across the notch. Primarily 45° fibres broke prior to final failure. Results of this study suggest that the 45° plies play a key role in the fatigue process of boron/epoxy laminates that contain them. Microscopic damage, intralaminar cracks and fibre breaks in both types of laminates occurred primarily in the 45° plies.

Stinchcomb *et al.*[132] conducted tensile fatigue studies of (0, ±45°) boron/epoxy composites, with two different stacking sequences at different frequencies. The tests were conducted on rectangular specimens with a centre hole. They found that the fractures contained both transverse and longitudinal damage but to different degrees, depending on the frequency. At lower frequencies, the damage mode was primarily transverse, but at the highest frequency, it was very predominantly longitudinal.

In each case, the damage nucleated as small longitudinal splits in the matrix from points on the edge of the hole but off the transverse axis.

3.3 Residual Strength Properties

Polymer composites often exhibit a change in static strength and stiffness properties as a result of fatigue loading. One fact not emphasised by many investigators is that any such change is dependent on the fibre orientation and the particular strength or stiffness parameter measured. Investigators often use just the term 'stiffness' or 'modulus', when reporting changes in 'axial tensile stiffness' or 'tensile modulus', as if this one parameter characterises the strain response under all conditions of loading. For example, as Nevadunsky *et al.*[127] point out, there are conditions under which, as a result of fatigue cycling, one might observe changes in axial tensile modulus but not torsional modulus, and vice versa. The same comments apply to strength measurements. Thus as a result of fatigue loading in the axial direction, a ($\pm 45°$) laminate would exhibit less change in the torsional modulus than in the axial modulus, whereas the reverse would be true for a (0°) laminate. Reporting the results as a change in modulus or stiffness alone without specifying the type or fibre orientation is inadequate and misleading and should be kept in mind in reviewing such data.

The investigations reported indicate that changes in strength properties of polymer composites as a result of cyclic loading depend on the configuration of the reinforcement, the cycling stress level, and the type of notch if used.

3.3.1 Glass Reinforced Composites

Mandell *et al.*[111] studied the effect of cyclic tensile loading on the monotonic strength of glass/epoxy laminates. The studies were conducted on unnotched and double-edge notched specimens having the configurations (0, 90°) and (0, $\pm 45°$). They found that for the (0, 90°) configuration, the residual toughness, K_Q, of the notched specimens increased up to 20% after cycling at several stress levels up to 10^6 cycles. The residual tensile strength of unnotched specimens decreased up to 20% for the same number of cycles.

For the (0, $\pm 45°$) material, the residual tensile strength of unnotched specimens decreased with cycling at three stress levels. However, the notched specimens also exhibited a decrease in net section fracture stress after cycling up to 10^6 cycles, which is contrary to the behaviour of the (0, 90°) material. The authors postulate that the increasing size of the damage zone for the (0, $\pm 45°$) material tends to extend the notch and weaken the material, whereas the extension of the damage zone in the (0/90°) material blunts the crack without greatly extending its effective length.

Underwood and Kendall[69] measured the residual tensile strength of single-edge notch specimens of (0, 90°) and (0, ±60°) glass/epoxy laminates after 10^6 cycles at various maximum stresses. They found that the strength increased with increasing values of fatigue load used in the prior cycling for both materials. The increases were attributed to the nature of the fatigue damage zone, which is primarily a region of delamination, with little fibre breakage. The high normal stresses and interlayer shear stresses which normally exist near a notch are relieved and spread over the larger area of the damage zone. Since the damage zones are larger at high fatigue loads, the residual strength is increased. The obvious limit to this increase in residual strength occurs when the fatigue damage zone proceeds far enough across the specimen to cause significant weakening or to cause fracture during cycling.

Hahn and Kim[34] conducted studies on tensile fatigue behaviour of (0, ±45, 90°) glass/epoxy laminates, using unnotched specimens. They found that a unidirectional laminate, fatigue loaded at what appears to be approximately 50–60% of the static ultimate tensile strength, did not exhibit any significant reduction in tensile modulus in 10^5 cycles, which was just prior to failure. Also, preloading the (0, ±45, 90°) laminate above the first ply failure stress has a negligible effect on the residual tensile strength after 10^6 cycles of fatigue at a stress slightly below the first ply failure stress.

Ply failure followed by partial delamination, as a result of tensile fatiguing for 100 cycles at a relatively high stress, resulted in a small decrease in the compression buckling strength.

Dharan,[117] using unnotched specimens, found that for a unidirectional glass/epoxy composite, fatigued at a maximum tensile stress of approximately one-half of the static strength, the tensile modulus decreased by approximately 50% after 10^5 cycles. There was little change in modulus for almost 10^5 cycles, after which it started to decrease at an increasing rate, reaching a value of about 60% of the initial after about $1{\cdot}3 \times 10^5$ cycles.

Tanimoto and Amijima,[119] also using unnotched specimens, measured the residual strength of specimens of satin weave glass cloth and polyester resin after cycling under tensile loading. Residual tensile strength and interlaminar shear strength were measured, and both showed similar trends. After several loading cycles, at stresses of 35% to 55% of the ultimate, cracks were formed in the specimens but there was little or no reduction in static strength, and possibly a slight increase. Then the residual strength decreased rapidly but only to about 20% in shear strength and 40% in tensile strength up to approximately half of the specimen life. At this point and until specimen failure, the residual strength does not change significantly.

This behaviour is in contrast to that reported by Mandell *et al.*[111] They measured the residual tensile strength of glass/epoxy laminates in both (0, 90°) and (0, ±45°) configurations. Unnotched specimens showed little change in residual strength until close to failure, at which point the strength decreased rapidly.

Sims and Brogdon[33] conducted tension-tension fatigue tests on glass/epoxy composites in a (±45°) configuration at a stress ratio, R, of 0·04 and a maximum stress of 40% of the ultimate. They found little reduction in specimen tensile stiffness after 10^3 cycles, approximately 10% reduction after 10^5 cycles, and then a 48% reduction at failure after 10^6 cycles.

Nevadunsky *et al.*[127] measured the residual strength and stiffness of (±45°) glass/epoxy and (0, ±45°) graphite/glass/epoxy tubular specimens after 10^7 cycles of axial fatigue. They found that both the residual tensile strength and the axial modulus decreased approximately 15%. The change in torsional modulus was much less than that for the axial modulus, due to the fibre orientation in the specimens. A change in stiffness due to fatigue damage can be detected very early in the total life up to fracture, depending on the magnitude of the fatigue load. The change occurs continuously over a large portion of the fatigue life.

3.3.2 *Graphite Reinforced Composites*

Ramani and Williams[121] studied the tensile fatigue behaviour of graphite/epoxy composite with various lay-ups. They found that for (0, ±45°) with a centre hole, the residual strength increased 40% over the static strength after 2×10^6 cycles, at 85% of the net section strength or 50% of the unnotched strength, whereas the modulus decreased 5%. If the fatigue stress level is increased to 65% of the unnotched strength for 10^6 cycles, there is now a 70% increase in residual strength over the static net section strength, but no further change in modulus. If the cycling stress is further increased to 90% of the unnotched strength for 10^6 cycles, there is further reduction in stiffness and indications that fatigue failure of the specimen would eventually occur through general unnotched failure mechanisms operating in the net section.

If a (0, ±45°) specimen with double-edge notch is cycled at 85% of the net strength or 60% of the unnotched strength for 10^6 cycles, the residual strength is increased by 30% with no change in modulus. If cycling continues at 85% of the increased residual net section strength, there are further increases in residual strength and eventually a stress level is reached where unnotched fatigue in the specimen net section occurs with attendant changes in specimen modulus of the order of 30% and more, which results in fatigue failure. (The presence of delamination crack suggests interlaminar weakening with degradation in transverse, compressive and shear properties.) When a centre hole specimen was precycled and then cycled at higher stresses, the residual strength increased further, but there were indications of much interlaminar weakening.

For (0, ±45, 90°) specimens with a centre hole, fatigued at 85% of the net section strength or 60% of the unnotched strength, the residual strength increased initially up to 15% at 10^5 cycles, then decreased sharply. The modulus decreased continuously, with the rate of decrease increasing after 10^5 cycles and achieving a 20% reduction at 10^6 cycles.

Sendeckyj *et al.*[125] measured the residual tensile strength of (0, ±45°) graphite/epoxy specimens with a partially through surface notch after spectrum fatigue loading. They found that the residual strength correlated with the size of the fatigue-induced matrix damage zone. If the zone is the same size or smaller than the delaminated region observed in static tests, then the residual strength equals the original static strength. However, if the fatigue-induced zone is much larger, then the residual strength is greater than the static strength. The extent of matrix damage and the residual strength increased with fatigue test temperature. The authors do not point out that this relation between damage zone size and residual strength applies in the case for tensile strength only. For example, the shear strength or compressive strength may be reduced by damage zones of this type.

The increase in residual strength with matrix damage was explained by considering the role of the matrix in the composite. The matrix provides the mechanism by which loads are transferred from fibres cut by the surface notch to adjacent intact ones. Load transfer, occurring by shear, results in load concentration in fibres adjacent to the notch. The magnitude of the load concentration dictates the strength reduction. As the load is applied, the matrix near the notch fails, the surface ply buckles and there is a reduction in load concentration.

Awerbuch and Hahn[98] derived a relation which indicated that a specimen whose static strength is above 85% of the average strength of that material shows an increase in residual strength over the original average static strength at 10^5 cycles. At $0{\cdot}5 \times 10^6$ and 2×10^6 cylces, however, the residual strength decreases slightly below the static strength, the amount of reduction increasing with the number of fatigue cycles.

Tests on unidirectional graphite/epoxy composites showed that the tensile modulus after fatigue for 2×10^6 cycles was higher in most cases than before. The reason for this unusual change was not clear.

Porter[124] studied the tensile fatigue behaviour of graphite/epoxy composite specimens with holes and slits, including partial and full penetrations in three configurations: (0, ±45, 90°), (0, ±80°), and (0, ±30°). The residual strength usually increased only slightly after fatiguing for up to approximately 10^6 cycles at stresses close to the static strength of the flawed specimen.

Ryder and Walker[126] found that the residual tensile strength and modulus of unnotched (0, ±45, 90°) graphite/epoxy specimens decreased slightly as a result of tension-tension and tension-compression fatigue, but the average strength and modulus increased with the severity of fatigue stress range, probably as a result of screening out of weaker coupons by early failure. Failure modes of the residual strength test coupons were similar to static failure modes and less similar to fatigue failure modes.

Sumison and Williams[130] conducted torsional and flexural fatigue tests on (0°) and (±45°) graphite/epoxy composites. Both the unidirectional and cross-ply laminates undergo fatigue damage as evidenced by reduction in torsional stiffness when subjected to cyclic torsional stress. For constant torsional deflection tests, the fatigue damage curves are characterised by: (1) an initial incubation period before damage begins; (2) an accumulation of damage as evidenced by a reduction in stiffness; and (3) a decrease in the rate of damage accumulation as evidenced by an approach to a constant stiffness value. Exposure to water during testing decreases the incubation period and increases the rate of damage accumulation; at 74°C, water also decreases the limiting stiffness. Cross-ply specimens are more prone to fatigue damage than unidirectional.

Reductions in torsional stiffness caused by stress cycling are accompanied by reductions in failure strength in bending. In contrast to torsion fatigue results, flexural fatigue studies showed a greater effect of water on unidirectional specimens than on cross-ply specimens.

3.3.3 Boron Reinforced Composites

Roderick and Whitcomb[131] studied the fatigue behaviour of boron/epoxy in two configurations, (0, ±45°) and (0, ±45, 90°). Specimens were fatigued at approximately two-thirds of the static ultimate tensile strength of the notched specimens. As a result of cycling up to 10^7 cycles, the tensile modulus decreased up to 20% for the (0, ±45°)

material and up to 15% for the (0, ±45, 90°) material, both unnotched and notched. There was no significant change in the residual tensile strength of the (0, ±45°) material and the (0, ±45, 90°) material exhibited a slight reduction in residual strength for notched specimens. The decrease in tensile modulus was apparently predominantly caused by matrix degradation.

Kulkarni *et al.*[115] reported on fatigue tests of (0, ±45°) boron/epoxy composite specimens with a centre hole. Tensile fatigue tests were conducted with maximum stresses of 80% and 67% of the ultimate static notched strength, R = 0·1, and frequency = 30 Hz. After 50 000 cycles, the residual tensile strength showed an increase of approximately 10%, and continued to increase slightly as the number of cycles increased further. The residual tensile modulus decreased slightly under the same conditions.

Durchlaub and Freeman[114] fatigue cycled (0, 90°) boron/epoxy composite specimens with a centre hole at stresses close to the ultimate net section stress. After 6×10^6 cycles, the material exhibited an increase in residual tensile strength of approximately 25%.

Reifsnider *et al.*[133] conducted tensile fatigue tests on (0, ±45°) boron/epoxy composite specimens with a centre hole at R = 0·1 and frequencies of 1–30 Hz. During cycling there is an early rapid decrease in tensile modulus corresponding to early damage development around the hole, followed by a more gradual decrease throughout the remainder of the test. The second stage of the decrease is, in most cases, much flatter for a strain-controlled test than for a load-controlled test. Most of the visible damage occurred around the hole in the early part of the test, sometimes within 10^5 cycles.

Stinchcomb *et al.*[132] conducted tension fatigue tests on two configurations of (0, ±45°) boron/epoxy composites, one symmetric and the other asymmetric. The tests were conducted at frequencies of 15, 30 and 45 Hz for 10^6 cycles, using rectangular specimens with a centre hole. They found that the static tensile stiffness decreased up to approximately 20%, depending on the cycling frequency. The dynamic stiffness decreased between 10 and 50%, depending on the laminate stacking sequence and the cycling frequency. The material with an asymmetric configuration suffered a dynamic stiffness reduction approximately 2·5 times that of the symmetric material. The asymmetric material also exhibited a reduction of 0–40% residual fatigue strength, whereas the symmetric material exhibited an increase of 0–20%, again depending on the frequency.

3.4 Frequency Effects in Fatigue

Fairly limited studies on frequency effects in polymer composites indicate that fatigue behaviour is influenced by cycling frequency. Results reported indicate that fatigue properties are usually improved at higher frequencies, except for one study which indicates that there is a maximum frequency beyond which properties decrease.

Reasons proposed for the improved fatigue properties at higher frequencies include: (1) less time-at-load at higher frequencies for a given number of cycles; (2) increase in monotonic fracture strength due to the increased loading rate at higher frequencies; and (3) increased local notch-tip heating at higher frequencies limiting the extent of the damage zone.

Underwood and Kendall[69] conducted tensile fatigue studies of (0, 90°) and (0, ±60°) glass/epoxy specimens at frequencies of 4 Hz and 30 Hz, using rectangular specimens with a centre hole. They found that the damage zone growth from the hole was faster for both materials at the lower frequency. They interpreted this as an indication that the local notch-tip heating, caused by the higher cyclic frequency, limits the extent of the damage zone and thus extends the life of the material. In addition, previously reported increases in monotonic fracture strength of composites due to increased loading rate could also account for the better fatigue properties observed at the higher frequency.

Hahn and Kim[34] conducted tension fatigue tests on a glass/epoxy composite in the (0, ±45, 90°) configuration at a stress ratio $R = 0{\cdot}1$, using a frequency of 4 Hz at high stresses and 10 Hz at low stresses. The fatigue strength did not appear to depend on frequency for the range investigated.

Using static fatigue data, the time-to-failure was converted to cycles-to-failure, assuming the frequency to be 4 Hz. The static fatigue data thus plotted closely resembled the cyclic fatigue data except for wider scatter.

Dharan[117] compared cyclic fatigue to static fatigue for unidirectional glass/epoxy composites by loading a specimen to 700 MN m^{-2} (10^5 psi) as in a creep test. The time to failure was greater than 10 h, which is much larger than the approximately 100 cycles required to cause failure at this stress level under cyclic fatigue.

Thornton and Kozik[134] conducted tensile fatigue tests of graphite/epoxy composites in five orientations: (0°), (0, ±15°), (0, ±30°), (0, ±45°), and (0, ±60°) at three different maximum loads and two frequencies, 3 Hz and 25 Hz. Cycling at 3 Hz was found to be more damaging than at 25 Hz in a limited number of tests.

Stinchcomb *et al.*[132] studied the fatigue behaviour of boron/epoxy composites in two different (0, $\pm 45°$) configurations, in axial tension under strain control at three frequencies—15, 30 and 45 Hz—using a specimen with a centre hole. They found that the applied stress (under strain control), the stiffness, and damping change during a fatigue test. The magnitude of the change depends on the frequency. Changes at 30 Hz were more severe than at 15 or 45 Hz, i.e. the response was a non-linear function of frequency. The residual tensile strength after $1{\cdot}5 \times 10^6$ cycles, determined from cyclic stress-strain tests, improved in tests conducted at 30 Hz and did not change at 15 or 45 Hz for one configuration, but decreased at both 15 and 30 Hz for the other configuration.

The type of damage that developed during fatiguing was also frequency dependent. There was transverse damage with some longitudinal damage at 15 Hz but this changed to predominantly longitudinal mode at 45 Hz. The damage nucleated as small longitudinal splits in the matrix at the edge of the hole but off the transverse axis at points of maximum strain energy density in the surface layers.

It was concluded that the influence of frequency on the behaviour of composites lies in the variation of the degree of contribution of several intermediate failure modes which combine to produce final failure. Presumably, these intermediate modes consist of fibre breakage, debonding, delamination and matrix cracking. It is conceivable, at least in boron/epoxy, that the local distribution of stress associated with each of these damage modes is time dependent. Therefore, it appears that the frequency of cycling can influence or even control the details of the selection, interaction and combination of these intermediate failure modes. The complexity of the process is demonstrated by the nonlinearity of the frequency dependence observed for various properties studied.

Reifsnider *et al.*[133] elaborated further, stating that the most likely explanation of the effect of frequency on fatigue damage is that time-at-load effects contribute more deformation during the longer time required for a given number of cycles at lower frequencies than at higher ones. Hence, the number of cycles required for a given amount of damage to occur is smaller at low frequencies than at high frequencies as observed for both strain-controlled and load-controlled tests. The matrix material, primarily, undergoes time-dependent deformation and stiffness reduction under the mean stress during a fatigue test and this degradation is thought to be the principal contributor to the frequency effects observed.

It may be possible to predict the effect of frequency by viscoelastic analysis based on proper data generated for the matrix material. It is not yet clear how the three-dimensionality of viscoelastic effects should be handled. A stress concentrator, such as a hole, complicates this type of pursuit, but 3-D finite element analysis can be used to approach the problem.

REFERENCES

1. *Composite Materials: Testing and Design* (Second Conference), ASTM–STP 497, American Society for Testing and Materials, Philadelphia, PA., 1972.
2. *Composite Materials: Testing and Design* (Third Conference), ASTM–STP 546, American Society for Testing and Materials, Philadelphia, PA., 1974.
3. *Fatigue of Composite Materials*, ASTM–STP 569, American Society for Testing and Materials, Philadelphia, PA., 1975.
4. *Fracture Mechanics of Composites*, ASTM–STP 593, American Society for Testing and Materials, Philadelphia, PA., 1975.
5. *Composite Reliability*, ASTM–STP 580, American Society for Testing and Materials, Philadelphia, PA., 1975.
6. Reifsnider, K. L. and Lauraitis, K. N., eds., *Fatigue of Filamentary Composite Materials*, ASTM–STP 636, American Society for Testing and Materials, Philadelphia, PA., 1977.
7. Sendeckyj, G. P., ed., *Composite Materials, Vol. 2, Mechanics of Composite Materials*, Academic Press, London, 1974.
8. Broutman, L. J., ed., *Composite Materials, Vol. 5, Fracture and Fatigue*, Academic Press, London, 1974.
9. Chamis, C. C., ed., *Composite Materials, Vol. 7, Structural Design and Analysis*, Part I, Academic Press, London, 1975.
10. Herakovich, C. T., ed., *Inelastic Behavior of Composite Materials, AMD Vol. 13*, American Society of Mechanical Engineers, New York, 1975.
11. Scala, E. *et al.*, eds., *Proceedings of the 1975 International Conference on Composite Materials*, The Metallurgical Society of AIME, New York, 1976.
12. Noton, B. R. *et al.*, eds., *Proceedings of the 1978 International Conference on Composite Materials*, The Metallurgical Society of AIME, New York, 1978.
13. Tsai, S. W. and Hahn, H. T., in *Inelastic Behavior of Composite Materials*, ed. Herakovich, C. T., AMD Vol. 13, American Society of Mechanical Engineers, New York, NY, 1975.
14. Sendeckyj, G. P., Richardson, M. D., and Pappas, J. E., *Composite Reliability*, ASTM–STP 580, American Society for Testing and Materials, Philadelphia, PA., 1975.
15. Yeow, Y. T. and Brinson, H. F., *Virginia Polytechnic Institute and State University Report* #VPI-E-76-21, Blacksburg, VA., 1976.
16. Chou, S., Orringer, O., and Rainey, J. H., *J. Comp. Mat.*, **11** (1977) p. 71.
17. McLaughlin, P. V., Huang, S. N., and Rosen, B. W., Materials Sciences Corp., TFR/7508, Naval Air Development Center, Warminster, PA., 1974.

18. FRANCIS, P. H., WALRATH, D. E., and WEED, D. N., Southwest Research Institute, AFML-TR-77-62, Air Force Materials Lab., Dayton, OH., 1976.
19. REIFSNIDER, K., in *Fracture Mechanics*, eds. Perrone, N., Liebowitz, H., Mulville, D., and Pilkey, W., Univ. of Virginia Press, Charlottesville, VA., 1978.
20. REIFSNIDER, K., *Proceedings of the 14th Annual Meeting of the Society of Engineering Science*, ed. Sih, G., Lehigh University Publishing, Bethlehem, PA., 1977.
21. SANDHU, R. S., AFFDL-TR-72-71, Air Force Flight Dynamics Lab., Dayton, OH., 1972.
22. SENDECKYJ, G. P., *Composite Materials: Testing and Design* (Second Conference), ASTM–STP 497, American Society for Testing and Materials, Philadelphia, PA., 1972.
23. WU, E. M., *J. Comp. Mat.*, **6** (1972) p. 72.
24. ROWLANDS, R. E., in *Inelastic Behavior of Composite Materials*, ed. Herakovich, C. T., AMD Vol. 13, American Society of Mechanical Engineers, New York, NY, 1975.
25. WU, E. M., in *Composite Materials, Volume 2, Mechanics of Composite Materials*, ed. Sendeckyj, G. P., Academic Press, New York, NY, 1974.
26. WU, E. M. and RUHMANN, D. C., *Composite Reliability*, ASTM–STP 580, American Society for Testing and Materials, Philadelphia, PA., 1975.
27. WU, E. M., Research Workshop on Mechanics of Composite Materials, Duke University, Durham, NC., 1978.
28. HASHIN, Z., Research Workshop on Mechanics of Composite Materials, Duke University, Durham, NC., 1978.
29. WU, E. M. and SCHEUBLEIN, J. K., *Composite Materials: Testing and Design* (Third Conference), ASTM–STP 546, American Society for Testing and Materials, Philadelphia, PA., 1974.
30. TENNYSON, R. C., MACDONALD, D., and NANYARO, A. P., *J. Comp. Mat.*, **12** (1978) p. 63.
31. GUESS, T. R. and GERSTLE, F. P., JR., *J. Comp. Mat.*, **11** (1977) p. 146.
32. NARAYANASWAMI, R. and ADELMAN, H. M., *J. Comp. Mat.*, **11** (1977) p. 366.
33. SIMS, D. F. and BROGDON, V. H., *Fatigue of Filamentary Composite Materials*, eds. Reifsnider, K. L. and Lauraitis, K. N., ASTM-STP 636, American Society for Testing and Materials, Philadelphia, PA., 1977.
34. HAHN, H. T. and KIM, R. Y., *J. Comp. Mat.*, **10** (1976) p. 156.
35. BELYTSCHKO, T. B., BURKE, J., and SCHULTZ, A. B., University of Illinois, AFML-TR-73-313, Air Force Materials Lab, Dayton, OH., 1974.
36. COLLINS, B. R. and CRANE, R. L., *J. Comp. Mat.*, **5** (1971) p. 408.
37. HAHN, H. T. and TSAI, S., AFML-TR-74-13, Air Force Materials Lab, Dayton, OH., 1974.
38. KANNINEN, M. F., RYBICKI, E. F., and BRINSON, H. F., *Composites*, **8** (1977) p. 17.
39. DURCHLAUB, E. and SACCO, P., Boeing Company, Vertol Division, AFFDL-TR-72-82, Air Force Flight Dynamics Lab, Dayton, OH., 1972.
40. OWEN, M. J. and BISHOP, P. T., *J. Comp. Mat.*, **7** (1973) p. 146.
41. SLEPETZ, J. M. and CARLSON, L., *Fracture Mechanics of Composites*, ASTM–

STP 593, American Society for Testing and Materials, Philadelphia, PA., 1975.
42. Pines, D. G., Fang, C. E., Mandell, J. F., and McGarry, F. J., Massachusetts Institute of Technology, AMMRC-CTR-75-15, Army Materials and Mechanics Research Center, Watertown, MA., 1975.
43. Barnby, J. T. and Spencer, B., *J. Mat. Sci.*, **11** (1976) p. 78.
44. Phillips, C. D., *J. Comp. Mat.*, **18** (1974) p. 130.
45. Spencer, B. and Barnby, J. T., *J. Mat. Sci.*, **11** (1976) p. 83.
46. Slepetz, J. M., *Proceedings of the Army Symposium on Solid Mechanics, 1976–Composite Materials: The Influence of Mechanics of Failure on Design*, AMMRC-MS-76-2, Army Materials and Mechanics Research Center, Watertown, MA., 1976.
47. Sani, M. L. and Stern, M., *Int. J. Fract.*, **12** (1976) p. 331.
48. Sih, G., *Handbook of Stress Intensity Factors*, Lehigh University, Bethlehem, PA., 1973.
49. Papaioannou, S. G., Hilton, P. D., and Lucas, R. A., *Eng. Fract. Mech.*, **6** (1974) p. 807.
50. Mandell, J., McGarry, F., and Wang, S. S., *J. Comp. Mat.*, **8** (1974) p. 106.
51. Atluri, S. N., Kobayashi, A. S., and Nakagaki, M., *Fracture Mechanics of Composites*, ASTM–STP 593, American Society for Testing and Materials, Philadelphia, PA., 1975.
52. Sih, G. C. and Chen, E. P., *J. Comp. Mat.*, **7** (1973) p. 230.
53. Sih, G., Chen, E., Huang, S., and McQuillen, E., *J. Comp. Mat.*, **19** (1975) p. 167.
54. Wu, E. M., *Failure Modes of Composite Materials with Organic Matrices and Their Consequences on Design*, AGARD 39th Meeting of the Structures and Materials Panel, Munich, Germany, 1974.
55. Tirosh, J. and Berg, C. A., *Composite Materials: Testing and Design* (Third Conference), ASTM–STP 546, American Society for Testing and Materials, Philadelphia, PA., 1974.
56. Morris, D. H. and Hahn, H. T., *J. Comp. Mat.*, **11** (1977) p. 124.
57. Smith, D. G. and Mullinix, B. R., T-78-6, U.S. Army Missile Research and Development Command, Redstone Arsenal, AL., 1977.
58. Whitney, J. M. and Nuismer, R. J., *J. Comp. Mat.*, **8** (1974) p. 253.
59. Nuismer, R. J. and Whitney, J. M., *Fracture Mechanics of Composites*, ASTM–STP 593, American Society for Testing and Materials, Philadelphia, PA., 1975.
60. Waddoups, M. E., Eisenmann, J. R., and Kaminski, B. E., *J. Comp. Mat.*, **5** (1971) p. 446.
61. Cruse, T. A., *J. Comp. Mat.*, **7** (1973) p. 218.
62. Konish, H. J., Jr. and Cruse, T. A., *Composites Reliability*, ASTM–STP 580, American Society for Testing and Materials, Philadelphia, PA., 1975.
63. Tirosh, J., *J. Appl. Mech.*, **44** (1977) p. 449.
64. Bhatia, N. M. and Verette, R. M., *Fracture Mechanics of Composites*, ASTM–STP 593, American Society for Testing and Materials, Philadelphia, PA., 1975.

65. WADDOUPS, M. E., *Polymer Eng. and Sci.*, **15** (1975) p. 160.
66. SENDECKYJ, G. P., *Fracture Mechanics of Composites*, ASTM–STP 594, American Society for Testing and Materials, Philadelphia, PA., 1975.
67. KUNZ, S. C. and BEAUMONT, P. W. R., *Fatigue of Composite Materials*, ASTM-STP 569, American Society for Testing and Materials, Philadelphia, PA., 1975.
68. MANDELL, J. F. and MEIER, U., *Fatigue of Composite Materials*, ASTM–STP 569, American Society for Testing and Materials, Philadelphia, PA., 1975.
69. UNDERWOOD, J. H. and KENDALL, D. P., *Proceedings of the 1975 International Conference on Composite Materials*, eds., Scala, E. *et al.*, Metallurgical Society of AIME, New York, NY., 1976.
70. BROUTMAN, L. J. and SAHU, S., *Composite Materials: Testing and Design* (Second Conference), ASTM–STP 497, American Society for Testing and Materials, Philadelphia, PA., 1972.
71. PIGGOTT, M. R., *J. Mech. Phys. Solids*, **22** (1974) p. 457.
72. SANCHEZ-MOYA, V. and PIPKIN, A. C., *Int. J. Solids and Struct.*, **13** (1977) p. 571.
73. CONRAD, N., Mechanics and Materials Research Center, Texas A & M University, MM 3168-76-10, College Station, TX., 1976.
74. MCCALL, I. R. and MORLEY, J. G., *J. Mat. Sci.*, **12** (1977) p. 1165.
75. ZWEBEN, C., *Eng. Fract. Mech.*, **6** (1974) p. 1.
76. WANG, S. S., MANDELL, J. F., and MCGARRY, F. J., *Eng. Fract. Mech.*, **9** (1977) p. 217.
77. KANNINEN, M. F., RYBICKI, E. F., GRIFFITH, W. I., and BROEK, D., Battelle Columbus Labs., NASA-CR-2689, National Aeronautics and Space Administration, Washington, DC., 1976.
78. NUISMER, R. J., Research Workshop on Mechanics of Composite Materials, Duke University, Durham, NC., 1978.
79. BEAUBIEN, L. A., MAST, P. W., MULVILLE, D. R., SUTTON, S. A., THOMAS, R. W., TIROSH, J., and WOLOCK, I., Memo Report 3433, eds. Bascom, W. and Lockhart, L., Naval Research Laboratory, Washington, DC., 1976.
80. BEAUBIEN, L. A., CLIFFORD, M. F., MAST, P. W., MULVILLE, D. R., SUTTON, S. A., THOMAS, R. W., TIROSH, J., and WOLOCK, I., Memo Report 3721, eds. Bascom, W. amd Lockhart, L., Naval Research Laboratory, Washington, DC., 1978.
81. SMITH, C. W., in *Inelastic Behavior of Composite Materials*, ed. Herakovich, C. T., AMD Vol. 13, American Society of Mechanical Engineers, New York, NY., 1975.
82. PIPES, B. and PAGANO, N., *J. Comp. Mat.*, **14** (1970) p. 538.
83. WANG, S. S. and CROSSMAN, F. W., *J. Comp. Mat.*, **11** (1977) p. 92.
84. HSU, P. W. and HERAKOVICH, C. T., *J. Comp. Mat.*, **11** (1977) p. 422.
85. HSU, P. W. and HERAKOVICH, C. T., Virginia Polytechnic Institute and State University, NASA-CR-148080, National Aeronautics and Space Administration, Hampton, VA., 1976.
86. WANG, S. S. and MANDELL, J. F., Massachusetts Institute of Technology, NASA-CR-135248, National Aeronautics and Space Administration, Washington, DC., 1977.

87. KULKARNI, S. V., RAMKUMAR, R. L., PIPES, R. B., and SCOTT, W. R., *Proceedings of the 1978 International Conference on Composite Materials*, eds. Noton, B. *et al.*, Metallurgical Society of AIME, New York, NY., 1978.
88. RYBICKI, E. F., SCHMUESER, D. W., and FOX, J., *J. Comp. Mat.*, **11** (1977) p. 470.
89. MOON, F. C., KIM, B. S., and FANG-LANDAU, S. R., Princeton University, NASA-CR-134999, National Aeronautics and Space Administration, Washington, DC., 1976.
90. CHAMIS, C. C., *Proceedings of the Army Symposium on Solid Mechanics, 1976–Composite Materials: The Influence of Mechanics of Failure on Design*, AMMRC-MS-76-2, Army Materials and Mechanics Research Center, Watertown, MA., 1976.
91. MALLICK, P. K. and BROUTMAN, L. J., *Eng. Fract. Mech.*, **8** (1976) p. 631.
92. WANG, S. S., *Proceedings of the 1978 International Conference on Composite Materials*, eds. Noton, B. *et al.*, Metallurgical Society of AIME, New York, NY., 1978.
93. PHOENIX, S. L., *Composite Materials: Testing and Design* (Third Conference), ASTM–STP 546, American Society for Testing and Materials, Philadelphia, PA., 1974.
94. HITCHON, J. W., MCCAUSLAND, W. H., and PHILLIPS, D. C., AERE-R8217, UK Atomic Energy Authority, Harwell, Oxfordshire, 1975.
95. MCMAHON, P., *Proceedings of the 19th SAMPE Symposium*, 1974, Society for the Advancement of Materials and Process Engineering, Azusa, CA.
96. SUN, C. T. and YAMADA, S. E., *J. Comp. Mat.*, **12** (1978) p. 169.
97. HARLOW, P. G. and PHOENIX, S. L., *J. Comp. Mat.*, **12** (1978) p. 195.
98. AWERBUCH, J. and HAHN, H. T., *Fatigue of Filamentary Composite Materials*, ASTM–STP 636, eds. Reifsnider, K. L. and Lauraitis, K. N., American Society for Testing and Materials, Philadelphia, PA., 1977.
99. HAHN, H. T. and KIM, R. Y., *J. Comp. Mat.*, **9** (1975) p. 297.
100. YANG, J. N., *J. Comp. Mat.*, **12** (1978) p. 19.
101. HALPIN, J. C., JOHNSON, T. A., and WADDOUPS, M. E., *Int. J. Fract. Mech.*, **8** (1972) p. 465.
102. HALPIN, J. C., *J. Comp. Mat.*, **6** (1972) p. 208.
103. ROTEM, A. and HASHIN, Z., *J. Comp. Mat.*, **9** (1975) p. 191.
104. MANDELL, J. F., WANG, S. S., and MCGARRY, F. J., *J. Comp. Mat.*, **9** (1975) p. 266.
105. IM, J., MANDELL, J. F., WANG, S., and MCGARRY, F. J., Massachusetts Institute of Technology, NASA-CR-935094, National Aeronautics and Space Administration, Washington, DC., 1976.
106. CHAPLIN, C. R., *J. Mat. Sci.*, **12** (1977) p. 347.
107. GARRETT, K. W. and BAILEY, J. E., *J. Mat. Sci.*, **12** (1977) p. 157.
108. STEVENS, G. T. and LUPTON, A. W., *J. Mat. Sci.*, **12** (1977) p. 1706.
109. GARRETT, K. W. and BAILEY, J. E., *J. Mat. Sci.*, **12** (1977) p. 2189.
110. PARVIZI, A., GARRETT, K. W., and BAILEY, J. E., *J. Mat. Sci.*, **13** (1978) p. 195.
111. MANDELL, J. F., MCGARRY, F. J., IM, J., and MEIER, U., *Failure Modes in Composites II*, eds. Fleck, J. N. and Mehan, R. L., The Metallurgical Society of AIME, New York, 1974.

112. SCOTT, J. M. and PHILLIPS, D. C., Atomic Energy Research Establishment, AERE-R7793, UK Atomic Energy Authority, Harwell, 1974.
113. HANCOX, N. L., *J. Mat. Sci.*, **12** (1977) p. 884.
114. DURCHLAUB, E. C. and FREEMAN, R. B., Boeing Co., AFML-TR-73-225, Air Force Materials Laboratory, Dayton, OH., 1974.
115. KULKARNI, S. V., MCLAUGHLIN, P. V., JR., and PIPES, R. B., Materials Sciences Corp., NASA CR-145039, National Aeronautics and Space Administration, Washington, DC., 1977.
116. LAURAITIS, K. N., *Fatigue of Filamentary Composites*, ASTM–STP 636, eds. Reifsnider, K. L. and Lauraitis, K. N., American Society for Testing and Materials, Philadelphia, PA., 1977.
117. DHARAN, C. K. H., *Fatigue of Composite Materials*, ASTM–STP 569, American Society for Testing and Materials, Philadelphia, PA., 1975.
118. KIM, H. C. and EBERT, L. J., *J. Comp. Mat.*, **12** (1978) p. 139.
119. TANIMOTO, T. and AMIJIMA, S., *J. Comp. Mat.*, **9** (1975) p. 380.
120. SMITH, E. W. and PASCOE, K. J., *Composites*, **8** (1977) p. 232.
121. RAMANI, S. V. and WILLIAMS, D. P., *Fatigue of Filamentary Composite Materials*, ASTM–STP 636, eds. Reifsnider, K. L. and Lauraitis, K. N., American Society for Testing and Materials, Philadelphia, PA., 1977.
122. CHANG, F. H., GORDON, D. E., and GARDNER, A. H., *Fatigue of Filamentary Composite Materials*, ASTM–STP 636, eds. Reifsnider, K. L. and Lauraitis, K. N., American Society for Testing and Materials, Philadelphia, PA., 1977.
123. STURGEON, J. B., *Composites*, **8** (1977) p. 221.
124. PORTER, T. R., *Fatigue of Filamentary Composite Materials*, ASTM–STP 636, eds. Reifsnider, K. L. and Lauraitis, K. N., American Society for Testing and Materials, Philadelphia, PA., 1977.
125. SENDECKYJ, G. P., STALNAKER, H. D., and KLEISMIT, R. A., *Fatigue of Filamentary Composite Materials*, ASTM–STP 636, eds. Reifsnider, K. L. and Lauraitis, K. N., American Society for Testing and Materials, Philadelphia, PA., 1977.
126. RYDER, J. T. and WALKER, E. K., *Fatigue of Filamentary Composite Materials*, ASTM–STP 636, eds. Reifsnider, K. L. and Lauraitis, K. N., American Society for Testing and Materials, Philadelphia, PA., 1977.
127. NEVADUNSKY, J. T., LUCAS, J. J., and SALKIND, M. J., *J. Comp. Mat.*, **9** (1975) p. 394.
128. PHILLIPS, D. C. and SCOTT, J. M., *Composites*, **8** (1977) p. 233.
129. PHILLIPS, D. C. and SCOTT, J. M., *Fibre Science and Technology*, **11** (1978) p. 23.
130. SUMISON, H. T. and WILLIAMS, D. P., *Fatigue of Composite Materials*, ASTM–STP 569, American Society for Testing and Materials, Philadelphia, PA., 1975.
131. RODERICK, G. L. and WHITCOMB, J. D., *Fatigue of Filamentary Composite Materials*, ASTM–STP 636, eds. Reifsnider, K. L. and Lauraitis, K. N., American Society for Testing and Materials, Philadelphia, PA., 1977.
132. STINCHCOMB, W. W., REIFSNIDER, K. L., WILLIAMS, R. S., and MARCUS, L. A., *Failure Modes in Composites II*, eds. Fleck, J. N. and Mehan, R. L., The Metallurgical Society of AIME, New York, 1974.

133. Reifsnider, K. L., Stinchcomb, W. W. and O'Brien, T. K., *Fatigue of Filamentary Composite Materials*, ASTM–STP 636, eds. Reifsnider, K. L. and Lauraitis, K. N., American Society for Testing and Materials, Philadelphia, PA., 1977.
134. Thornton, H. R. and Kozik, T. J., Texas A & M University, NASA CR-134691, National Aeronautics and Space Administration, Washington, DC., 1975.

Chapter 9

FRACTURE OF POLYMERS IN DENTAL APPLICATIONS

M. BRADEN and B. E. CAUSTON
The London Hospital Medical College, London UK

SUMMARY

Both elastomeric and glassy polymers are used in dentistry, the latter including composite materials. Elastomers are used as impression materials and soft prostheses, glassy polymers for denture bases and other oral appliances, and composites as filling materials. In all these applications, it is desirable that the component does not fail. Hence fracture mechanics is an essential study both of the intrinsic properties of the materials used, and of their response to the rather hostile aqueous environment of the mouth. Tearing energy studies of elastomers, using the methods of Rivlin and Thomas, reveal marked differences between the materials used in dentistry which explain the differences in clinical performance. Similarly classical fracture mechanics and fracture toughness studies reveal differences between materials and their degradation in the oral environment, and suggest possible mechanics of wear in dental composites. Such studies are a valuable pre-requisite to the development of new clinical materials.

1 INTRODUCTION

Polymers comprise a very important and growing part of dental materials as a whole. Elastomers are used as impression materials, soft lining materials for dentures, and maxillo-facial prostheses materials. Glassy polymers are used for denture bases and artificial teeth, orthodontic appliances, temporary crown and bridge materials, and as the basis for composite filling materials. In all of these applications, it is essential that the materials do not undergo fracture, or at least that the conditions under which they fail are well-understood. On the basis of such understanding, it is easier to predict clinical performance, and sometimes to produce improved clinical materials. The fact that the British Department of Health and Social Security undertakes over one million denture repairs per annum indicates the magnitude of one particular problem.

2 FRACTURE OF ELASTOMERIC MATERIALS USED IN DENTISTRY

2.1 Impression Materials

A major use of elastomeric materials in dentistry is as impression materials. Although they are ancillary materials in the sense that they are only used in the process of making an appliance, the successful completion of, say, a multiple-unit bridge costing several hundred pounds relies heavily on a good impression.

Most materials are fluid in the uncross-linked state, with viscosities in the range of 10–10^5 N s m^{-2} and hence of relatively low molecular weight.[1] They are, of course, usually non-Newtonian, so these figures are only a rough guide.

Prior to taking an impression, the material is mixed to incorporate the cross-linking reagent, and loaded into an impression tray. This is then placed in the patient's mouth, where it flows into the interstices of the teeth and oral mucosa. Under the influence of the enhanced temperature of the mouth (30–37 °C) the setting reaction is accelerated and the material cross-linked, typically three to eight minutes after introduction into the mouth. The impression is then removed, strains being induced where the rubber has set into undercut areas. The recovery from such strains is important, but not relevant to the present discussion. More important, the material must not tear, otherwise the impression is spilt; the tear of course may not be easily visible to the clinician, and not apparent until the model is cast.

The materials for impression taking are:

(i) *Alginates based on sodium/potassium alginate polyelectrolytes.*

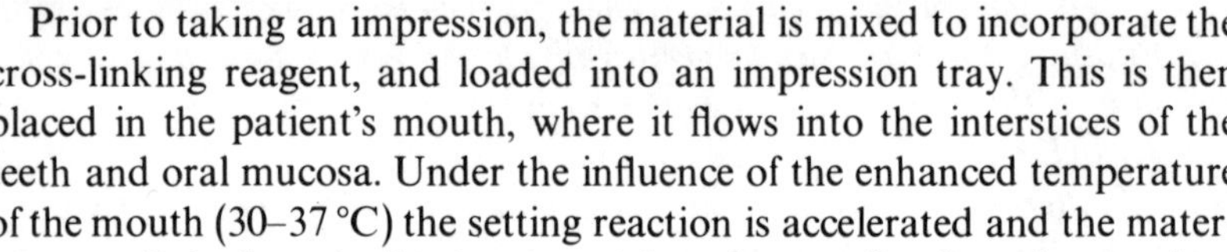

These are supplied as a powder containing a filler, calcium sulphate, and sodium phosphate. On adding water, a paste is formed which ultimately sets to a gel when the calcium ions from the calcium sulphate cross-link the alginate macromolecules.

(ii) *Silicone rubbers.*
These may be α-ω-hydroxy-terminated dimethyl siloxanes[2] which are cross-linked by a tetra-alkyl silicate activated by an organo-tin compound such as dibutyl laurate.

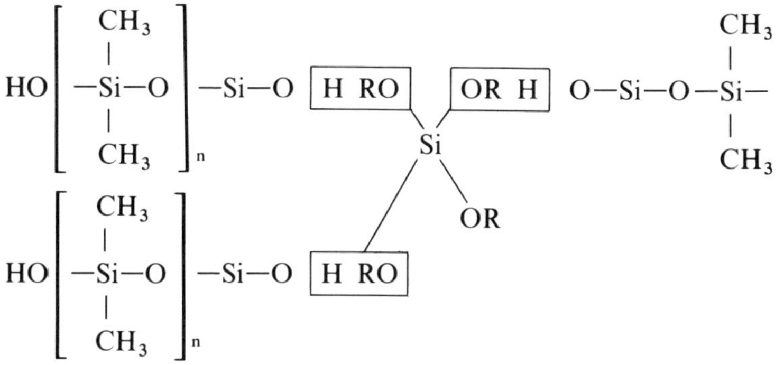

Alternatively, vinyl-terminated silicones are used, which are cross-linked by means, for example, of hydro-silicones activated by a platinum salt.[3]

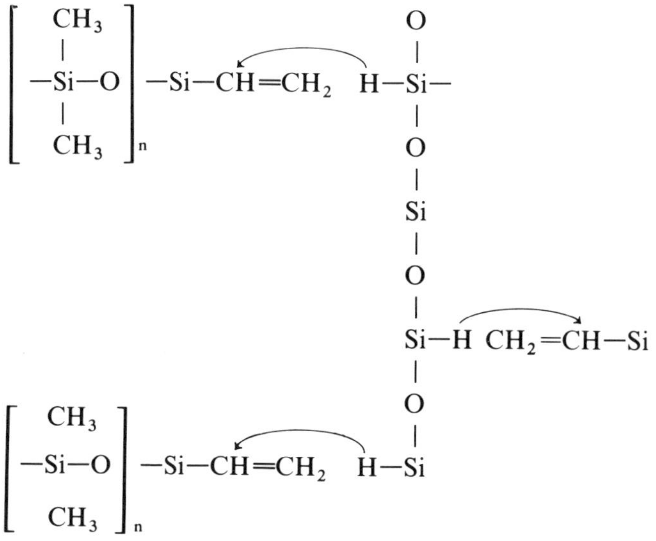

In both cases cross-linking is at chain ends, so the cross-linked network is markedly different from conventional vulcanisates. In each case strength is conferred by the incorporation of suitable fillers.

(iii) *Polysulphides.*

These are based on Thiokol LP–2:[4]

$$HS{-}(CH_2{-}CH_2{-}O{-}CH_2{-}O{-}CH_2{-}CH_2{-}S{-}S)_{23}$$
$$-C_2{-}H_4{-}O{-}CH_2{-}O{-}C_2H_4{-}SH$$

This has a molecular weight of about 4000; in addition to the terminal —SH groups shown, there is 2 mol % branching to afford cross-linking sites.

Chain lengthening and cross-linking is achieved by an oxidising agent such as lead dioxide, PbO_2:

$$\begin{array}{c} O \text{ (from oxidising agent)} \\ + \\ \sim\sim S\ H \quad H\ S\sim\sim \longrightarrow \sim\sim S{-}S\sim\sim \\ + \; H_2O \end{array}$$

(iv) *Imine-terminated polyether.*[5]

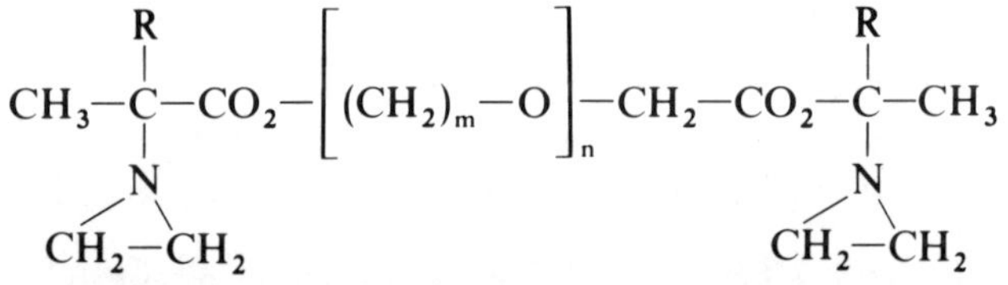

Cross-linking is achieved by a carbonium ion (say from an aromatic sulphonic acid ester) which opens the imine ring to give a cationic polymerisation process, i.e

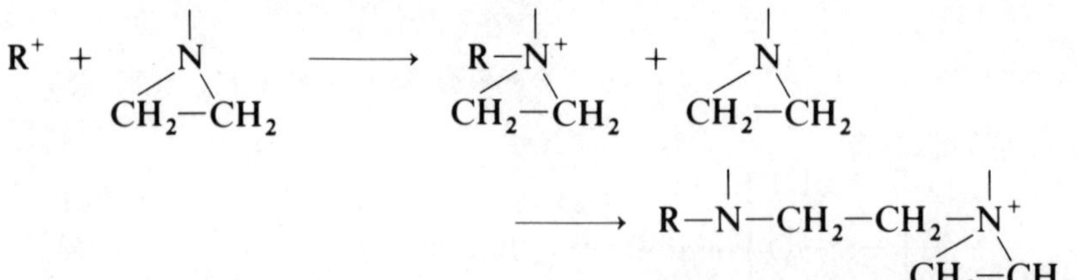

There is a further compliant impression material, based on agar-agar, which is a thermally reversible gel. On heating it becomes fluid, so that an impression can be taken, whilst on cooling it gels.

2.2 Soft Lining Materials for Acrylic Dentures, and Maxillo-Facial Prostheses

Some patients cannot tolerate a hard acrylic denture, and so a denture is prescribed with a soft elastomeric lining, to spread occlusal load.

Silicone rubbers, such as those described previously, can be used with a suitable adhesive. Alternatively, soft acrylic materials are used, which are usually made up from a poly(ethylmethacrylate) powder and a higher methacrylate monomer mixed with a plasticiser.[6] The latter materials have good adhesion to the acrylic denture, but harden due to the loss of plasticiser in the oral environment.

Silicones, soft acrylics and polyurethanes are also used for maxillo-facial prostheses (ears, noses, etc.). Clearly, in both applications, but particularly in the case of maxillo-facial prostheses, tear resistance is extremely important, because the patient has to remove prostheses for cleaning purposes.

2.3 Measurement of Tear Resistance of Dental Elastomers

The fundamentals of the tearing behaviour of elastomers were established by the now-classic work of Rivlin and Thomas.[7] This work was based on the Griffith criterion, namely that as fracture involves the formation of new surfaces (δA) the basic parameter is the energy per unit area ($\mathscr{T}$) necessary to produce these surfaces.

If $\mathscr{E}$ is the total stored energy in the specimen, then the criterion is:

$$\mathscr{T} = -\frac{\partial \mathscr{E}}{\partial A} \tag{1}$$

Thomas devised a series of test-pieces, with which $\mathscr{T}$ could be determined experimentally and which gave consistent results.[8] For one such test-piece, the so-called trousers test specimen, it can be shown that:

$$2\mathscr{T} = \frac{2F\lambda}{h} - 2bW_0 \tag{2}$$

where F is the applied force, h the specimen thickness, b the leg width, λ is the extension ratio in the legs, and W_0 the stored energy per unit volume in the legs; usually $hbW_0 \ll 2F/h$. The quantity $2\mathscr{T}$ corresponds to the tearing energy T originally defined by Rivlin and Thomas. Greensmith and Thomas[9,10] showed for a wide range of elastomers that $\mathscr{T}$ is dependent on the rate of tear and temperature.

This method has been applied to both impression materials (by Braden[11] and Inglis[12]) and to soft prostheses materials by Goldberg and Craig.[13] All these materials exhibit steady, as distinct from stick-slip, tearing.

Figure 1 shows a plot of tear distance against time, where it is seen that linear behaviour obtains, so that for a given value of F (hence $\mathscr{T}$) there is

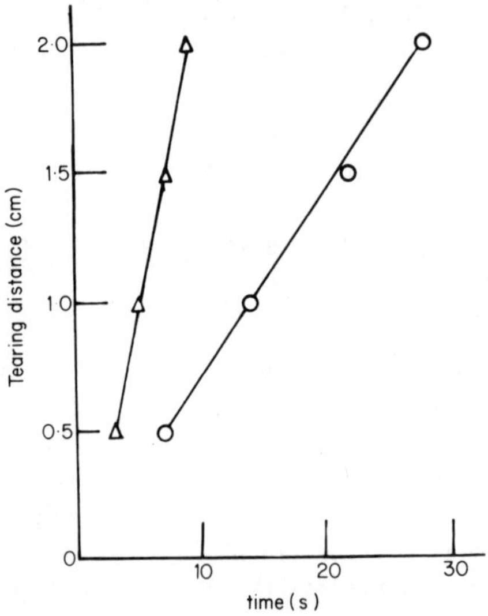

FIG. 1. Tearing distance as a function of time for:
(a) alginate impression material △;
(b) polysulphide impression material ○.

an associated tear-rate R. (As with the original work of Rivlin and Thomas, measurements are facilitated by lightly scoring the centre-line of the specimen to ensure linear tear propagation).

Figure 2 plots $2\mathscr{T}$ as a function of tear-rate for the range of impression materials listed previously. In agreement with the work of Greensmith and Thomas on non-crystallising rubbers, the tearing energy increases monotonically with rate. With these particular materials, within the limits

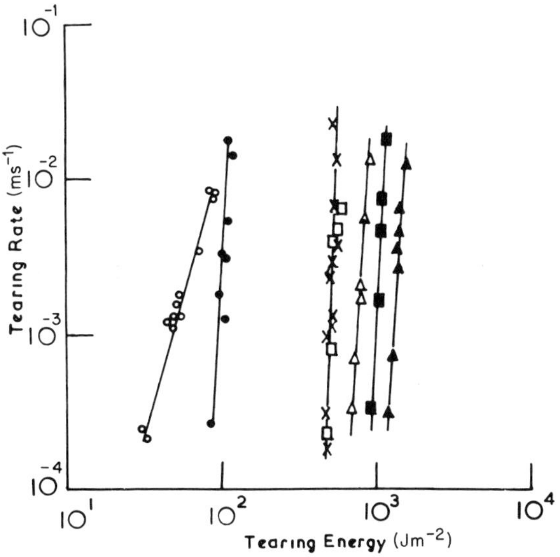

FIG. 2. Tearing rate as a function of tearing energy.

- ○ reversible hydrocolloid
- ● alginate
- × imine-terminated polyether
- □ silicone
- △ low-viscosity polysulphide
- ■ medium-viscosity polysulphide
- ▲ high-viscosity polysulphide

of experimental error, a linear relationship between log $\mathscr{T}$ and log rate obtains, indicating that over the limited range studied,

$$R \propto \mathscr{T}^n \tag{3}$$

The index n clearly varies with the material, lying in the range 4 to ≈ 20. It is of interest to note that clinicians have, for many years, empirically recommended the withdrawal of impressions by a swift pull rather than a slow drag.

The maximum tearing energies shown in Fig. 2 are comparable to the values reported by Greensmith for unfilled non-crystallising rubbers. The elastomers used as impression materials are (a) of low molecular weight and (b) cross-linked at chain ends in some cases. Even so they contain fillers, and it is interesting to speculate what low values of $\mathscr{T}$ would be achieved with unfilled polymers of these types. Hydrocolloids containing large amounts of water are, as may be expected, weak. However, it is not clear why agar-agar materials are weaker than alginates.

Since it is well known that alginates commonly tear in practice, but silicones rarely do so, it is tempting to try to establish a tearing energy criterion for impression materials. However, this is not feasible because the stiffness (modulus) varies widely between the materials shown (Table 1). The clinical condition imposed on the material is clearly the strain involved in removing the impression, which is in turn dictated by the anatomy of the tooth or teeth involved. Hence the imparted elastic energy for a given degree of undercut will increase with stiffness. For example, the polyether material (Fig. 2) has what may appear to be reasonable tearing energies; however, its high modulus (Table 1) compared with other materials means that in clinical practice it will sometimes tear.

TABLE 1

TYPICAL SHEAR MODULI OF IMPRESSION MATERIALS

Material	*Shear Modulus* ($MN\ m^{-2}$)
Reversible hydrocolloid	0·28
Alginate	0·20
Silicones	0·43–1·07
Polysulphides	0·26–0·64
Imine-terminated polyether	2·27

Tearing energies of impression materials are in general temperature-dependent, in agreement with the work of Greensmith and Thomas, although the silicone materials are hardly affected by a temperature change from 20 to 37 °C.

Figure 3 shows the data for silicone and soft acrylic materials used in soft lining and maxillo-facial prostheses. The soft acrylics have the highest tearing energies. Since analysis shows these to be unfilled, it seems likely that these high energies are consequent on high mechanical loss factors.[14] Tan δ values up to unity are often obtained, δ being the loss angle between stress and strain. Goldberg and Craig[13] report that polyurethanes have the highest tearing energies. It should be emphasised that certain silicones, with hydrophillic fillers, suffer dramatic reduction in $\mathscr{T}$ (at least one order of magnitude) on immersion in water.

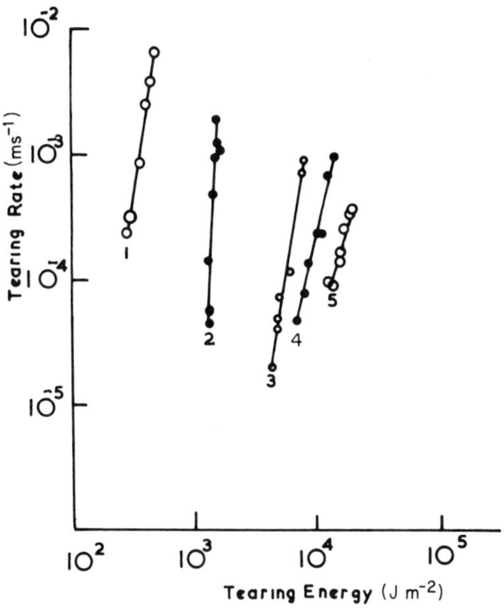

FIG. 3. Tearing energies of soft lining materials:
(1) silicone;
(2) silicone/acrylic co-polymer;
(3) vinyl co-polymer;
(4), (5) soft acrylics.

3 THE FRACTURE OF GLASSY POLYMERS USED IN DENTISTRY

3.1 Denture-Base Polymers

The great majority of denture bases are made of poly(methylmethacrylate). A conventional denture-base acrylic is made from a mixture of beads of poly(methylmethacrylate) with methyl-methacrylate monomer to form a dough which is moulded and then polymerised. The fracture characteristics of denture-base acrylic will depend therefore on several factors, e.g. the molecular weight and molecular structure of the two phases, the size and distribution of the phases and the additives added to the basic formula to aid ease of handling, weathering or aesthetic appeal.

The bead phase is either poly(methylmethacrylate) to which has been added plasticiser (e.g. dibutyl phthalate) to aid gelling, or a co-polymer of methyl methacrylate and ethyl methacrylate. The viscosity average molecular weight of the bead polymer ranges from $\overline{M}_v \approx 5 \times 10^5$ to $\overline{M}_v \approx 6 \times 10^6$ and bead size varies from 10 to 200 μm in diameter. The monomer is methyl methacrylate to which a cross-linking agent such as ethylene glycol dimethacrylate has been added to reduce crazing during service. The monomer contains 10 ppm of hydroquinone inhibitor to extend shelf life. Removal of the inhibitor prior to curing the dough has no significant effect on the final tensile strength, which is fortunate when it is considered that few dental laboratories are equipped to carry out such a chemical separation.

Denture-base acrylic is cured by one of three techniques, of which the most commonly used is the heat cure process. When the mixture of polymer beads and monomer is made, doughing involves the diffusion of monomer into the polymer beads, plasticising and partially dissolving them. The beads contain about 0·6% of benzoyl peroxide which dissolves in the monomer and diffuses out into the liquid phase. The dough is packed under pressure into plaster moulds, whose surface has been sealed with alginate, and heated. Heat cycles are often left to the personal preference of the technician, though in general they involve a slow rise in temperature for one hour to about 100 °C and then a slow cooling overnight to room temperature. The strength of the final product is very dependent on the heating cycle. Too rapid an initial rise results in porosity due to monomer boiling, whilst too slow a cure gives large amounts of free monomer, which plasticises the final polymer with adverse effects on mechanical properties.

The other two techniques used in denture production are the so-called 'cold-cure' methods. An aromatic tertiary amine (e.g. *N,N*-dimethyl *p*-toluidine) is incorporated in the monomer; this causes the benzoyl peroxide to break down into radicals at room temperature and hence initiate the radical polymerisation of the monomer. The whole process takes about 20 minutes from mixing the dough to complete cure. The difference between the two cold-cure techniques lies in the mode of gelling. In one technique a conventional dough is formed and packed into a plaster mould, in the other a polymer of a higher molecular weight that gels more slowly is used; whilst still in its liquid state the monomer/bead mixture is poured into an agar-agar mould (hence this is known as the 'pour technique'). In both cold-cure techniques a considerable exotherm develops; to prevent monomer evaporation, the moulds are placed in pressure vessels at $3{\cdot}45 \times 10^5$ Pa. This pressure of $3{\cdot}45 \times 10^5$ Pa is sufficient to prevent porosity due to monomer evaporation and any air bubbles incorporated during mixing are compressed. The cold-cure denture bases have two major disadvantages compared to those that are heat-cured. Firstly, the degree of polymerisation is lower and branching greater, hence the strength of the polymer polymerised in the mould is lower. Secondly, whereas the monomer in a heat-cured denture base diffuses towards the centre of the beads prior to polymerisation and forms an interpenetrating polymer network, the monomer diffusion in the cold-cure technique takes place at room temperature and occurs in less than 20 minutes. Consequently, monomer penetration of the beads is restricted to the outer coat.[15]

The structural differences between heat- and cold-cured acrylics leads to differences in their mode of fracture. When heat-cured acrylic fractures, the cracks travel in one plane throughout. However, in cold-cured acrylics there is a tendency for the crack to deviate around the beads, indicating a structural weakness at the bead/interstitial polymer interface.[16]

It will be evident from the preceding paragraphs that denture-base acrylic is a very variable entity, the composition of the final product being dependent on both manufacturer and dental technician.

Over a period of time, two methods of fracture mechanics have been used to evaluate denture-base resins. These are the earlier method of Berry,[17] characterising fracture in terms of a Griffith fracture surface energy and an inherent flaw size, and the subsequent method employing fracture toughness and stress intensity factors. A brief description of the importance of each component in cured denture-base acrylic with regard to its strength will be given.

The importance of the heterogeneity of denture-base acrylic compared to homogeneous poly(methylmethacrylate) can be seen in Table 2. The lower tensile strength and fracture surface energy of the denture-base resin may be due in part to a disparity in the Young's moduli of the two phases present. The fracture surface of a heat-cured denture-base acrylic shows two distinct phases (Fig. 4): a tough phase associated with the IPN that is predominantly linear 'bead' polymer, and a brittle interstitial phase. The Young's moduli of the two phases are such that the tough phase can transfer stress to the stiff brittle phase. When the difference in Young's moduli of the two phases is reduced, there is a subsequent improvement in tensile strength (Table 2 and Fig. 5) showing maximum improvement when there is the greatest similarity in the Young's moduli of the two phases.

Raising the molecular weight of the polymer beads increases the tensile strength of the cured denture-base acrylic (Table 3), as would be expected

TABLE 2

FRACTURE PROPERTIES OF DENTURE-BASE RESINS

	Tensile strength (MN m^{-2})	*Impact strength* (*Hounsfield*) (N m^{-1})	*Young's modulus* (GN m^{-2})	†*Fracture surface energy* (J m^{-2})	‡K_{Ic} (MN $m^{-3/2}$)
Poly(methylmethacrylate) $\bar{M}_v$ 6 × 10^6	78	0·06	2·9	2·0	2·71
Conventional heat-cured denture-base acrylic	55	0·046	2·6	1·4	2·53
Denture-base acrylic having beads and interstitial polymer of the same Young's modulus	76	0·047	2·88	—	—
'Pour-type' denture-base acrylic[39]	63	0·045	2·3		1·76
High-impact acrylic	64	0·123	2·27	6·0*	1·78*

* Ductile failure.
† Tapered cleavage specimen.
‡ Single-edge-notch specimen.

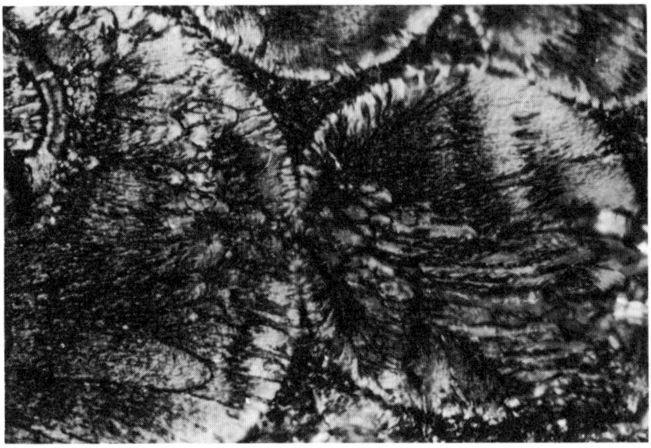

FIG. 4. Fracture surface of a heat-cured denture-base resin (note original PMM bends in structure.)

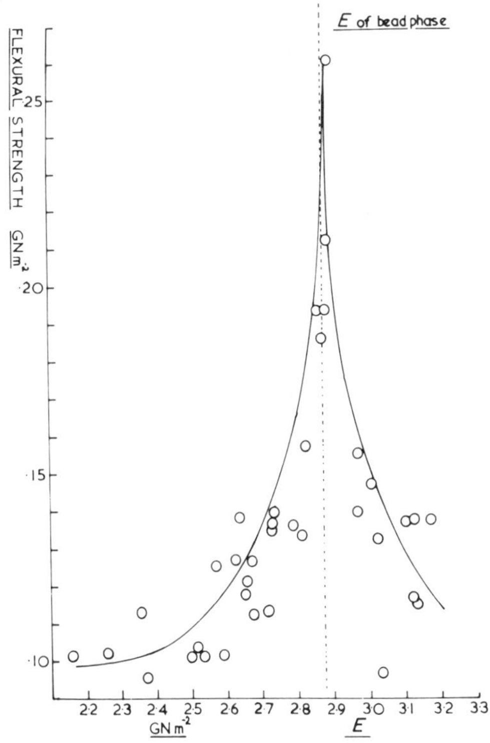

FIG. 5. Flexural strength as a function of Young's modulus, *E*.

TABLE 3

DEPENDENCE OF TENSILE STRENGTH OF DENTURE-BASE RESINS ON MOLECULAR WEIGHT.

	$\bar{M}_v$ *of bead polymer*	*Tensile strength* ($MN\ m^{-2}$)
Conventional heat-cured denture-base acrylic	$4{\cdot}95 \times 10^5$	55
Conventional heat-cured denture-base acrylic	$1{\cdot}5 \times 10^6$	78·5
'Pour-type' denture-base acrylic[39]	$6{\cdot}0 \times 10^5$	66

from Berry's work on the effect of molecular weight.[18] The bead polymer in the 'pour-type' resins often has a higher molecular weight than that of the heat-cured resins. As a result pour-type resins often have a higher tensile strength, thus illustrating the importance of the bead phase in determining the tensile strength of the cured resin because the improvement is achieved despite the presence of 8% residual monomer in the interstitial phase. (The latter, however, has marked deleterious effects on the glass-transition temperature, T_g).

The importance of structural differences between heat-cured and 'pour-type' denture-base acrylics is also evident from their stress intensity factors (K_{Ic}). The 'pour-type' resins show a considerable reduction in K_{Ic} values (Table 2) and hence a greater tendency to crack propagation. The clincal significance of such a reduction is discussed later.

The service life of a denture-base acrylic may be reduced by one or more of three major factors. First and most obvious is accidental impact. The second and most common is fatigue along the mid-line of the denture due to masticatory forces (Fig. 6). Finally, the oral environment is capable of crazing acrylic either by enzyme and aqueous means or *via* dietary sources, e.g. peppermint from toothpaste and sweets, or chloroform from medicines.

To improve impact resistance a dispersed rubber phase is sometimes added to the bead polymer (Fig. 7); this reduces the Young's modulus (and in some earlier products reduced the T_g to unacceptable levels) but the impact strength is improved 10-fold (Table 4). The incorporation of

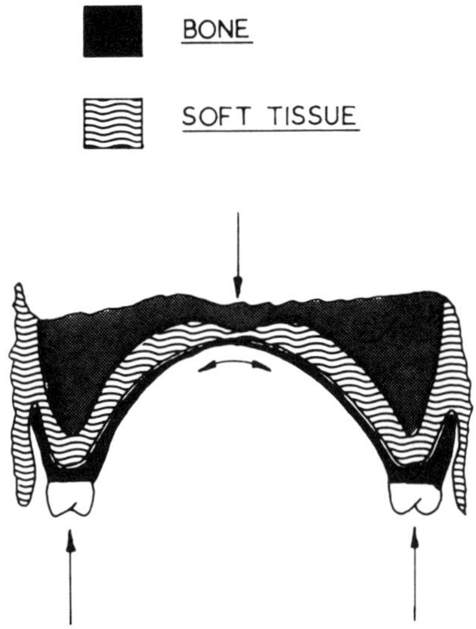

FIG. 6. Mid-line fracture of a denture base.

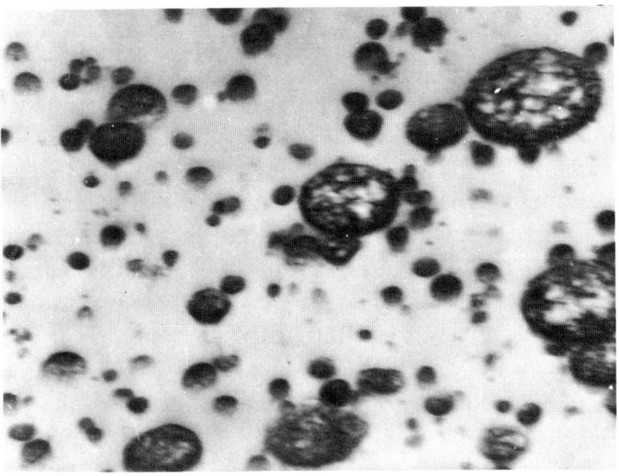

FIG. 7. Dispersed rubber phase in a denture-base polymer.

TABLE 4

COMPARISON OF IMPACT STRENGTH AND FATIGUE LIFE BETWEEN HIGH-IMPACT AND CONVENTIONAL DENTURE-BASE RESINS (AFTER REFERENCE 17).

	Impact strength (Nm) (*Hounsfield*)	*Fatigue*† *life* (*cycles*)
Conventional heat-cured denture-base acrylic	0·046	479 000
High-impact denture-base acrylic	0·123	1 444 000

† notched bend specimen 37 °C

the rubber phase also greatly improves fatigue resistance. Stafford has shown a three-fold increase in fatigue life of impact acrylic over conventional heat-cured acrylic.[17] The high-impact resins, not being brittle, can give anomalous results in fracture toughness experiments; however, a value for G_{Ic} of 6·0 MN $m^{-3/2}$ was obtained from a tapered cleavage specimen, illustrating the much-improved resistance to crack propagation in these resins.

The fatigue resistance of all conventional denture-base acrylics, whether 'pour-type' or heat-cured, is about the same. Only the impact acrylics, as stated previously, show significant improvement. One millon denture repairs per year, the vast majority of which were made necessary by fatigue failure, are carried out in the UK by N.H.S. technicians alone, excluding the private sector. The impact acrylics are too expensive for general use and hence there is an urgent economic need for a conventional, inexpensive alternative.

Water is the major environmental component that a denture base has to cope with. The effects of water on the denture-base acrylics are, however, very difficult to monitor because the dough is cured in a wet mould, be it plaster of paris or agar-agar, and therefore water is incorporated in its matrix during cure. Removal of the water results in internal strain which causes a fall in fracture toughness, whereas specimens equilibrated fully with water at 37 °C show an improved fracture surface energy (Table 3). Alternatively, in terms of the Berry approach the increase in fracture surface energy $\mathcal{T}$ (denoted γ by Berry) is accompanied by a compensating increase in the inherent flaw size c_0. This being so, Waters' formula relating c_0 and fatigue life would predict a shortened fatigue life for wet denture-base acrylic.[19] However, in practice the rise in $\mathcal{T}$ appears to compensate and, at the small strain rates experienced in fatigue experiments, water appears to act as a plasticiser. Stafford has found that Perspex shows a 10-fold improvement in fatigue life when tested in water and denture-base acrylics show only a two-fold improvement[20]. The lower improvement in fatigue life of denture-base acrylic can be attributed in part to the incorporation of water during the cure cycle.

To test the effects of the oral environment on fracture toughness a notched three-point bend specimen has been used. The solution to be tested is placed in the notch of the specimen (Fig. 8) and the specimen broken. The stress intensity factor (K_{Ic}) was calculated from Freitag and Cannon's equation,[21] using Brown and Srawley's shape correction factor:[22]

$$K_{Ic} = y\frac{Ma^{1/2}}{bw^2} \tag{4}$$

where:

$$y = 6\left[1{\cdot}93 - 3{\cdot}07\left(\frac{a}{w}\right) + 14{\cdot}53\left(\frac{a}{w}\right)^2 - 25{\cdot}11\left(\frac{a}{w}\right)^3 + 25{\cdot}8\left(\frac{a}{w}\right)^4\right]$$

$$\frac{S}{w} = 8;$$

a = crack length

b = specimen thickness

w = specimen width

M = bending moment at notch = $SP/2$

S = distance between supports

P = force at break

The effects of whole saliva, peppermint oil in water, and chloroform water on ease of crack propagation have been tested in this way. Also a comparison of the effect on crazing of first cross-linking the acrylic with 10% EGDM has been studied by the same method.[23] It can be seen from Fig. 9 that the solutions have little or no effect in general on the cross-linked dental resin, whereas uncross-linked PMM showed a considerable

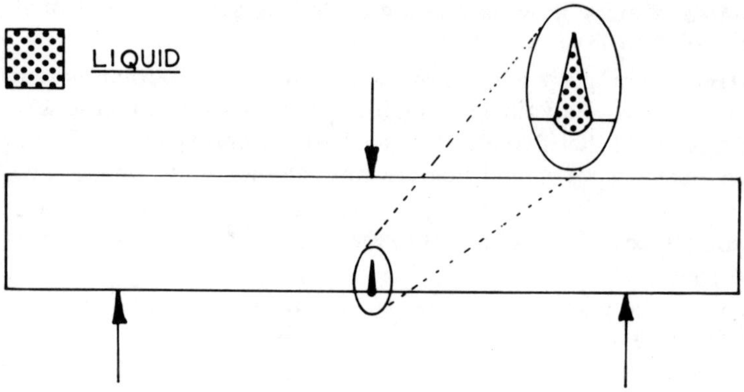

FIG. 8. Liquid sited in fracture notch. (see eqn. 4)

fall in K_c (instability). The effect on fracture toughness characteristics of increasingly long exposure to the solutions of peppermint oil and chloroform water show an initial increase in K_{Ic} due to broadening of the crack tip. However, the reduction in the tan δ of the resin, 100 s after exposure to the test solution, reduces the K_{Ic} value. The system is equilibrated within 10 000 s, the uncross-linked resin having a lower K_{Ic} than before application of the crazing agent, and the cross-linked resin a higher K_{Ic} value (Fig. 9a). The lower graph in Fig. 9 shows the solvent dependence of the stress intensity factor at the point of catastrophic failure of the specimen; this shows a logarithmic time dependence, the K_c values of the uncross-linked resin falling faster than that of the cross-linked resin.

Fracture mechanics is playing an ever-increasing role in predicting the clinical performance of denture-base acrylics.[23,24] However, with the diversity of test methods available, seemingly anomalous behaviour can be observed and erroneous conclusions drawn as to clinical serviceability. The crack speed attained in any given test can give rise to such anomalous behaviour, for it is known to affect such parameters as the K_{Ic} value.[25] This must be considered when comparing the results from a tapered cleavage experiment, where the crack speed is slow, with the results obtained from a three-point loaded specimen, where crack speeds are very high by comparison. Perspex and conventional heat-cured acrylic show little change in relative crack susceptibility between the two tests. However, high-impact denture-base acrylic gives very high G_{Ic} values (Table 2) when a tapered cleavage specimen is used where the time for plastic flow to occur is considerable; whereas in the three-point load test

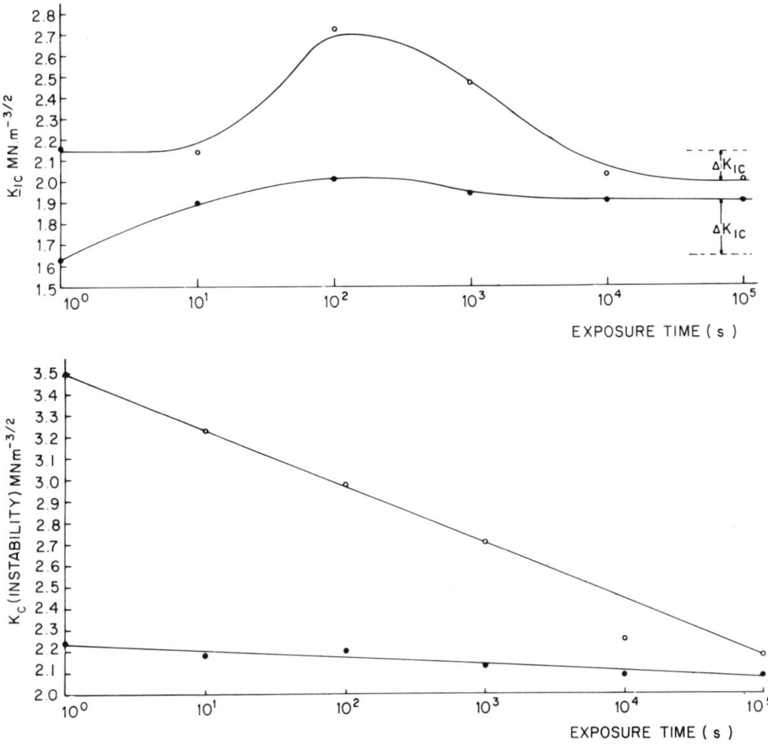

FIG. 9. (a) K_{Ic} (stress intensity factor, in MN m$^{-3/2}$) and (b) K_c (instability) as a function of exposure time to 1% chloroform in water at 25 °C.
○ Bulk polymerised PMM (Perspex.)
● Cross-linked denture-base acrylic.

high-impact acrylic has a lower K_{Ic} value than Perspex and some of the conventional denture-base acrylics. This indicates that fracture toughness testing of denture-base acrylics should be carried out at both high and low crack speed. In this way not only would the ductile nature of each material be indicated, but it might also be possible to relate the low crack speed to the performance of the acrylic during mastication and the high-speed crack data to the traumatic situation.

3.2 Composite Restorative Materials

The composite restorative filling materials are primarily aesthetic materials designed to replace the silicate cements in anterior restorations. In

TABLE 5

FRACTURE PROPERTIES OF COMPOSITE FILLING MATERIALS.

	Compressive strength†[40] (Mn m^{-2})	*Tensile strength*†[40] (MN m^{-2})	*Transverse strength*†[40] (MN m^{-2})	*Young's modulus*†[40] (GN m^{-2})	K_{Ic} (MN $m^{-3/2}$)
Amine/ peroxide-cured composite	243	43	105	13	0·3–>0·5
Light/α diketone-cured composite	266	44	106	14	—

† After 1 month in water at 37·4 °C.

such uses a composite of acrylic resin and silane-treated quartz is adequate both mechanically and aesthetically. However, their success in replacing silicate cements has led to an extension of their use to more testing applications such as restoring incisal edges and replacing the occlusal surfaces of posterior teeth. This shift in their use means that high strength and good weatherability are of greater importance and a more rigorous investigation of their strength characteristics is required.

The majority of composite restorative materials are based on an aromatic dimethacrylate, *bis*-GMA:

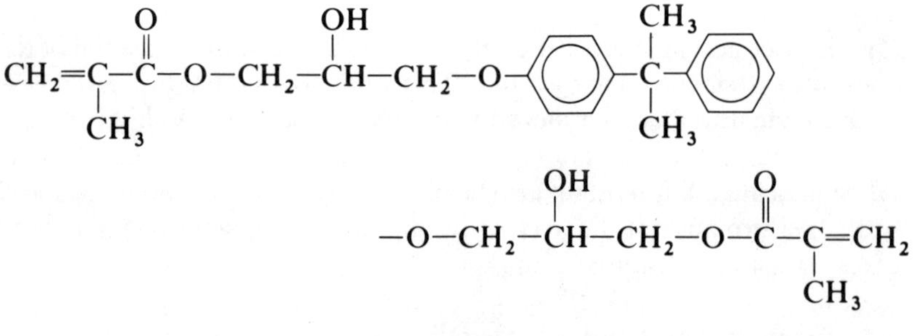

Bis-GMA is first thinned with approximately 25% of a glycol dimethacrylate and then mixed with 75% by weight of crushed silane-treated quartz

(5–150 μm), forming a thick translucent paste. By mixing two such pastes, one containing 0·5% benzoyl peroxide, and the other 1–2% of an aromatic tertiary amine, the *bis*-GMA is cured in about seven minutes to give a hard translucent composite. Other curing systems involve the use of UV light and benzoin methyl ether or blue visible light and α diketones. However, provided that the thickness of the composite cured by light-activated catalysts is not excessive, there is no evidence to show any significant differences in fracture behaviour attributable to the curing system used (Table 5). The volume of resin used per filling rarely exceeds 0·1 ml, and when such a small volume cures there is very little exotherm. If, however, a compression or fracture toughness specimen is made with a volume of 0·5 ml, there will be a large rise in temperature during setting and hence a higher degree of cross-linking will take place. Therefore one dilemma in the testing of composite restorative materials is the design of test specimens of comparable size to the fillings used in clinical practice without introducing large errors when measuring the specimen dimension.

Once a test specimen has been designed, relating the results to the clinical situation is still difficult. A filling material is subjected to many affronts, often simultaneously, and a single figure for abrasion may not be relevant unless combined with data indicating enzyme attack or temperature cycling. The assessment of abrasion resistance is a good example in many ways. Early tests involved a hardness measurement and a toothbrush abrasion test. The *in vitro* conclusion was that abrasion resistance of dental composites was very good; in practice they were worse than unfilled resins. The *in vitro* tests lacked several factors present in the mouth and illustrate the need for a comprehensive knowledge of the total stress and environmental conditions present before assessing individual measurements.

In the mouth there are three stress cycles operating; incisal edge restorations[26] are subjected to large bending moments and also shear forces along the margin between enamel and restoration. Restorations out of occlusion, i.e. on the enamel/gingival margin have no stress except the force of a toothbrush or piece of food passing over them. Finally the occlusal surfaces of the molars are subjected to grinding and cracking forces; these are the greatest forces in the mouth and are repeated continually during meals and during periods of psychological stress. The environmental factors vary from patient to patient; a 'mouth breather' will dry out his incisal etch restoration between meals, and marginal restorations may be covered in a thick layer of acid-excreting plaque for months at a time, whereas occlusal surfaces are kept clean by the tongue and opposing teeth and will only be covered in saliva.

The point of weakness of incisal edge restorations is the bond between enamel and composite. Attempts to promote bonding by first treating the enamel surface with a vinyl silane failed because of the rigours of thermal cycling that such bonds experience. The answer was to etch the enamel with 50% phosphoric acid; this produces a retentive surface and enamel/composite bond strengths of 20 MN m^{-2} in tension are possible.[38] These bond strengths are adequate since it has been calculated by Farah[27] that the shear stress on the bond will be approximately 10 MN m^{-2}. In the occlusal situation of posterior restorations, bulk fractures are rare in practice but marginal failure is very common. Marginal failure is thought to occur because unsupported enamel prisms shear off leaving an unsupported composite edge; this in turn breaks off, leaving a large vee-shaped gap that can fill with bacteria. The marginal seal may also be lost and oral fluid will penetrate the dentine causing secondary caries (Fig. 10). Recently modifications of margin design have proved[28] that bevelling the edge and etching the surrounding enamel reduces margin fracture by 30%.

In occlusal situations, wear and abrasion are the factors limiting the use of composite filling materials. Of the many modes of wear and abrasion known, three are thought to be relevant to the dental occlusal case. The first, brittle fracture wear,[29] is thought possible because the tensile strengths of dental composites are less than one-third of their compressive strengths. However, single-pass wear experiments carried out on several filled composites show that the mode of surface failure is ductile in

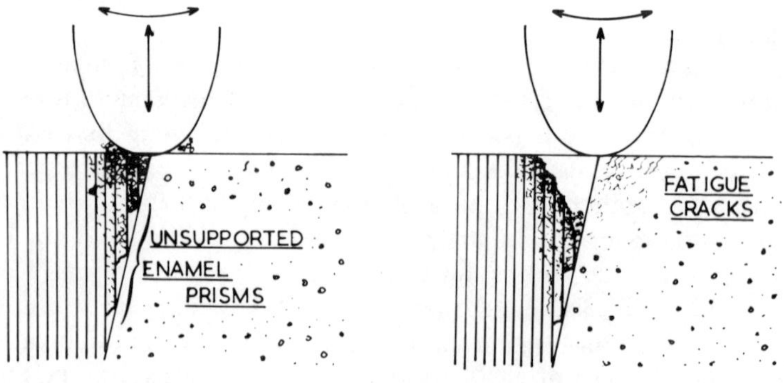

FIG. 10. Marginal failure of composite filling material.

nature. The addition of silane coupling agent to the filler reduced wear; however, despite a stronger bond between filler and organic matrix the surface failure remained ductile.[30] Most abrasive wear is of the three-body type, an abrasive being trapped between opposing enamel and the restoration. Abrasive wear is rarely seen except in steel workers and miners, where abrasive particles such as iron oxide are ingested on a daily basis. Finally, adhesive wear may occur as a result of strong adhesive forces between enamel and the opposing composite formed during sliding contact.[31–33] This wear occurs whether the surface is lubricated or not and will be dependent on the crack resistance of the weaker adherent surface.[34] Simple two-body abrasion tests using an abrasive belt and three-body tests using a slurry and brush technique[35] predict that composite filling materials will be more abrasion-resistant than dental amalgam, whereas the reverse is true in practice. Hence these simple tests fail to combine all the components of wear and abrasion present in the mouth. Powers *et al.*,[36] using an irrigated sliding-wear machine, have succeeded in predicting wear behaviour and ranking the abrasion resistance of amalgam and composite correctly.[36] However, the machine is complex and expensive and requires a careful running-in procedure. For routine testing a fracture mechanics approach may suffice. Powers' sliding-wear machine demonstrates the importance of adhesive wear on the total wear process; hence an accurate measure of the crack susceptibility of a material would be expected to rank materials according to their adhesive-wear tendency.

Roberts *et al.*[37] have reported a method of accurately making small notch fracture toughness specimens from dental composite material and have determined G_{Ic}, K_{Ic}, and fracture toughness data for three materials. The data correlate well with wear data produced by Powers *et al.*[30] using a single-pass wear test. Therefore it may be possible with further work to relate such fracture mechanics data to the clinical situation and compare composites with materials known to have adequate abrasion and wear resistance, e.g. dental amalgam or inlay gold alloys.

REFERENCES

1. BRADEN, M., in *Scientific Aspects of Dental Materials*, ed. A. J. von Fraunhofer, Butterworths, London and Boston, 1975, chap. 13.
2. BRADEN, M. and ELLIOTT, J. C. *J. Dent. Res.*, **46** (1966) p. 1016.
3. BRADEN, M. *J. Dentistry.*, **4** (1976) p. 1.
4. BRADEN, M. *J. Dent. Res.*, **45** (1966) p. 1065.

5. BRADEN, M., CAUSTON, B. E. and CLARKE, R. L. *J. Dent. Res.*, **51** (1972) p. 889.
6. WRIGHT, P. S. *J. Dentistry.*, **4** (1976) p. 247.
7. RIVLIN, R. S. and THOMAS, A. G. *J. Polymer Sci.*, **10** (1953) p. 291.
8. THOMAS, A. G. *J. Polymer Sci.*, **3** (1960) p. 168.
9. GREENSMITH, H. W. and THOMAS, A. G. *J. Polymer Sci.*, **18** (1955) p. 189.
10. GREENSMITH, H. W. *J. Polymer Sci.*, **21** (1956) p. 175.
11. BRADEN, M. *Dent. Pract.*, **14** (1963) p. 67.
12. INGLIS, A. T. Ph.D. thesis University of London, 1977.
13. GOLDBERG, A. J. and CRAIG, R. J. *J. Dent. Res., Special Issue A.*, **56** (1977) p. 523.
14. MULLINS, L. *Trans. I.R.I.*, **35** (1959) p. 213.
15. CAUSTON, B. E. Ph.D. thesis, University of London, 1972.
16. KUSY, R. P. and TURNER, D. T. *J. Dent. Res.*, **53** (1974) p. 520.
17. BERRY, J. P. *J. Polymer Sci.* **50** (1961) p. 107.
18. BERRY, J. P. *J. Polymer Sci.* **A2** (1964) p. 4069.
19. WATERS, N. E. *Dental Pract.*, **18** (1968) p. 389.
20. STAFFORD, G. D. Ph.D. thesis., University of Wales, 1978.
21. FREITAG, T. A. and CANNON, S. L. *J. Biomed. Mater. Res.*, **10** (1976) p. 805.
22. BROWN, W. F. and SRAWLEY, J. E. ASTM STP 410, p. 13.
23. STAFFORD, G. D., CAUSTON, B. E., HUGGETT, R. and HANDLEY, R. W. *Proc. IADR., British Division* 1978.
24. CAUSTON, B. E. *J. Dent. Res.*, **54** (1975) p. 339.
25. MARSHALL, G. P., CULVER, L. E. and WILLIAMS, J. G. *Crack and Craze Propagation in Polymers: A Fracture Mechanics Approach. Crack Growth in Polymethylmethacrylate in Air* (1969) pp. 75–87.
26. FARAH, J. W. and CRAIG, R. *J. Dent. Res.*, **53** (1974) p. 859.
27. FARAH, J. Doctoral Dissertation, Ann Arbor, University of Michigan, 1972.
28. OILO, G. and JORGENSEN, K. D. *J. Oral Rehab.*, **4** (1977) p. 305.
29. RABINOWICZ, E. *Friction and Wear of Materials*, John Wiley & Sons, New York, 1965.
30. POWERS, J. M., ROBERTS, J. C. and CRAIG, R. G. *J. Dent. Res.*, **55** (1976) p. 432.
31. BOWDEN, F. P. and TABOR, D. *Friction and Lubrication*, Methuen, London, 1956.
32. MARIN, J. *Mechanical Behavior of Engineering Materials*, Prentice Hall, Englewood Cliffs, N.J., 1962.
33. MERIAM, J. L. *Statistics, Mechanics Part I*, Wiley N.Y., 2nd. ed, 1959, pp. 1–101.
34. BURWELL, J. T., Jr. *Wear*, **1** (1957) p. 119.
35. SWARTZ, M. L., PHILLIPS, R. W., NORMAN, R. D. and OLDHAM, D. F. *J. Am. Dental Assoc.* **67** (1963) p. 367.
36. POWERS, J. M., PHILLIPS, R. W. and NORMAN, R. D. *J. Dent. Res.*, **54** (1975) p. 1183.
37. ROBERTS, J. C., POWERS, J. M., and CRAIG, R. G. *J. Dent. Res.* **56** (1977) p. 748.
38. YOUNG, K. C., HUSSEY, M. GILLESPIE, F. C. and STEPHEN, K. W. *Proct. Int. Symp. Acid Etch Tech.*, North Central Publishing Co., St. Paul, Minnesota, 1975 p. 50.
39. M. BARSBY, private communication, 1978.
40. *International Symposium on Fotofil Dental Restorative* 1978 p. 8 and 16.

INDEX